The Marsh Tit
and the
Willow Tit

The Marsh Tit
and the
Willow Tit

RICHARD K. BROUGHTON

T & AD POYSER
LONDON · OXFORD · NEW YORK · NEW DELHI · SYDNEY

T & AD POYSER
Bloomsbury Publishing Plc
50 Bedford Square, London, WC1B 3DP, UK
29 Earlsfort Terrace, Dublin 2, Ireland

BLOOMSBURY, T & AD POYSER and the T & AD Poyser logo are
trademarks of Bloomsbury Publishing Plc

First published in the United Kingdom 2025

A catalogue record for this book is available from the British Library.

Library of Congress Cataloguing-in-Publication data has been applied for.

ISBN: HB: 978-1-4729-8031-1
PB: 978-1-4729-8032-8
ePDF: 978-1-4729-8029-8
ePub: 978-1-4729-8030-4

2 4 6 8 10 9 7 5 3 1

Design by Susan McIntyre
Printed and bound in India by Replika Press Pvt. Ltd.

MIX
Paper | Supporting
responsible forestry
FSC
www.fsc.org   FSC™ C016779

Cover: A Willow Tit displays territorial behaviour towards a Marsh Tit. (© Darren Woodhead)
Frontispiece: A first-year Marsh Tit in Norfolk, England. (© David Tipling)

Contents

An adult Willow Tit, northern Europe. (© Henri Lehtola/Shutterstock)

Preface

My aim in writing this book is to summarise what is known about two small woodland birds that are part of our common natural heritage. Marsh Tits and Willow Tits are an important part of forest ecosystems across Eurasia, and they are fascinating and rewarding birds to observe. They have also dominated much of my research career.

To see Marsh Tits and Willow Tits can sometimes take a little effort, entering the ancient woodlands, old forests and tangled shrublands that they inhabit. They occur at relatively low densities and, despite being vocal and bold, they are good at blending into their surroundings. Nevertheless, both species are widespread, resident and sedentary, and so they can be found throughout the year from Great Britain to Japan. With a little patience they can readily be seen in suitable places, and will often allow a close approach. In some areas Marsh Tits and Willow Tits also live alongside us in leafy villages, large parks or suburbs, becoming regular visitors to garden bird-feeders. Watching them for any length of time reveals a social life of strong bonds and a deep attachment to their chosen patch of woodland. Individuals can survive for a decade or more, and so we can get to know them as they live out their lives. They are resourceful, resilient and intelligent birds, and these attributes make them a real joy to study.

Quite a lot has been written about Marsh Tits and Willow Tits since the 19th century, although this is dwarfed by the literature on the more familiar Blue Tit and Great Tit. Regardless, Marsh Tits and Willow Tits are increasingly considered as model species in ecology and evolutionary science, along with their close relatives, the North American chickadees. Much has been learnt from these species about social organisation, spatial memory and communication, and these are some of the ways that Marsh Tits and Willow Tits help us to understand the natural world. Importantly, both species are also sentinels for forest ecosystems. Being resident throughout the year, Marsh Tits and Willow Tits are wholly reliant on what happens within and around their woodlands. If the habitat deteriorates, then they will struggle, and this alerts us to something going seriously wrong in the woods. In many parts of Europe, and especially in Great Britain, this is exactly what is happening. The alarming population declines of both species over the past 50 years have triggered great concern, but so far only modest action.

These declines are an unavoidable sign of the wider biodiversity crisis that dominates our era. In my own lifetime I have seen the loss of Marsh Tits and Willow Tits from places where they were once familiar. Willow Tits were regular birds in my childhood garden and local birding patch in northern England. By the time I had started my research projects in the 2000s they had all gone. Now, the Marsh Tits in my long-term study at Monks Wood, in eastern England, are also dwindling away; the study has changed from monitoring a functioning population to documenting its collapse on the downward spiral towards local extinction. These losses can feel personal and profound, and similar stories are not uncommon elsewhere in Europe.

Researching and preparing the book has therefore felt timely, and bringing together what we know about both species may help to inform their conservation. I have tried to summarise the major research projects from Great Britain and elsewhere in Europe and Asia. The published literature, postgraduate theses and unpublished data have provided a wealth of information to draw upon. The coverage is inevitably biased towards Europe, however, as this reflects where much of the research has been carried out so far.

Working on the book has been a pleasure, but has been set against the backdrop of difficult times. The global pandemic of coronavirus affected everybody, and during the crisis many people found solace in the nature around them. Fieldwork in the woodlands, when possible, was one of my own ways of maintaining a sense of normality, and I particularly appreciated my study time with the Marsh Tits during this period.

War has also returned to Europe, and Russia's invasion of Ukraine continues to cause humanitarian and ecological devastation that ripples throughout the region. The conflict has impacted one of the iconic sites for Marsh Tit research, the unique Białowieża National Park in Poland, which borders Belarus. Escalating militarisation and an engineered migrant crisis through the forest have created an unfolding tragedy that damages the fragile ecosystem, and also impacts the researchers who study there.

During this difficult period, the research community also lost one of the foremost authorities on Marsh Tits, Professor Tomasz Wesołowski, a friend and colleague from the Białowieża studies. I felt this loss when visiting the forest to search for Marsh Tits and Willow Tits, and when referencing his many scientific papers. Tomasz is remembered and appreciated throughout the book.

While writing this preface I visited important places for woodland research, and for myself, at Monks Wood and the Białowieża National Park. At both locations I could hear Marsh Tits just outside as I wrote. I took inspiration from these places and also from some fantastic earlier monographs of woodland birds, notably Susan Smith's *The Black-capped Chickadee*, Ian Newton's *The Sparrowhawk* and Erik Matthysen's *The Nuthatches*, which convey much of the pleasure of research and discovery. On reading through all of the previous literature, and also my own notes and data spanning over 35 years, I am struck by just how rewarding science and natural history can be. Marsh Tits and Willow Tits are such wonderful subjects, and immersing yourself in the woodlands to follow their lives is a special experience. I hope that this book inspires others to look more closely at these birds and their habitats, whether for research or just for the personal pleasure of connecting with nature. I also hope that the information gathered here can be used to better understand these birds, their habitats and our impacts upon them, so that we may conserve them into the future.

Richard K. Broughton
Monks Wood and Białowieża, 2024

CHAPTER 1

Marsh Tits and Willow Tits – similar yet different

The Marsh Tit and the Willow Tit are small, brown and buff-coloured birds with black-and-white heads, and are closely related members of the tit family, the Paridae (Figures 1.1 and 1.2). Both species are agile, vocal, intelligent and bold, and they are familiar birds in some parts of their ranges, but are scarce and elusive in others. They live in wooded habitats across Eurasia, where they are non-migratory and resident all year round. Willow Tits have a more northerly distribution than Marsh Tits, and are able to survive the northern winter in the boreal forests. Willow Tits also have a substantial overlap with Marsh Tits in temperate woodlands across the milder parts of their range. Where Marsh Tits and Willow Tits occur together in the same habitats they can be an identification challenge owing to their similar appearance, particularly in Great Britain. Some distinctive calls and songs help to distinguish between the two species, and, with practice, they can be readily separated on plumage and physical characteristics.

Despite Marsh Tits and Willow Tits being relatively well studied, with major research projects in several parts of Europe, there remain many significant knowledge gaps regarding their ecology, behaviour and conservation, especially in much of Asia. Of particular concern are catastrophic declines of both species in some parts of Europe, leading to local extinctions and range contractions. The major causes of these declines are obvious for some populations, but in others they are poorly understood. This is alarming, as Marsh Tits and Willow Tits are important indicators of the forest and woodland ecosystem, so their largely unexplained declines show how little we really know about the functioning of these habitats.

Discovery and naming

The convoluted history of how Marsh Tits and Willow Tits got their names helps to explain why they seem a little confusing, or even misleading. Small songbirds with brown backs and black caps were first documented in Europe by the Swiss polymath Conrad Gessner (sometimes Gesner) in his 1551 *Historia Animalium*, where he named them *Parus palustris*, literally the 'small marsh bird'. This was based on birds seen in swampy woodlands, but it was not yet realised that two species were involved. The English name Marsh Titmouse was formalised in *The Ornithology of Francis Willughby*, published in 1678. The name 'titmouse' derives from Middle English, meaning small bird, and was used for European species in the Paridae family until the 20th century, when the abbreviated 'tit' took over. Meanwhile, the scientific name *Parus palustris* was cemented for the Marsh Titmouse by Carl Linnaeus in his 1758 *Systema Naturae*, which formed the basis of today's taxonomic nomenclature.

The Willow Tit had to wait until 1827 to be recognised by another Swiss naturalist, Thomas Conrad von Baldenstein ('Conrad'), in his *Neue Alpina*. Conrad objected to the name of Marsh Tit and suggested renaming them 'Grey Tits' (*Parus cinereus*) to reflect their wider habitat use, but this didn't catch on. Conrad described two subspecies of his Grey Tits that differed in their songs, calls and appearance. One subspecies (*communis*) was the 'common'

Figure 1.1. An adult Marsh Tit in Northamptonshire, England, February 2018. (© www.garthpeacock.co.uk)

Figure 1.2. A first-year Willow Tit at Pennington Flash, England, August 2020. (© Philip Schofield)

form that favoured the wooded valleys, which were the birds that we know as Marsh Tits. The other subspecies were distinctive birds in the higher mountain forests that Conrad named *montanus*, the 'mountain' form, which were actually Willow Tits. This is how a bird that is often associated with lowland habitats in many parts of Europe got the scientific name of *montanus*, as it was first described from the Swiss Alps. Soon afterwards, in the 1831 *Handbuch der Naturgeschichte aller Vögel Deutschlands*, German ornithologist Christian Ludwig Brehm described another distinctive tit found in willows along streams and rivers, which he named *Parus salicarius*, the 'Willow' Tit. Brehm did not realise that this was the same species already described by Conrad from Switzerland.

Subsequent authors eventually unravelled that there were only two species involved, which were both widespread across Europe. The English name of Willow Tit was finally adopted, but awkwardly retaining the scientific name of *Parus montanus*. Meanwhile, the English and scientific names of Marsh Tit and *Parus palustris* unfortunately persisted, leaving us with one of the most misleading bird names in European ornithology: a woodland bird named for a marshland habitat that is more commonly occupied by a similar-looking species named after the mountains! Whilst the English name of Willow Tit is not a bad one, it would be sensible to rename the Marsh Tit as something else, perhaps the Wood Tit or Forest Tit, similar to the Wood Warbler or Forest Dormouse in the same habitat (see Appendix 1 for the scientific names of species mentioned in the book). This is more than a trivial academic niggle, as names and labels matter when communicating clearly to policy makers and non-specialists. It can be difficult to explain to people why woodlands are of critical importance for the conservation of a bird that we call the Marsh Tit.

Current taxonomy

Marsh Tits and Willow Tits remain within the family Paridae, which encompasses all of the tits and chickadees of Eurasia, North America and Africa. Both species were previously placed in the genus *Parus*, alongside Great Tits, Blue Tits, Coal Tits and most other tits and chickadees, but this was reviewed in 2005 based on the nucleotide sequences in the mitochondrial cytochrome-*b* gene (Gill *et al.* 2005). The result was that a group of Eurasian tits and North American chickadees were moved into the new genus *Poecile*, giving us the current scientific names of *Poecile montanus* for the Willow Tit and *Poecile palustris* for the Marsh Tit. The *Poecile* species all share a similar template of brownish body, pale underparts, dark cap and white cheeks, with a similar vocabulary structure (see Chapter 2). The *Poecile* species also have similar habits of living year-round in large territories and caching food (see Chapters 4 and 5).

Marsh Tits and Willow Tits diverged more than 4 million years ago and are not each other's closest relative (Figure 1.3), but they are in neighbouring clades within the *Poecile* genus

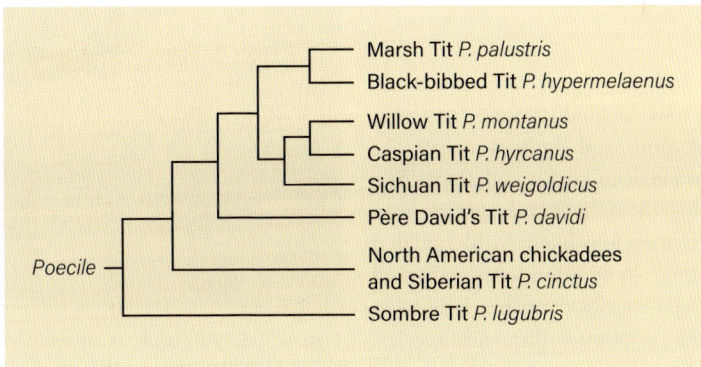

Figure 1.3. A simplified phylogeny showing the relatedness of species in the *Poecile* genus, focusing on the Eurasian species. Modified from Tritsch *et al.* (2017)

Figure 1.4. A Black-capped Chickadee collecting seeds in Nova Scotia, Canada. For the first half of the 20th century the Black-capped Chickadee of North America was erroneously considered to be the same species as the Willow Tit of Eurasia. (© J. W. Ross MacKinnon)

(Johansson *et al.* 2013, Tritsch *et al.* 2017). As such, both species are more closely related to each other than to any of the American chickadees. This shows the value of genetics, as for the first half of the 20th century the Willow Tit was thought to be the same species as the Black-capped Chickadee of North America (Figure 1.4). This notion was eventually contested by David Snow (1956), based on what now seem like obvious discrepancies in appearance, although Siegfried Eck was still arguing for combining Willow Tits and Black-capped Chickadees into one species as late as 1980 (Eck 1980). Snow (1956) was vindicated by the later genetic studies, which conclusively showed that Willow Tits and Black-capped Chickadees are totally different species.

SUBSPECIES

Taxonomy and phylogeny are constantly evolving as new techniques arise, concepts change over time and authors revisit the relationships between bird populations. Any current phylogeny of the *Poecile* species, as in Figure 1.3, is therefore likely to be updated before very long. In addition to the variation between *Poecile* species there has long been interest in the variation within each species, leading to the classification of numerous subspecies in the 19th and 20th centuries (Vaurie & Snow 1957, Eck 1980).

Figure 1.5. An adult northern Willow Tit at Dombås in Norway. Willow Tits in this region are attributed to the subspecies *P. m. borealis*. (© Robert Fredagsvik)

Among Willow Tits there are obvious differences in the size, colouration and vocalisations between regions. Large, pale Willow Tits in Norway (Figure 1.5) look quite different to the small, brown ones in Great Britain (Figure 1.2), and Willow Tits in the high Alps sing a different song to those in Sweden (see Chapter 2). Based on such attributes, perhaps 14 subspecies of Willow Tit are widely recognised (Tritsch *et al.* 2017, Gosler *et al.* 2020a), including an endemic island subspecies in Great Britain (*P. m. kleinschmidti*). Similar variation in size and plumage has led to around nine subspecies of Marsh Tit also being recognised, although the birds in Great Britain (*P. p. dresseri*) are not currently regarded as an endemic form (Gosler *et al.* 2020b). The subspecies in the recent literature are summarised in Tables 1.1 and 1.2.

Table 1.1. Subspecies of the Willow Tit that are currently recognised, based on Harrap & Quinn (1995), Tritsch *et al.* (2017) and Gosler *et al.* (2020a).

Willow Tit subspecies	Approximate distribution
P. m. kleinschmidti	Great Britain
P. m. rhenanus	Western Europe
P. m. montanus	Central European uplands, including the Alps
P. m. salicarius	Central European lowlands
P. m. borealis	Northern and eastern Europe, including Fennoscandia
P. m. uralensis	Western Siberia to northern Kazakhstan
P. m. baicalensis	Eastern Siberia, Mongolia, northern China to Korea
P. m. anadryensis	Northeast Siberia
P. m. kamtschatkensis	Kamchatka
P. m. sachalinensis	Sakhalin, possibly Japan (Hokkaido)
P. m. restrictus	Japan
P. m. songarus	Eastern Kazakhstan, Tien Shan, northwest China
P. m. affinis	Central-northern China
P. m. stoetzneri	Northeast China

Table 1.2. Subspecies of the Marsh Tit that are currently recognised, based on Harrap & Quinn (1995) and Gosler *et al.* (2020b).

Marsh Tit subspecies	Approximate distribution
P. p. dresseri	Great Britain, western France
P. p. palustris	Western, central and northern Europe
P. p. stagnatilis	Eastern Europe, southeast Europe, northern Turkey
P. p. italicus	Southern Alps and Italy, including Sicily
P. p. kabardensis	Western Caucasus
P. p. brevirostris	Siberia, Mongolia, northeast China, northern Korea
P. p. hellmayri	Eastern China, southern Korea
P. p. ernesti	Sakhalin
P. p. hensoni	Northern Japan (Hokkaido)

Figure 1.6. An adult Marsh Tit near Trondheim in Norway, in fresh plumage in mid September. Scandinavian birds are assigned to the nominate subspecies *P. p. palustris*, and tend to be slightly larger and paler than Marsh Tits in western Europe. (© Robert Fredagsvik)

Despite the long history of subspecies classification there is no universally agreed taxonomy, as each author tends to include, exclude or combine apparent subspecies based on their own individual assessment of specimens. For example, the subspecies recognised by Eck (1980), Vaurie & Snow (1957) or Svensson (2023) differ from each other, based on their own interpretations of different source material. Regardless of which classification is preferred, recent analyses suggest there is very little genetic basis for it. Across the Willow Tit's very large geographical range, studies by Kvist *et al.* (2001) and Salzburger *et al.* (2002) found little or no genetic differentiation between almost all of the claimed subspecies, despite differences in appearance and songs. There are no detailed genetic studies of Marsh Tits as yet, but most of the currently accepted subspecies appear even less visually or vocally distinctive than the Willow Tit subspecies (Figure 1.6).

Most of the original subspecies classifications are well over a century old (e.g. Hartert 1910) and were often based on very few specimens. For example, Carl Hellmayr's (1900) description of the endemic British Willow Tit (*P. m. kleinschmidti*) was initially based on perhaps only two birds. Leonhard Stejneger (1886) also proposed classifying an endemic British subspecies of the Marsh Tit, based on subtle colouration of a few preserved skins. Vaurie & Snow's (1957) influential review disagreed with Stejneger and lumped British Marsh Tits with those in northern France, yet retained Hellmayr's British Willow Tit. However, Vaurie and Snow's own assessments were also based on rather limited evidence, typically involving just 5–34 individuals examined from each region, and sometimes only a single bird. With so few specimens it seems questionable whether this could account for the variation of size

Figure 1.7. A Willow Tit in the Netherlands (ostensibly subspecies *P. m. rhenanus*). These birds in western Europe appear to be essentially indistinguishable from those in Great Britain on the basis of plumage and morphology, undermining the current classification of British birds as an endemic island subspecies (*P. m. kleinschmidti*). (© Han Onderwater)

and colouration within a population, which we now know can vary with age and sex, as well as individually (Kniprath 1967, Broughton *et al.* 2016). The lack of genetic differentiation between claimed subspecies of the Willow Tit underlines how these subspecies classifications that we still use today are built on rather shaky foundations.

Increasingly I take the view that most of the recognised subspecies of Marsh Tits and Willow Tits are not very meaningful. The variation within each species is clinal and not categorical, with only a subtle gradation of size and colour along latitude, longitude and altitude. In colder climates in northern regions, and at higher mountain elevations, both species become a little larger in size, greyer on the upperparts and whiter on the underparts (Figures 1.5 and 1.6). Meanwhile, towards the milder and humid climates of western Europe and southeast Asia, both species become incrementally smaller, browner and darker (Figures 1.1 and 1.2). Vaurie & Snow (1957) and Svensson (2023) acknowledged the subtle gradation along these clines, with no clear boundaries and a difficulty in assigning many birds to one subspecies or another. Categorising populations into subspecies along these gradients therefore seems a bit subjective, unless there is a clear genetic basis. Indeed, I find the argument for an endemic British subspecies of Willow Tit to be unconvincing, as with the Marsh Tit. There seems to be an almost complete overlap in biometrics and colouration between Willow Tits from Great Britain and those from adjacent parts of the continent, just on the other side of the North Sea (Cramp & Perrins 1993). If a Dutch or Belgian Willow Tit ever arrived in Great Britain, I very much doubt that anyone would distinguish it from British birds (Figure 1.7).

15

Distribution and population size

Marsh Tits and Willow Tits both have vast ranges across Eurasia. The Willow Tit has a continuous distribution from Great Britain through to Japan (Figure 1.8). The Marsh Tit's distribution also extends from Great Britain to Japan, but with a large gap separating two disjointed regions in Europe and east Asia (Figure 1.9).

The current distribution of both species may not necessarily reflect their natural ranges, as these have been affected by human impact. For example, both are curiously absent from the island of Ireland, which may be a true quirk of biogeography or possibly a result of very early deforestation that caused their extinction in prehistory. Even more recently Willow Tits have undergone a rapid range contraction in Great Britain over the last century, with both species now having a rather patchy distribution (Figure 1.10). Despite being restricted to a small area of Scotland today, Willow Tits probably occurred over much of the country when it had

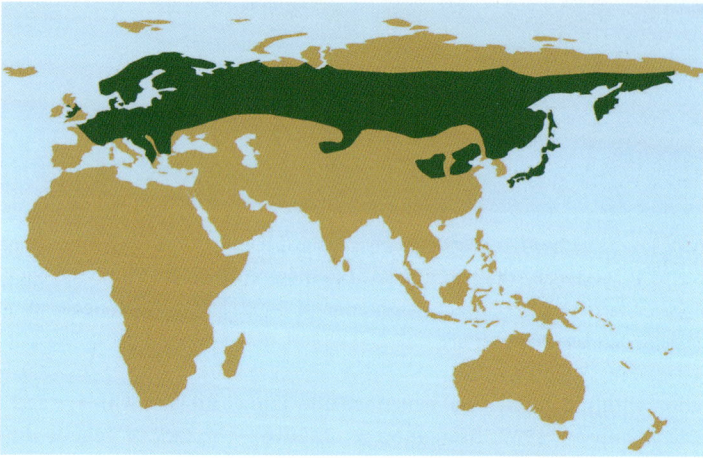

Figure 1.8. The current global distribution of the Willow Tit in Eurasia.

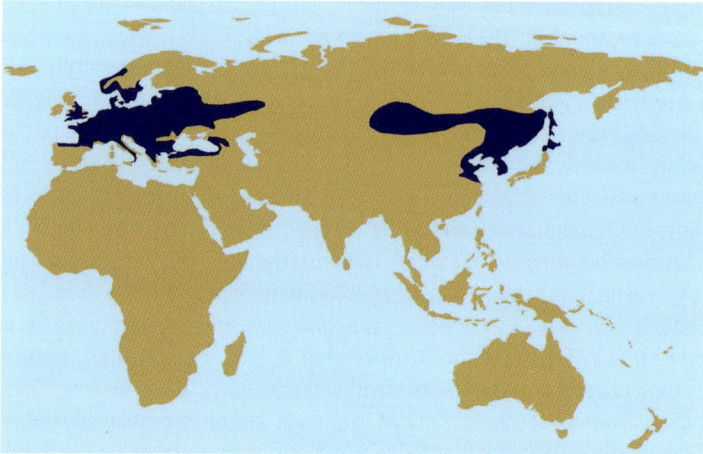

Figure 1.9. The current global distribution of the Marsh Tit in Europe and east Asia.

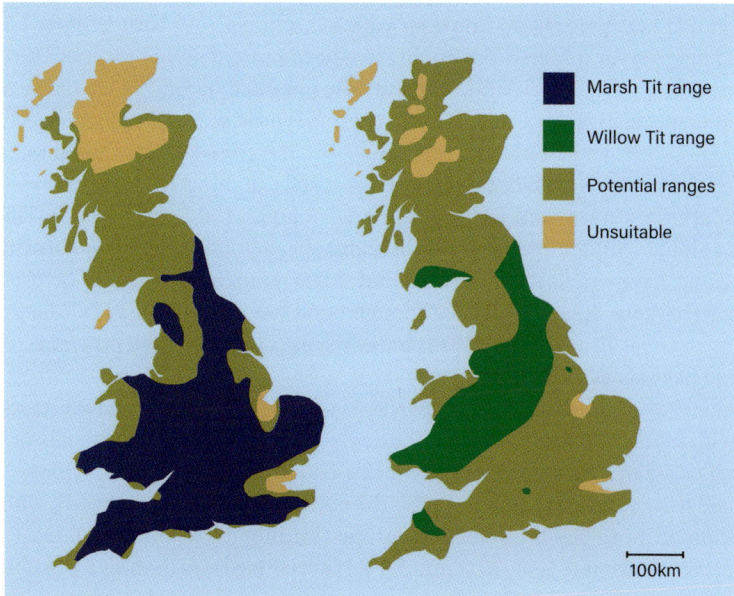

Figure 1.10. The approximate current range of the core populations of Marsh Tits (left) and Willow Tits (right) in Great Britain. Also shown is the potential natural range of each species during the woodland maximum of the Holocene epoch, around 6,000 years ago, before widespread deforestation. Current range estimates are extrapolated from Balmer *et al.* (2013), Amaral-Rogers (2021), Pinder & Carr (2021) and numerous published reports from British regions. Potential ranges are modified from woodland habitat extents in Rackham (1976) and Smout *et al.* (2007).

plentiful native forest. Willow Tit bones are frequent in Mesolithic or Neolithic (11,500–4,500 years ago) archaeological deposits from Oban in western Scotland (Bartosiewicz *et al.* 2010), suggesting they were common before deforestation. Even in the early 20th century Willow Tits were still quite widespread in Scotland, breeding as far north as Inverness and Wester Ross (Andrews 2014). William Brunsdon Yapp (1962) reported resident Willow Tits in Abernethy Forest, probably in the mid 20th century, and Eric Simms (1971) found Willow Tits in Spey Valley birchwoods as late as 1957. A steady decline meant that by the early 21st century Willow Tits were restricted to a small area around Dumfries and Galloway in Scotland's southwest. From the 1980s Willow Tit populations also began disappearing from southern and eastern England, a process which is continuing apace into the north and west – with the result that any recent distribution map for Willow Tits in Great Britain is quickly out of date. Nationally, the abundance of British Willow Tits declined by 86 per cent between 1995 and 2020 (Harris *et al.* 2022). Declines and range contractions are also occurring elsewhere in Europe, such as the Netherlands, France and Finland.

The Marsh Tit's distribution in Great Britain is also not static, with an ongoing range contraction over the past 50 years and a 46 per cent decline in abundance between 1995 and 2020 (Balmer *et al.* 2013, Harris *et al.* 2022). The losses of Marsh Tits have not yet been as severe as for Willow Tits (Figure 1.10), but the downward trend shows no sign of improvement and there is no knowing where it will end. As with Willow Tits, before widespread deforestation Marsh Tits could have been present across most of Great Britain, possibly including much of lowland Scotland. The potential natural range of the Marsh Tit in Scotland is speculative, however, as there are no records or physical evidence away from southern regions below the Firth of Forth. However, the climate and original forest habitat should have been very suitable, particularly in forests dominated by oaks and Common Hazel in southern, central and eastern Scotland (Smout *et al.* 2007). Marsh Tits are still present in a similar climate and habitat in Scandinavia, but intense deforestation in lowland Scotland could have eradicated them around 2,000 years ago or more, before anybody knew they were there. It's an intriguing thought, and the natural ranges of Marsh Tits and Willow Tits in Great Britain are relevant today when considering the potential restoration of woodland habitats and species.

The national or international population sizes of Marsh Tits and Willow Tits are difficult to assess with much certainty, as they are usually extrapolated from estimates or densities within smaller areas, rather than from accurate counts. Worldwide, for Willow Tits there are estimated to be very approximately 175–253 million territorial individuals, with perhaps 61–88 million of those being resident in Europe (BirdLife International 2023). For Marsh Tits, estimates are around 10.6–20.9 million territorial birds, including a possible 2.9–5.7 million in Europe. These are little better than rough estimates, however. Within Great Britain, a more accurate dedicated survey in 2019–2021 produced a total of just under 5,700 pairs of Willow Tits, with 76 per cent of these in England, 21 per cent in Wales and 3 per cent in Scotland (Amaral-Rogers 2021). Estimates for the British Marsh Tit population are far more uncertain, as there has not been any dedicated survey. Based on broad extrapolations from limited sampling, the best estimate is 28,500 Marsh Tit territories in Great Britain in 2016 (Woodward *et al.* 2020).

Identification

Marsh Tits and Willow Tits are easily told from most other species, but they can be notoriously difficult to distinguish from each other in some regions. The two species have a broadly similar structure and proportions, and although posture and fluffing up of the plumage can significantly change their shape, this is not a structural difference (Figure 1.11). Willow Tits do have somewhat fluffier or looser plumage than Marsh Tits, which is because their body feathers have longer barbs and more barbules (Figure 1.12), and this may be an adaptation for better insulation in their more northerly distribution.

Much has been written about separating Marsh Tits and Willow Tits, especially in Great Britain, but for many years too great an emphasis was placed on unreliable features in popular field guides, such as cap gloss, size of the throat patch and presence or absence of pale fringes on the secondary wing feathers. Several studies have tested the identification features on a range of live birds and museum skins, primarily but not exclusively for British birds (Scott 1999, Broughton *et al.* 2008a, Broughton 2009, Broughton & Alker 2017, Konno 2018). For all populations examined, the most consistent and objective physical differences between the two species are the 'bill spot' and the gradation of the tail. Another very strong feature is the pattern on the cheek patches, while the colour of the greater coverts and fringes of the secondary wing feathers can also be useful.

Figure 1.11. Willow Tits can sometimes appear rounded when they fluff up their dense plumage, and they have been described as looking more 'bull-necked' than Marsh Tits, but this is just a temporary posture. This bird is from northern England. (© Philip Schofield)

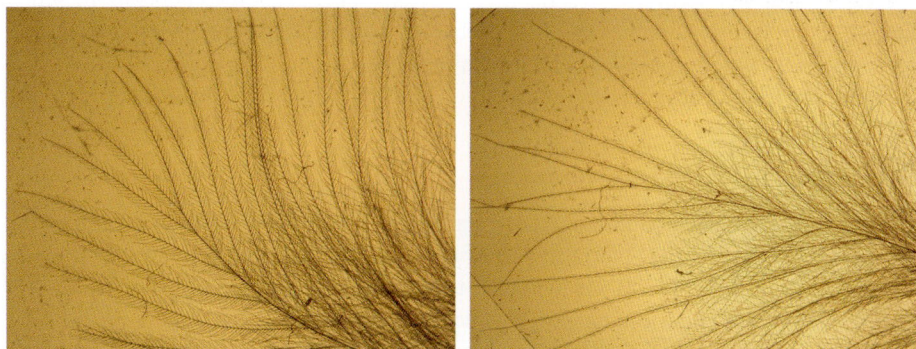

Figure 1.12. Magnification of a flank feather of a British Marsh Tit (left) and Willow Tit (right), showing the relatively longer, looser, wispy barbs and denser barbules that give the Willow Tit's plumage a fluffier appearance. (© Richard K. Broughton)

THE MARSH TIT'S BILL SPOT

This small but significant identification characteristic was first proposed by Dewolf (1987), who noticed that Marsh Tits have a whitish spot at the base of the upper mandible of the otherwise dark bill. On Willow Tits the bill is all dark and inky black (Figure 1.13). Dewolf didn't provide any numbers to support the bill-spot criterion, so colleagues and I tested it on 199 live Marsh Tits and 27 Willow Tits from England, along with 57 Marsh Tit skins from England, Poland and Scandinavia, and 112 Willow Tit skins from England, Switzerland,

Figure 1.13. The all-dark bill of a Willow Tit (left) compared with the distinctive white spot visible at the base of a Marsh Tit's bill (right). (© Richard K. Broughton)

Austria and Scandinavia, all held in the British Natural History Museum (Broughton *et al.* 2008a). Pooling the live birds and the skins, we found that 99 per cent of Marsh Tits and 94 per cent of Willow Tits were correctly identified using only the presence or absence of a whitish bill spot. Of the 6 per cent of Willow Tits that did show some kind of bill spot, only half showed clear marks on both sides of the bill, with most of the other marks attributed to a scratch or abrasion. As such, applying a threshold of the presence or absence of clear bill spots on both sides of the bill allowed the correct identification of 99 per cent of Marsh Tits and 97 per cent of Willow Tits overall. This criterion seems to apply to populations across Europe and in Japan (Konno 2018), so is probably valid across the species' ranges. The bill spot is not very practical to see in the field, although it can be surprisingly visible at close quarters, and is often obvious on photos and on birds in the hand during ringing.

TAIL GRADATION

Differences in the relative lengths of the tail feathers provide a useful distinction between the species. On Marsh Tits the outermost pair of tail feathers are shorter than all the others by a typical 3–4mm (Figure 1.14). On Willow Tits the outermost pair are also shorter than the longest tail feathers, and this difference is typically 5–6mm. In both species this difference is greater for adults than for juveniles in summer or first-years in their first autumn and winter after the post-juvenile moult.

In Willow Tits, however, there is also a gradual increase in the length of the tail feathers from the outer pair inwards, with the innermost pairs of tail feathers being the longest. This creates a stepwise sequence of increasing length visible on the underside of the folded tail (Figure 1.14), visible in the hand or on good photos in the field. These differences in tail gradation and feather lengths hold true in Great Britain, Switzerland and Japan (Amann 1980, Abe & Kurosawa 1984, Scott 1999, Broughton *et al.* 2016), so are probably applicable everywhere. In the British studies the measured difference between the shortest and longest tail feathers, and the number of stepped feathers visible on the tail underside, were significantly different between the species. There is some overlap in tail gradation, however, and some birds' tails can resemble the other species.

Willow Tit first-year

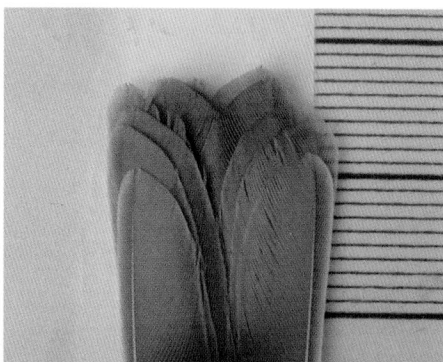

Willow Tit adult

Marsh Tit first-year

Marsh Tit adult

Figure 1.14. Differences in the tail gradation of Willow Tits and Marsh Tits, viewed on the underside of the tail. Note the more pointed feather tips of first-years compared to the rounded feathers of adults. Also the stepwise increase in tail-feather length in Willow Tits, compared to only the shorter outer feathers in Marsh Tits, although first-year tails are less distinctive than those of adults. (© Richard K. Broughton)

OTHER PLUMAGE FEATURES

Older literature promoted several plumage features for separating Marsh Tits and Willow Tits, especially in Great Britain. Among these was a supposedly glossier black cap in Marsh Tits compared to a dull matt cap in Willow Tits, and a smaller and neater black throat patch (or 'bib') in Marsh Tits. These plumage features have been tested in British studies and found to have so much overlap that they are of no practical use (Scott 1999, Broughton 2009). The cap of both species can have a glossy bluish sheen on adult males, particularly on Marsh Tits and in bright sunlight, but on females and summer juveniles there is usually no gloss and sometimes even a faint brownish tint (Figure 1.15). The black throat patch is even more subjective and variable, and is heavily reliant on posture, so is of no real use.

A key identification feature emphasised in earlier field guides was a 'pale wing panel' on Willow Tits, formed by whitish or buff fringes on the secondary feathers of the closed wing. This feature can be useful as a supporting characteristic in identification, but it requires some caution with respect to the angle of viewing. A pale wing panel can be very pronounced on some Willow Tits (Figure 1.2) but much less obvious on juvenile and first-year birds. Marsh

Figure 1.15. A juvenile Marsh Tit in Sussex, southern England, in July–August. The bird has begun its post-juvenile moult (note the missing inner greater coverts on the wing) but has not yet moulted the head. The clean white cheeks and dull blackish cap (with a brownish tint) are typical of juvenile Marsh Tits and Willow Tits. (© Adam Nicolson)

Tits can sometimes show quite pale fringes on the secondaries, although they are often only slightly paler than the rest of the upperparts. Two British studies showed that the pale secondary fringes are reliable in only around half of Marsh Tits and Willow Tits (Scott 1999, Broughton 2009), although in Japan it was useful for 100 per cent of Willow Tits and 84 per cent of Marsh Tits (Konno 2018).

A plumage feature useful in western Europe and Japan at least, which was largely overlooked in earlier field guides, is the pattern of the pale cheek patches on the birds' heads, including the ear coverts and onto the nape. On Willow Tits in Great Britain and Japan this whole area of the cheek patch is whitish or warm creamy buff, often with a golden tint where the cheek patch meets the shoulder and mantle (Figure 1.16). On Marsh Tits there is a clearer distinction between the white ear coverts and the cool greyish or brownish tint behind the ear onto the nape, giving a 'two-tone' cheek pattern (Figure 1.17). The cheek pattern can be quite noticeable in the field, and an assessment of skins and live birds in Great Britain found that almost 9 out of 10 birds were consistent with expectations (Broughton 2009). In Japan 96 per cent of Willow Tits and 73 per cent of Marsh Tits were correctly identified using the cheek pattern (Konno 2018). An exception is that the cheek pattern feature does not apply to fledglings and juveniles before they have completed their post-juvenile moult in September. Before this, during the summer, unmoulted juveniles of both species have similar clean white cheeks (Figure 1.15).

A relatively new identification feature that we recently tested for British Marsh Tits and Willow Tits is the pattern on the greater coverts on the wing. Again, this is only really useful for birds in the hand or on good photographs of birds in the field. In Willow Tits, each greater covert has a dark blackish central shaft, a dark greyish centre and a contrasting brown fringe to

Figure 1.16. An adult British Willow Tit showing a typical cheek pattern of a clean white ear covert that extends towards the nape, where a subtle warm buff tint creates a sharp contrast with the brown mantle and shoulder. (© Philip Schofield)

Figure 1.17. A British Marsh Tit showing a typical cheek pattern of a clean white ear covert on the side of the head, but a dull grey-brown wash behind the ear towards the nape. This results in a 'two-tone' cheek patch that blends into the brown mantle and shoulder. Note the atypical white feather in the black cap. (© www.garthpeacock.co.uk)

the outer web of the feather, often with a paler tip (Figure 1.18). In Marsh Tits, the central shaft is a less obvious brown, and the whole outer web of each covert is a more uniform brown, with no contrasting fringe, except for usually a slightly paler tip. The whole effect is that the greater coverts of Marsh Tits are more uniform, while Willow Tits show a more contrasting pattern (Figure 1.18). We trialled this feature using an online survey, where 140 British ringers were asked to identify the species (Marsh Tit or Willow Tit) based only on photos of the greater coverts of 18 birds (Broughton & Alker 2017). Overall, 84 per cent of respondents identified the birds correctly, using just this feature.

Key identification criteria

Individually, the bill spot, tail gradation, cheek pattern or greater covert pattern can each correctly identify 80–98 per cent of birds, with the bill spot alone identifying almost all birds. If a couple of these features are combined, perhaps also considering pale fringes on the secondary wing feathers as a supporting characteristic, then essentially all Marsh Tits and Willow Tits can be readily identified. This may vary with populations, as the cheek pattern will probably be less useful

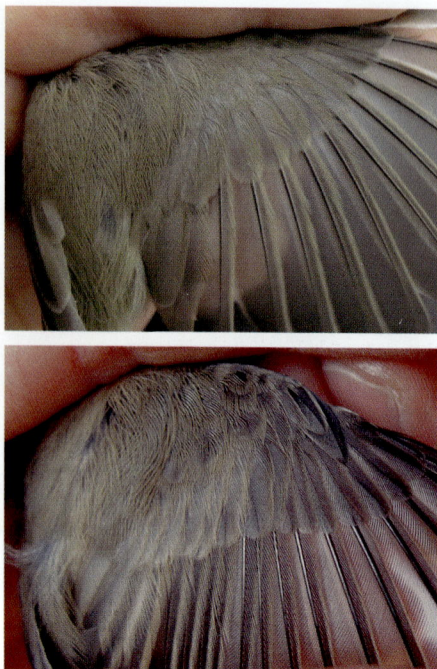

Figure 1.18. Greater coverts of an adult Marsh Tit (top) and Willow Tit (bottom), showing the subtle yet distinctive difference in colour and pattern. Marsh Tit greater coverts are rather uniform brown, whereas Willow Tit has a more contrasting pattern of dark centres to each feather with a paler brown fringe. (© Richard K. Broughton)

in Fennoscandia, but the bill spot and tail gradation appear to be highly reliable across both species' ranges, wherever they have been tested. Each species also has distinctive vocalisations that can enable identification, and these are discussed in Chapter 2.

Moult

Both species are initially in juvenile plumage after fledging, but they later undergo a partial post-juvenile moult in their first summer and autumn, when they become first-years, and then a full annual moult the following summer to become full adults. Juveniles begin their first moult about six weeks after fledging, when they have settled in a home-range. In our study in England the first juvenile Marsh Tits began their post-juvenile moult in mid July, with the last ones not completing until early October, although most had finished by mid to late September (Broughton *et al.* 2008b). The post-juvenile moult involves the replacement of all the body and head feathers, but not the primary or secondary wing feathers and usually not the tail feathers. Also retained are the primary coverts, alula and usually several greater coverts on the wings, marking them out as first-year birds until the following summer.

Adults start their full moult at or near the end of the breeding period, which in our study of English Marsh Tits was from around late May, and birds continued moulting until mid to late August (Broughton *et al.* 2008b). The moult begins with the shedding and replacement of tail feathers and primary wing feathers, as well as body contour feathers. In June the adults can look very ragged (Figure 1.19). The primary feathers are moulted sequentially from the innermost to the outermost, followed by the secondaries. The approximate duration of primary moult and complete replacement in England was 67 days, and by mid August only some of the head and body feathers were still being replaced. For British Willow Tits a similar duration of approximately 70 days is reported for primary moult, lasting over a similar period of mid May until mid September, although many adults had completed the moult by the end of August (Ginn & Melville 1983). A study of Willow Tits in Finland during the 1970s found that the adult moult overlapped with the latter stages of breeding, commencing around mid June and finishing by the end of August, altogether lasting around 77 days (Orell & Ojanen 1980).

Figure 1.19. A moulting adult Marsh Tit at Monks Wood in mid June 2013. Note the extensive feather loss on the head, the missing primary wing feathers and the short tail, which is regrowing. (© Richard K. Broughton)

For first-year birds, the juvenile tail feathers retained after the post-juvenile moult show a characteristic tapered shape towards a pointed tip (Figure 1.14). In an English population only 25 per cent of 616 juvenile Marsh Tits had replaced any tail feathers by the autumn (Broughton *et al.* 2008b, and additional data). Just one of these birds had replaced the whole tail, and most (70 per cent) had replaced only one or two feathers. Only 11 birds (7 per cent) showed any kind of symmetrical replacement of tail feathers, which would have indicated genuine moult. Instead, it looked as if most of the first-years had replaced tail feathers that had been lost during accidents, attacks or fights, possibly during the dispersal and settling period. Information for British Willow Tits suggests that they have a similar strategy of retaining most or all of the juvenile tail feathers throughout their first year (Ginn & Melville 1983). This is very useful, as it means that birds can readily be aged during ringing, providing reliable demographic data for the proportions of adults and juvenile recruits in a sampled population.

Greater coverts retained after the post-juvenile moult are typically very slightly longer than the adjacent greater coverts that have been moulted, and usually have paler tips than adult feathers (Figures 1.20 and 1.21). This creates a subtle yet distinctive *moult limit* on the greater coverts, which was present on the great majority of first-year Marsh Tits and Willow Tits that I have examined in England (Broughton *et al.* 2008b). Many birds commonly retain 3–4 greater coverts after the post-juvenile moult. In the Monks Wood study, for example,

Figure 1.20. Moult limit in the greater coverts of a first-year Willow Tit at Combe Wood in southern England. The five unmoulted greater coverts (marked) are slightly longer than the inner moulted greater coverts, creating a distinct moult limit. (© Richard K. Broughton)

Figure 1.21. Moult limit in the greater coverts of a first-year Marsh Tit at Monks Wood in eastern England, with three unmoulted greater coverts (marked) that are slightly longer and paler-tipped than the moulted inner greater coverts. (© Richard K. Broughton)

80 per cent of 610 first-year Marsh Tits had retained 3–4 unmoulted greater coverts, and almost all (97 per cent) had retained at least one. For Willow Tits, 59 per cent of 41 first-years that I have examined in England had retained at least one greater covert, and again the most frequent number was 3–4 (39 per cent of birds).

Ageing, sexing and biometrics

In the hand, and often on good photographs in the field, it is possible to distinguish adults from juvenile and first-year birds up to one year old, based on the moult limit formed by unmoulted greater coverts and the shape of the tail feathers, as outlined earlier. Among all studies the more pointed shape of the juvenile tail feathers is the most widely used feature for ageing birds as first-years or older adults, as young birds appear to almost never replace all of the tail during the post-juvenile moult (Laaksonen & Lehikoinen 1976, Amann 1980, Broughton *et al.* 2008b, Svensson 2023). For Marsh Tits at Monks Wood, checking for unmoulted juvenile tail feathers and greater coverts allowed every bird to be aged as a first-year or older adult.

Males average slightly larger than females, especially in Marsh Tits (Table 1.3) where wing length can reliably sex around 95 per cent of birds (Amann 1980, Nilsson 1992, Broughton *et al.* 2008b, Aparisi *et al.* 2018). However, the critical cut-off measurement in wing length between males and females varies between regions due to the clinal variation in body size. For wing lengths of live birds measured as the maximum chord (where the closed wing is flattened and straightened on a ruler; see Svensson 2023), a measurement of 63mm or less indicated a female and 64mm was a male in an Austrian population, and this accurately sexed all birds (Aparisi *et al.* 2018). For the slightly smaller birds in Great Britain a measurement of 62mm or less indicates a female and 63mm or more indicates a male for 93 per cent of individuals

(Broughton *et al.* 2016). When age is also taken into account, as first-year or adult, then wing length can be even more accurate for sexing Marsh Tits. In Switzerland 95 per cent of birds were sexed correctly using wing-length criteria of 64/65mm (female/male) for first-years and 65/66mm for adults (Amann 1980). We found a similar 95 per cent accuracy for a sample of 559 sexed Marsh Tits in England, using wing-length separation criteria of 62/63mm for first-years and 63/64mm for adults (Broughton *et al.* 2016). Other biometrics are all slightly larger for male Marsh Tits, such as tail and tarsus length and body mass, but the differences are less distinct than for wing length. Detailed reference biometrics for Marsh Tits from the Monks Wood study and affiliated data are given in Appendix 3.

Unfortunately it seems that the simple wing-length biometric cannot be used to reliably sex Willow Tits in Great Britain, because there is too much overlap between males and females (Table 1.3; Broughton *et al.* 2016). This limits the scope of population studies, as birds can only be sexed by colour-ringing and field observations of behaviour, or by genetic sampling. Appendix 4 gives reference biometric data for Willow Tits in England, derived from reliably sexed and measured birds. There seems to be a greater ability to sex Willow Tits using biometrics in populations of larger birds elsewhere. For example, Markovets (1992) proposed a sexing method using a discriminant equation based on wing and tail length for birds in the southern Baltic region.

Table 1.3. Wing-length measurements (maximum chord, mm) of Marsh Tits and Willow Tits of known sex from different European populations.

Region	Female				Male			
	Mean	s.d.	Range	No. of birds	Mean	s.d.	Range	No. of birds
Marsh Tits								
England	61.2	0.9	59.0–65.0	272	64.1	1.0	61.0–67.0	312
Austria	62.6	0.9	60.5–64.0	40	67.1	0.8	64.5–69.0	65
Switzerland	63.6	1.0	61.0–65.0	80	66.5	0.9	64.0–68.0	106
Germany	63.0	1.2	62.0–64.5	13	66.7	1.5	64.0–70.0	18
Willow Tits								
England	59.1	1.4	56.0–63.0	33	61.0	1.4	58.0–63.0	45
Germany	61.8	1.4	60.0–62.5	10	64.2	1.6	61.5–67.0	19
Curonian Spit (Baltic)	61.9	0.7	61.0–63.0	34	65.1	1.2	63.0–68.0	31
Norway	62.8	0.8	61.0–64.0	43	66.6	1.0	65.0–69.0	48

Means are derived from pooled adults and first-years, except for Willow Tits in Norway, where only first-years are included.

Sources: Ludescher 1973, Amann 1980, Markovets 1992, Hogstad 2011, Broughton *et al.* 2016 plus additional data, Aparisi *et al.* 2018.

Figure 1.22. A colour-ringed Marsh Tit at Monks Wood in eastern England (left: © www.garthpeacock. co.uk), and a Willow Tit at Oulu in northern Finland (right: © Anne Laine). Colour-ringing with unique colour combinations allows the identification and monitoring of individuals throughout their lives without the need to recapture them.

Studies of Marsh Tits and Willow Tits

Both species are quite well studied, although the research is heavily biased towards Europe. Marsh Tits and Willow Tits are relatively straightforward to catch and ring at certain times of the year, using baited traps or mist-nets, and their nests and nestlings are usually quite accessible, allowing the marking and monitoring of individuals throughout their lives. The use of colour-ringing to monitor individuals within discrete populations has been in use since the first half of the 20th century, providing insights on breeding, territoriality, dispersal, foraging, habitat selection and social organisation (Figure 1.22). Being resident within territories or home-ranges all year round, individuals are also relatively easy to find, which means that their disappearance can be used to confidently infer mortality. The first detailed studies of Marsh Tits and Willow Tits began in the 1930s and 1940s, and there have been multiple major research projects up to the present time, some of them lasting for decades. A general chronology of significant studies is outlined below.

BAGLEY AND WYTHAM WOODS, OXFORD, ENGLAND

The Edward Grey Institute of Field Ornithology (EGI) at the University of Oxford has produced several important studies of Marsh Tits and Willow Tits in Great Britain. Averil Morley's 10-year study at Bagley Wood, initially with Mick Southern, began in 1937 and provided the first detailed information on the ecology and behaviour of Marsh Tits. The study focused on 4–5 annual territories across 20ha of ancient woodland (Morley 1949, 1950, 1953, Southern & Morley 1950). Morley's published papers are full of ecological detail but also clearly convey the sheer pleasure of being immersed in the woods and discovering the lives of Marsh Tits. Averil Morley died in 1957 at just 43 years of age. In 2013 I revisited Morley's study site in Bagley Wood to try and colour-ring Marsh Tits for a new project (Figure 1.23), but I found only two pairs remaining.

Figure 1.23. Bagley Wood at Oxford, in southern England, the site of the first major ecological study of Marsh Tits by Averil Morley, between 1937 and 1947. The photo was taken in November 2013 in Morley's former study area. (© Richard K. Broughton)

Perhaps the first detailed study of Willow Tits took place nearby, at Oxford's Wytham Woods during 1948–1950, where John Foster and Christina Godfrey (1950) documented breeding, foraging and territorial behaviour in two territories over 22ha of woodland. Although focusing on only a few pairs, this study remains important for its detail and its setting in mature woodland, a habitat where Willow Tits now rarely occur in western Europe.

More recently at Wytham Woods, Jane Carpenter (2008) completed a major doctoral thesis on Marsh Tit ecology and habitat associations over 250ha of habitat during 2004–2008, supervised by Andrew Gosler and Shelley Hinsley. Later, in 2021–2023, Marta Maziarz analysed the Marsh Tit data from Wytham Woods' innovative PIT tagging study system. The PIT tags are small passive transponders attached to leg rings that are activated at a network of feeding stations to log the presence and movements of individual Marsh Tits, Blue Tits and Great Tits throughout Wytham Woods. The results of the analysis by Maziarz *et al.* (2023) provide important insights into the potential negative effects of interspecific competition on Marsh Tits.

BASEL, SWITZERLAND

Between 1947 and 1954 Fritz Amann studied a population of Marsh Tits in a core study area of 70ha in fragmented woodland, just outside Basel in Switzerland. However, the results were not published until decades later (Amann 1997, 2003, 2007, Schaub & Amann 2001).

The research was wide-ranging and especially detailed, focusing on territoriality, survival, dispersal distances and juvenile settling. Based on a much larger population than the one studied by Morley at Oxford, Amann's work represented a significant advance, and the study remains highly relevant and useful today.

PFRUNGER RIED, GERMANY

The four-year study by Fritz-Bernd Ludescher (1973) of Marsh Tits and Willow Tits in a 50ha birch forest at Pfrunger Ried was a landmark piece of research. The study ran from 1967 to 1970 and is unique in its multi-year comparison of the breeding and spatial ecology of around 10 pairs of each species, both living in the same place. No other study since has directly compared the two species in such detail, and it is still the most comprehensive account of the behaviour of lowland Willow Tits in central Europe.

LUND, SWEDEN

In 1982 a study of Marsh Tits breeding in nest-boxes and natural cavities was established in fragmented woodland over an area of 64km² near Lund, in southern Sweden. In a series of classic papers, Jan-Åke Nilsson and Henrik Smith investigated the causes, mechanisms and consequences of natal dispersal and settling behaviour (Nilsson & Smith 1985, 1988a, Nilsson 1989a, 1989b). Other important papers have focused on nesting ecology (Nilsson & Smith 1988b, Nilsson 1991, Smith 1993a, 1993b, Nilsson & Nord 2017). The wider project is still ongoing in the same study system, and the long-term data are providing important insights into the impacts of climate change on Marsh Tits and other species (Andreasson *et al.* 2023). Professor Nilsson's Marsh Tit papers were very influential in my own studies, and they are essential reading for anyone interested in dispersal.

OULU, FINLAND

Long-term studies of colour-ringed Willow Tits breeding in boreal forests were begun in the Oulu region of central Finland in 1969, by Markku Orell (Figure 1.24) and Mikko Ojanen. Other collaborators provided major contributions, including Kari Koivula, Kimmo Lahti, Jussi Laukkala, Veli-Matti Pakanen, Seppo Rytkönen and Emma Vatka. The research has continued to the present time and has produced a large number of important papers, particularly on breeding

Figure 1.24. Professor Markku Orell inspecting a Willow Tit nest with an endoscope, part of the Oulu study in Finland's boreal zone. (© Anne Laine)

ecology, dispersal and the effects of climate change and forest management (e.g. Orell & Ojanen 1983, Vatka *et al.* 2011, Pakanen *et al.* 2016a, Kumpula *et al.* 2023). The Oulu study has been fundamental to our understanding of northern Willow Tits in Europe, and also as a comparison for Willow Tit and Marsh Tit studies elsewhere in Eurasia.

BIAŁOWIEŻA NATIONAL PARK, POLAND

From the 1980s, Tomasz Wesołowski (Figure 1.25) led annual spring research into the ecology of Marsh Tits in multiple large plots spread across the Białowieża National Park, the last extensive remnant of primeval lowland temperate forest in Europe, set within the wider Białowieża Forest complex (Wesołowski 2007a). This iconic site (pronounced *Bee-awo-vee-eysha* for anglophones) is an example of the type of forest that likely existed in many parts of Europe before deforestation, and so represents the natural habitat that Marsh Tits occupied after the last ice age (see Chapter 3). The Białowieża research has produced dozens of published papers and has provided an essential benchmark for studies elsewhere in Europe and Asia, by showing how Marsh Tits function in undisturbed natural conditions (e.g. Wesołowski 1996, 2015, 2023, Wesołowski & Neubauer 2017). This information tells us what is 'normal' for Marsh Tits, such as their breeding behaviour, population densities, predation levels and dispersal capabilities, and this has been essential for interpreting the results from all other studies.

Figure 1.25. Professor Tomasz Wesołowski at the entrance gate to the Białowieża National Park in Poland, in June 2013. Tomasz's trusty fixed-gear bicycle dates from the Communist era and was still in regular use for fieldwork in the 2020s. (© Richard K. Broughton)

31

Figure 1.26. Dr Marta Maziarz catching a Marsh Tit at a nest in the Białowieża Forest in Poland, in 2015. (© Richard K. Broughton)

The four decades of intensive Marsh Tit fieldwork and analyses at Białowieża have also involved other key researchers, especially Marta Cholewa, Monika Czuchra, Grzegorz Hebda, Marta Maziarz (Figure 1.26), Grzegorz Neubauer and Patryk Rowiński. This research has included an annual census of the entire bird community of the forest for over 45 years (Wesołowski *et al.* 2022). The censuses provide essential context for the Marsh Tit studies with regard to natural densities of their competitors and predators, such as Blue Tits, Great Tits and Great Spotted Woodpeckers. I have spent lots of time in this forest to visit the Marsh Tit study, to search for Willow Tits, and also to collaborate on other woodland bird research (e.g. Maziarz *et al.* 2019, 2020, Broughton *et al.* 2022a). Professor Tomasz Wesołowski died in 2021, and we lost one of the world's authorities on Marsh Tits, the Białowieża Forest and woodland ecology in general.

SIBERIA (MAGADAN)

One of the key studies of Willow Tits in Asia is the paper by Vladimir Pravosudov and Elena Pravosudova (1996), based on four years of fieldwork on colour-ringed birds breeding in natural cavities in northeast Siberia. The research gave a comprehensive account of breeding biology and nestling diet for 22 nests, and is perhaps the most important study from this vast part of the Willow Tit's range in northern Asia. The team later relocated to the United States of America, where Professor Pravosudov leads innovative long-term research on the cognition and behavioural ecology of Mountain Chickadees in California.

MONKS WOOD, CAMBRIDGESHIRE, ENGLAND

In the early 1990s Shelley Hinsley (Figure 1.27) and Paul Bellamy began colour-ringing Marsh Tits in a group of fragmented woodlands centred on the 160ha Monks Wood National Nature Reserve, in eastern England. I joined the project in 2002 and led more intensive colour-ringing and nest-monitoring each year. These data were used to build a spatial database of

Figure 1.27. Dr Shelley Hinsley on an academic visit to the Białowieża National Park in 2007. (© Richard K. Broughton)

Figure 1.28. The author using an endoscope to inspect a Marsh Tit nest low down in a Field Maple tree at Monks Wood, May 2013. (© Richard K. Broughton)

Marsh Tit movements within a GIS (geographical information system) to investigate social organisation, dispersal, breeding success and habitat selection (Broughton *et al.* 2006, 2010, 2011, 2015a). The habitat analyses used novel combinations of Marsh Tit territory locations and remote sensing datasets, particularly lidar (light detection and ranging) and multispectral data (Broughton *et al.* 2012a). Data collection is ongoing and the project is now a long-term study that has run for over 22 years, allowing continued analyses of survival, longevity and other aspects of demography. Monks Wood has been the main focus throughout the project, but the study encompasses several neighbouring woods in the core area. At its maximum extent the monitoring has included up to 67 Marsh Tit territories across 16 local woodlands within a 7km radius of Monks Wood, totalling 518ha of wooded habitat.

By 2024 the Monks Wood study had accumulated 11,200 individual records of Marsh Tit movements in the spatial database, involving over 1,690 colour-ringed individuals and 200 nest-sites (Figure 1.28). As well as colour-ringing observations, we have also undertaken radio-tracking to monitor ranging behaviour in winter (Broughton *et al.* 2014). The Monks Wood research has involved a number of collaborators, particularly Ross Hill, Marta Maziarz and Jane Carpenter, whose EGI doctoral fieldwork was partly based at Monks Wood (Carpenter 2008). The Monks Wood study was also extended to another ancient woodland at Bradfield Woods in Suffolk during 2014–2017, collaborating with Daria Dadam, which involved colour-ringing and territory mapping of around 10 pairs per year.

WIGAN AND ELSEWHERE IN NORTHERN ENGLAND

There have been various studies of Willow Tits in northern England in recent decades, largely as a response to the species' alarming decline in Great Britain and the dearth of basic ecological information for British Willow Tits. Studies led by the RSPB and carried out by Alex Lewis, Elisabeth Charman and Arjun Amar during 2005–2006 provided some information on habitat associations and breeding success in South Yorkshire, Nottinghamshire and Derbyshire (Lewis *et al.* 2007, 2009a, 2009b). These data were also combined in a more wide-ranging doctoral thesis by Finn Stewart (2010), which focused on Willow Tit breeding success and habitat associations in multiple regions across Great Britain, including southern Scotland and central and southern England.

Between 2017 and 2021 the RSPB and the Yorkshire Wildlife Trust also delivered the Willow Tit Recovery Project, led by Sophie Pinder as part of the 'Back From the Brink' initiative (Pinder & Carr 2021). This project was centred on the Dearne Valley in South Yorkshire, and involved trials of habitat management as well as collection of valuable field data, particularly radio-tracking of Willow Tits during winter and early spring. Parallel work was also carried out by the RSPB at its Lake Vyrnwy reserve in north Wales.

The Wigan study was initiated by John Gramauskas in 2012 and later progressed by Wayne Parry, and it has focused on Willow Tits in 596ha of the Flashes of Wigan and Leigh National Nature Reserve, in northwest England. The study has involved annual colour-ringing and nest-monitoring in around 36 Willow Tit territories, generating extensive data on breeding density and nesting success. I collaborated on the project during 2018–2021 to analyse the data for breeding productivity, nest failure rates and spatial analyses of the nest-sites in relation to lidar models of the habitat (Parry & Broughton 2018, Broughton *et al.* 2020). The papers from the Wigan study are perhaps the most comprehensive assessment of British Willow Tits to date, particularly in mosaics of early-successional woodland, wetland and grassland, which has become a key habitat refuge for the declining population in Great Britain (Figure 1.29).

Summary

Marsh Tits and Willow Tits have very large distributions across Eurasia, from Great Britain to Japan. The Willow Tit has a continuous and more northerly distribution than the Marsh Tit, which has a break in its range across central Asia. Both species are curiously absent from Ireland and most of Scotland, although the latter at least could be due to habitat loss caused by deforestation. Their ranges in Great Britain remain unstable, with both suffering serious range contractions over the last century, particularly the Willow Tit.

Figure 1.29. A first-year British Willow Tit at Pennington Flash in northwest England, part of the Flashes of Wigan and Leigh National Nature Reserve that also contains the Wigan study area, home to an important metapopulation of the species. Note that in its post-juvenile moult this bird has moulted one central tail feather, which has a rounded adult-type shape. (© Philip Schofield)

The taxonomy of Marsh Tits and Willow Tits has also been unstable, with genetic studies currently classifying them within the *Poecile* genus alongside other dark-capped, brown tits and chickadees. Numerous subspecies have been proposed, although many appear to have a weak basis that is not supported by genetic analysis. The case for endemic subspecies in Great Britain appears to be quite tenuous, and this requires clarification. Nevertheless, there is a clinal variation of both species across their ranges, with larger and paler birds in northern latitudes and smaller, darker birds in the milder and wetter regions, such as western Europe.

Distinguishing Marsh Tits and Willow Tits from each other can be challenging, but differences in the bill markings, tail gradation, cheek pattern, greater coverts and secondary wing feathers can together identify essentially all birds by appearance alone. Differences in the tail shape and biometrics can also be used to age and sex Marsh Tits with a very high degree of accuracy, although reliable sexing of Willow Tits is difficult, at least in Great Britain.

Both species are quite well studied, with several major and long-term population studies in Europe, some lasting for decades. There are far fewer studies from Asia, and much remains to be learnt about these species across their full ranges.

CHAPTER 2

Communication

Effective communication relies on a signal being exchanged between a sender and a receiver, which might then elicit a reaction. Communication signals mostly operate between members of the same species, and they can involve positive or negative interactions. Communication also happens between different species, such as in mixed flocks or in response to alarm calls.

Marsh Tits and Willow Tits employ a range of vocalisations, postures and displays in their communication systems. Both species have a very complex array of calls and songs, not all of which are fully documented, and their body postures and displays are even less well known. However, there is a lot more research into communication among the closely related Black-capped Chickadee and Carolina Chickadee, and this provides insights into analogous vocalisations and postures of Marsh Tits and Willow Tits.

Some major vocalisations are distinctive enough that we can use them for identification of Marsh Tits and Willow Tits, and to separate them from each other. There is added complexity in that Marsh Tits and Willow Tits show local variation in calls and songs, so they can sound a little different from one region to another. The postures and displays are virtually identical in both species and all populations, however, and are probably common to all *Poecile* tits and chickadees.

Vocalisations

A vocalisation is a single note or a string of notes that are grouped together as a coherent unit. Jack Hailman (1989) reviewed the major vocalisations of tits and chickadees and listed three major groupings: the *chick-a-dee* call, the *gargle* call, and *song*. There are many other calls in the species' vocal repertoires, but these three groupings make up many of the distinctive vocalisations that we hear in the field. The songs of tits and chickadees are quite simple, whereas the chick-a-dee and gargle calls are far more complex and are still being decoded by researchers.

Most information and sonograms for Marsh Tit and Willow Tit vocalisations come from a fruitful era of research from the 1960s to 1990s (Thönen 1962, Romanowski 1978, 1979, Haftorn 1993a, Martens & Nazarenko 1993, Martens *et al.* 1995). Good summaries are also given by Cramp & Perrins (1993) and Harrap & Quinn (1995), but there are few recent studies despite advances in technology for recording and analysis. There are also increasingly large and useful repositories of sound recordings and sonograms from around the world that are open to all (e.g. xeno-canto.org). Much of the following information is based on these sources and my own additional recordings.

CHICK-A-DEE CALLS

The primary vocalisations of Marsh Tits and Willow Tits are based on the chick-a-dee call system, which is shared by all *Poecile* species and was first outlined in detail for Black-capped Chickadees and Carolina Chickadees (Smith 1972, Ficken *et al.* 1978). Chick-a-dee calls are

Figure 2.1. A Willow Tit in the Oulu study in Finland, calling at researchers near its nest. Chick-a-dee calls form the basis of nest-defence and mobbing behaviour. (© Anne Laine)

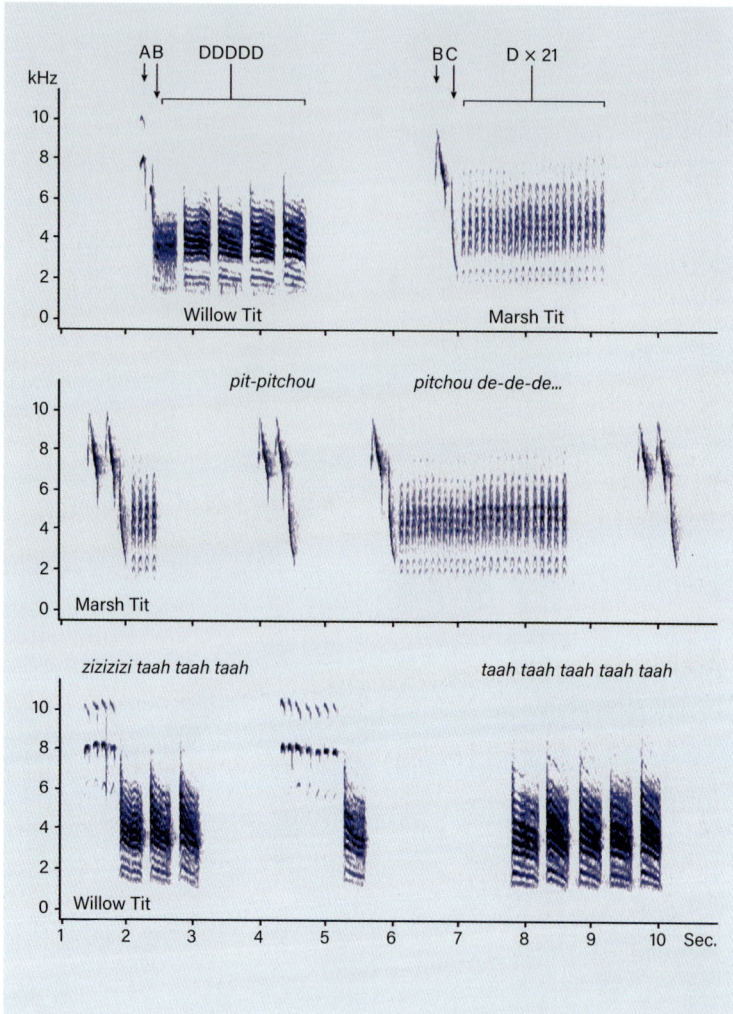

Figure 2.2. Top: sonograms of Willow Tit and Marsh Tit chick-a-dee calls, showing the notes of the A-B-C-D call system. The form of the notes (e.g. the length of the D notes) varies between species. Middle: typical string of Marsh Tit calls, containing *pitchou* notes (B-C or B-B-C) followed by rapid *dee* (D) notes. Bottom: typical Willow Tit calls, with high *zi* (A) notes followed by nasal *taah* (D) notes, or just a series of D notes.

used as contact calls and social calls, and in situations of annoyance or alarm, such as when mobbing a predator (Figure 2.1). The chick-a-dee system is a way of constructing calls using basic rules built around four notes, which were first identified in Black-capped Chickadee calls. These notes have been labelled in the literature as A-B-C-D, and they have a clear parallel in Marsh Tits and Willow Tits (Figure 2.2).

The A notes are simple, peaked chevrons that represent a high *si, zi* or *tsi* sound. B notes are similar but stronger chevrons, peaking at a lower frequency and sounding like a firm *sip*, *tsit* or *pit*. C notes have a lower peak frequency and harder delivery, like a sharp *chik*, *chit* or *chou*. Finally, the very different D notes have a wideband frequency and sound like a buzzing or nasal *dee*, *daa* or *taah*. The A-B-C-D notes can be combined and recombined into unique calls by repeating or omitting certain elements, like words in a sentence. Each unique string of notes encodes information to signal to other birds. For example, a call may involve the string A-A-D-D or A-C-D-D-D, but notes always occur in the same sequential order, where an A note comes before any B or C note, and any D note always comes last.

The chick-a-dee call system shares several elements with human language, including recombinant grouped units, a basic syntax and the economical encoding of information. The system is also generative, where the vocabulary expands with ever-increasing combinations. Hailman *et al.* (1985) recorded a vocabulary of 362 different calls in Black-capped Chickadees, all constructed from just the four A-B-C-D notes, but the potential number of unique calls is essentially limitless.

What the different variations of chick-a-dee call actually mean is still unclear, but Lucas & Freeberg (2007) have reviewed how some patterns are beginning to be decoded. For example, the number of D notes and the rate of delivery in mobbing calls seems to correlate with the degree of the threat. The A notes tend to be neutral and associated with movement, while C notes can denote aggression but also the location of food. Chick-a-dee calls seem to encode information on the identity of the individual, its flock or its population, thereby distinguishing between partners, neighbours and strangers. Call rates might even convey the bird's physiological condition, such as its stress levels and body mass.

Marsh Tits and Willow Tits are less studied than North American chickadees, but the close evolutionary relationship between them means that their basic vocabularies are very similar. The Marsh Tit's chick-a-dee call is composed of a combined B-C unit (or possibly A-C), which makes up the distinctive *pitchuu* or *pitchou* call, followed by a variable number of D notes (Figure 2.2). The full chick-a-dee call may be something like *pitchou de-de-de* (B-C D-D-D), but can include ten or more D notes. The D notes are rarely given without a preceding B-C unit, but often the *pitchou* element is given without any D notes. Sometimes just a *pit* (B note) or *chou* (C note) is given on its own. The D notes are typically brief, each lasting around 0.1 to 0.15 seconds.

The Willow Tit's chick-a-dee call is quite different from the Marsh Tit's, usually comprising a combination of A-D or B-D notes (Figure 2.2), or just the D notes on their own. The full call has been transcribed as *si-si tchay*, or *zi-zi taah*, or *sisi tää* (A-A-D). The initial A notes can be repeated in a rapid, high-pitched string, but are often inaudible from a distance, and so only the D notes are heard. The D notes are typically delivered in a string of two or three, but sometimes seven or more, and are much longer in duration than those of Marsh Tits, at around 0.3–0.5 seconds. This gives the Willow Tit's chick-a-dee call a more nasal quality than the shorter D notes of Marsh Tits.

Marsh Tits can occasionally call with more drawn-out and nasal-sounding D notes like Willow Tits, and Willow Tits can utter a string of shorter D notes that sound a bit like Marsh Tits, but usually the chick-a-dee calls are very distinctive. Furthermore, Willow Tits do not have a call that sounds anything like the distinctive *pitchou* of the Marsh Tit, and so this is reliable for identification. Sooner or later during an encounter a foraging Marsh Tit will give a *pitchou* call, or can be induced to do so with playback of recorded calls. Both species do show an interest in playback of each other's vocalisations, but respond most vigorously to their own.

GARGLE CALLS

The gargle call is a rapid, complex burst of sounds given by *Poecile* tits and chickadees over short distances, usually in specific circumstances of intense excitement or aggression. Gargles have been called 'fighting calls' or 'supplanting calls' (Baker & Gammon 2007), and Averil Morley (1953) referred to Marsh Tit gargle calls as the 'battle-cry'. Gargle calls were named in early studies of Black-capped Chickadees, referring to the complexity of the liquid jumble of bubbling, buzzing and slurred notes produced by the birds. They include a variety of high- and low-frequency sounds, sometimes with short trills, with the full call usually lasting less than a second (Figure 2.3). Individual Black-capped Chickadees have an average repertoire of 7–8 different gargle calls, and there is local variation indicating that neighbours copy or adopt similar versions (Baker *et al.* 2000, Baker & Gammon 2007). Chickadees within each population construct their gargle calls from a pool of 50–60 individual syllables or sounds. This is probably similar for Marsh Tits and Willow Tits, and Haftorn (1993a) recorded a Willow Tit giving three different gargles in rapid succession during a territorial dispute.

Gargle calls are difficult to describe. Ludescher (1973) transcribed a Marsh Tit gargle as *pi-te-wue-wue*, and Morley (1953) wrote another as *ter-tew-ti tu-ti tup-yup-tup*. These attempts give an idea of the complex variability of the calls. Nevertheless, gargles have a distinctive pattern of rapid frequency modulation that swings from high to low frequencies. The Willow Tit's gargle sometimes contains a short burst of a very fast, staccato trill, which can sound almost electronic, and which I have never heard from Marsh Tits.

Gargles are overwhelmingly given by males, although apparently versions have been documented from female Black-capped Chickadees (Hailman 1989). Haftorn (1993a) also recorded female Willow Tits uttering gargle calls when harassed by hungry fledglings. Gargles may be given by male Willow Tits during aggressive encounters, or just prior to copulation, and sometimes when foraging alone or in groups. Birds can give a string of gargles with only short intervals between them. For Marsh Tits and Willow Tits I have most commonly heard gargles between two males involved in a skirmish at their territory borders. I always found these observations fascinating, with the birds full of intense energy and giving lots of calls. The action is so quick that it is often difficult to keep up.

In both Marsh Tits and Willow Tits, gargles are frequently given in response to playback of vocalisations, and the bird will agitatedly flit back and forth around the source of the sound.

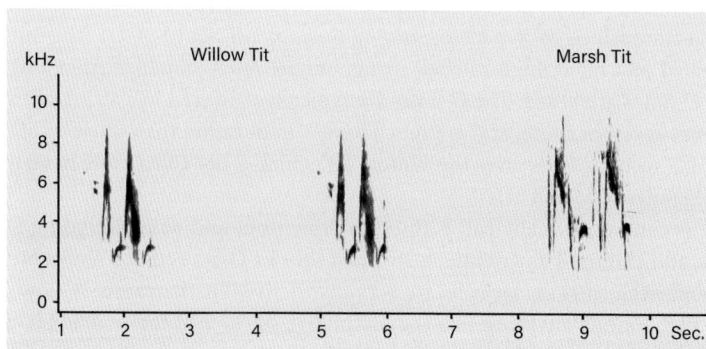

Figure 2.3. Sonogram of two examples of Willow Tit gargle calls, and a single Marsh Tit gargle, containing short bursts of complex notes with rapid frequency modulation.

In Black-capped Chickadees, gargles have also been induced by introducing two unfamiliar males in an aviary setting, where they quickly confront each other to establish dominance. Other circumstances eliciting gargle calls include males meeting in close quarters at bird-feeders, and especially between competing juvenile males during the post-dispersal settling phase in summer.

From these observations, the main function of gargle calls seems to be to test and assert dominance between (usually) males in contentious situations (Baker & Gammon 2007). Gargle calls also seem to encode dominance status within their structure, with subordinate juveniles giving imperfect and plastic gargles compared to dominant adults, and immigrants giving unfamiliar gargles to established residents, using their own local variants. The winner of an aggressive encounter is apparently the male that gives the most gargle calls with the most varied repertoire.

Songs

What we refer to as *song* is really just a distinctive type of vocalisation that is used to advertise or assert the bird's presence in a stylised way, often given from a prominent perch (Figure 2.4). In Marsh Tits and Willow Tits the songs are overwhelmingly given by males, but females do sometimes produce song-like vocalisations. The main type of song structure produced by

Figure 2.4. A male Marsh Tit singing in Wennington Wood, eastern England. (© Richard K. Broughton)

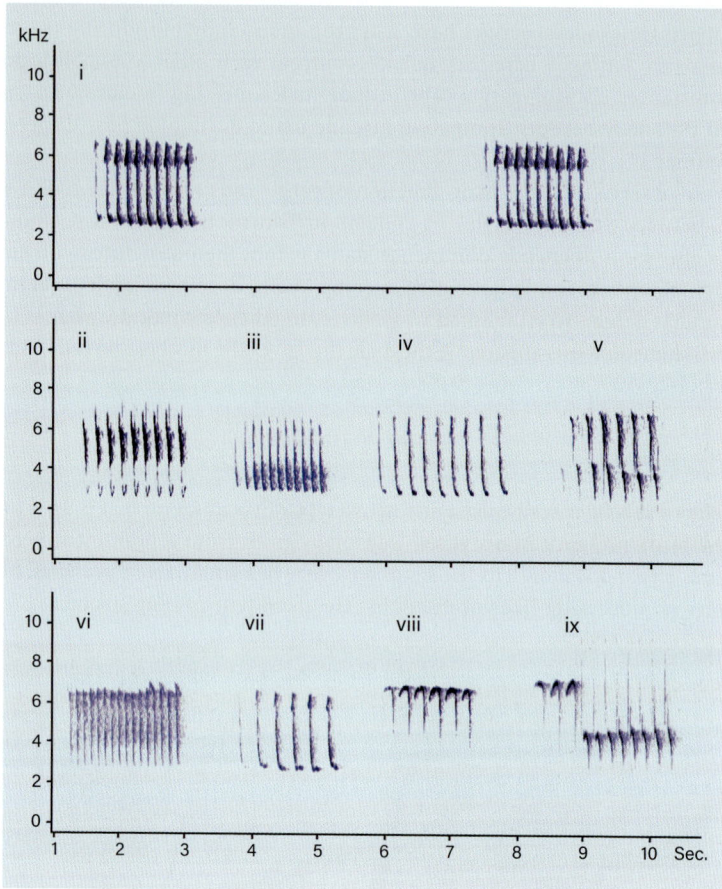

Figure 2.5. Sonograms of Marsh Tit songs from England. Song (i) shows two strophes of a typical monosyllabic song (*schip-schip-schip* …). Strophes (ii)–(iv) are other common monosyllabic songs that vary in tempo; (v) is a disyllabic variant (*wichu-wichu-wichu* …); (vi) is a rapid monosyllabic variant; (vii) is a slow song from a male separated from the female (*tu tu tu* …); (viii) is a rare high-pitched variant (*tse-tse-tse* …); (ix) switches between high and low notes in the same strophe (*tse-tse-tse-chup-chup-chup* …).

males of both species is a loud and simple note or short syllable that is repeated several times in groups, or strophes. These strophes are then repeated after short pauses, together making up a song bout.

Marsh Tits and Willow Tits both have multiple song types, but Marsh Tits have a much wider repertoire. Romanowski (1978) identified 37 different local song variants for Marsh Tits in southern Germany, but only four minor variations for Willow Tits. In the same study individual Marsh Tits were each found to have up to five songs in their repertoire, but I suspect they have more than this. Some Marsh Tit songs are quite similar, but others can be very different in structure and sound, and across their range there must be a very large number of song types.

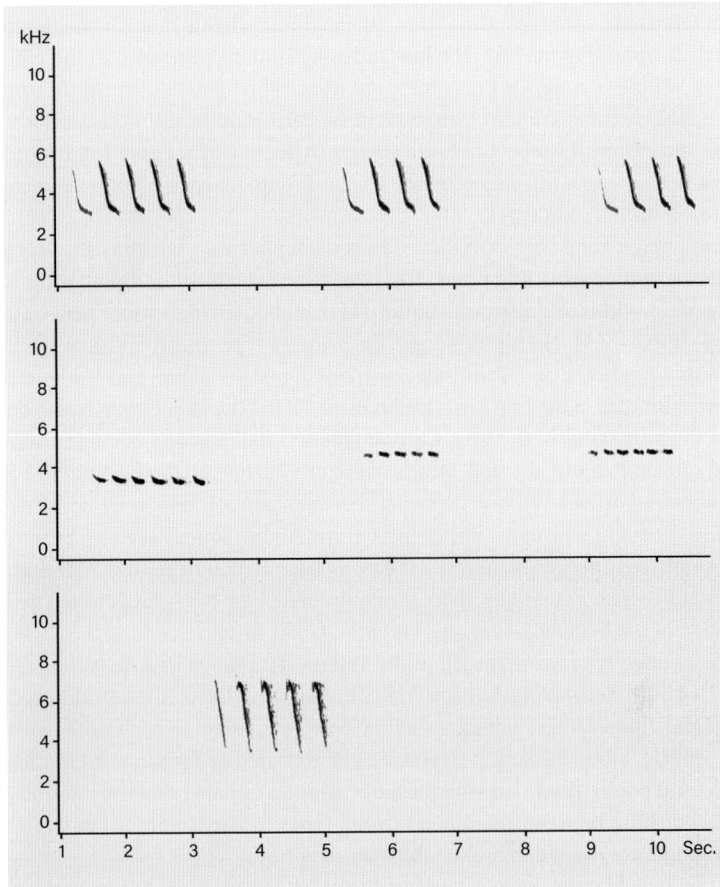

Figure 2.6. Sonograms of Willow Tit songs. Top: a lowland song from England (*tiuu tiuu tiuu* …). Middle: an alpine song from Białowieża, showing one low-frequency strophe (*duu-duu-duu* …) followed by two high-frequency strophes (*hee-hee-hee* …). Bottom: a slow Marsh Tit song variant that resembles the Willow Tit's lowland song, with a similar tempo and a downward-slanting form.

The most common Marsh Tit song is a rapidly repeated monosyllabic note, often described as a ringing rattle, such as *schip-schip-schip-schip*, or *tchip'chip'chip'chip*, or *chi'chi'chi'chi* (Figure 2.5). Each note is usually repeated around 5–12 times in a strophe, at a rate of 6–8 notes per second, but can involve longer, faster or slower strophes. Other common songs involve a repetition of more complex units, such as a disyllabic *wichu-wichu-wichu* or a trisyllabic *chi'pi'we-chi'pi'we-chi'pi'we*. There are many variations on these songs, but they all have a familiar ring to them, somewhere between the tone of a Coal Tit and a Great Tit.

In spite of its extensive distribution across the Palearctic, the Willow Tit produces only four major song types. The *lowland song* is sung by birds across almost all of the species' range from Great Britain to east Asia, although it is unclear if it occurs in Japan. The lowland song is characterised by a descending, slow, melancholic whistle that is usually repeated 2–7 times in each strophe (Figure 2.6). The lowland song is often compared to the piping whistle of the Wood Warbler's song, and is commonly transcribed as a mournful *tiuu tiuu tiuu*, or *tsiu tsiu tsiu*. This is the dominant song type in western Europe and is also very common in Scandinavia and eastern Europe, but becomes less frequent further east. Interestingly, Willow Tits in Great Britain appear to sing only the lowland song, whereas all other populations sing one or more additional song types.

The second major song type of Willow Tits is the *alpine song*, which is a repeated series of around 3–8 even-toned whistling notes. There are two variants of the alpine song, with one at a frequency of 3.5–4kHz and the other about 1kHz higher, giving a more penetrating quality to the whistle (Figure 2.6). The alpine song is the primary type sung by Willow Tits in the Alps, and is the only song heard above altitudes of 1,700m, but the alpine and lowland songs both occur at lower altitudes in the pre-Alps (Knaus *et al.* 2018). The alpine song is also dominant or common in the mountains of the Jura, Carpathians and Balkans, and occurs widely alongside the lowland song in Fennoscandia, central and eastern Europe. In the Białowieża Forest I have heard the same individual Willow Tit switch between the lowland song and then both the high- and low-frequency variants of the alpine song. Alpine songs have also been recorded as far west as south-central France, and as far east as Lake Baikal in central Asia and the Sea of Japan near Vladivostok. Essentially, then, the alpine song type has been detected across almost the entire span of the Willow Tit's range, except in Great Britain.

A dominant song type across much of the Willow Tit's range in Asia is the *Siberian song*, which has also been detected in Europe. The Siberian song looks like an intergrade between the lowland and alpine songs, having a shallow downward slant to each note that terminates in an even whistle. Siberian song types have been detected as far west as Germany and the Netherlands, and populations that sing the Siberian song can also sing the lowland and alpine songs. Distinguishing between the different song types by ear is not always straightforward, and so sonograms are needed to look at the note structure.

Finally, the *Sino-Japanese song* is perhaps the most complex and enigmatic of the Willow Tit's song types. It is similar to the alpine song but consists of alternating high- and low-frequency notes within the same strophe (e.g. *si'pee-si'pee-si'pee*). This song occurs only along the southern edge of the species' Asian range through the Altai mountains, from Kazakhstan and across northern Mongolia, and also in Japan (Martens *et al.* 1993, Harrap & Quinn 1995, Fijen 2015).

Almost all of the Marsh Tit's songs are very different from those of the Willow Tit, except for one. There is a slow variant of the Marsh Tit's monosyllabic song that can sound quite similar to the Willow Tit's lowland song in tempo, frequency and the downward slant of each note (Figure 2.6). This variant seems to be sung by Marsh Tits only in specific circumstances, and in Monks Wood I have heard it from males that are unpaired and trying to attract a female in the spring. It is worth being aware of this occasional variant as a potential pitfall in identifying birds by song alone.

For both species the song bouts may last for only a few strophes, or they may go on for hours as the bird sings on and off while feeding and moving around its territory. In some parts of the day the bird may sing very frequently, especially in the morning, and at others it will sing only occasionally. Males of both species sing throughout the year, but the major song period coincides with late winter and early spring, which in temperate regions (such as Great Britain) peaks in March. Singing is uncommon once breeding has begun in April or May, and it is most infrequent from June to December. An exception is the juvenile dispersal and settling period from June to July, when conflicts between young birds can involve jumbles of song and gargle calls, and may also stimulate a local adult to sing.

One of the primary contexts for male song is to signal and assert territory ownership, which they do by singing loudly to neighbouring territory-holders. The neighbouring males often sing back, and this *counter-singing* can be at close quarters over a few metres across the territory boundary, or at much greater distances of several hundred metres. Sometimes three or more males will engage in counter-singing, with each bird singing in the pauses left by the other birds, showing that they are listening to the information in their rivals' songs rather than just trying to drown them out.

The other main context for male song is advertising to females, and unpaired males sing loudly all over their territory. This kind of singing behaviour becomes increasingly intense as the nesting period approaches in the spring and if the male still hasn't been joined by a female. The males will sing as they forage, pausing to eat a food item, but all the time listening for a responding call from a female that might be dispersing through the area.

Females of both species can give vocalisations that sound a lot like songs, but these are often quieter and given in different situations, so they might not always mean the same thing as a male's song. On several occasions at Monks Wood I have seen females giving soft versions of simple songs when they are separated from their male and want to find him. This usually happens as the female leaves the nest after incubation and the male is not around. The female utters a soft *tu-tu-tu-tu-tu* and the male immediately comes to her (Broughton 2008). The soft song is clearly analogous to the *faint fee-bee* vocalisation of Black-capped Chickadees, which is a subdued version of song that is used in exactly the same context of mate separation (Ficken *et al.* 1978). In Marsh Tits it is probably also the same soft song used by males when separated from their mate (Figure 2.5vii).

Only on three occasions at Monks Wood have I seen a female singing loudly during territorial disputes, engaging in counter-singing with males or with playback of recorded song. One female sang repeatedly over several minutes against a neighbouring male, while her accompanying mate only called but did not sing. The female's song was a little less strident than a male's, and sometimes had a slurred element such as *tu-tu-tu-tu-tsup* (Broughton 2008).

Foster & Godfrey (1950) reported singing by a female Willow Tit in the late spring and summer, which may have involved locating a male or territorial disputes, but this was not clear. In Scotland it took Jimmy Maxwell several years of fieldwork before confirming that female Willow Tits can produce the lowland song, but it was clearly quite rare and the context unclear. Haftorn (1993a) also found that only on rare occasions do females sing like males.

OTHER CALLS

Marsh Tits and Willow Tits have other distinctive calls that are used in recognisable situations, and these are analogous to calls in the published repertoires of other chickadees. Both species have a *hawk alarm* call that is given in response to an aerial predator, such as an attacking Eurasian Sparrowhawk. For Marsh Tits, Morley (1953) described this alarm call as an explosive *pitz* note, sometimes extended to *pitz-itz-itz*, and that's also how it sounds to me. For Willow Tits, Foster & Godfrey (1950) described the hawk alarm call as a sharp low *zut* note, and this seems to be the same as the *spitt* alarm call described by Haftorn (1993a), which Thönen (1962) called the 'scare call'.

The hawk alarm call is accompanied by the bird and any companions within earshot diving straight down into low cover. The birds drop like stones from wherever they are in the tree canopy into dense vegetation near the ground, to hide from possible attack. Birds might also respond to the hawk alarm call by freezing motionless in the tree canopy, sometimes for several minutes, only blinking or tilting their heads to scan for the danger. This tactic relies on having not already been spotted by the predator. During fieldwork in Monks Wood when I have flushed Eurasian Woodcocks these have also elicited hawk alarm calls from Marsh Tits, perhaps because of the woodcock's sudden break from cover and rapid flight through the trees. Marsh Tits obviously find this very frightening. Averil Morley (1953), who was studying Marsh Tits in Bagley Wood at Oxford during the Second World War, commented that Marsh Tits did not respond to formations of Allied bomber aircraft roaring low over the wood, which Averil herself found quite alarming.

Another alarm call used by both species is the *high zee*, which warns of a potential predator approaching or nearby (Figure 2.7). The high zee is a single high-pitched note that is repeated several times, and its acoustic structure makes it difficult for predators to localise the sound, giving the caller some degree of protection while warning its companions. The high zee can have a similar effect to the hawk alarm call of sending birds diving into cover or freezing motionless (Ficken & Witkin 1977, Haftorn 1993a). Jimmy Maxwell (unpublished) noted the high zee given by Willow Tits in response to Eurasian Sparrowhawks, and I heard it many times from Marsh Tits at Monks Wood in similar scenarios to those described for Willow Tits and chickadees (Smith 1991, Haftorn 1993a). The high zee can be given in response to avian or mammalian predators, and it may vary in pitch and duration, which for Black-capped Chickadees was related to different predators. Haftorn (1993a) commented on how Willow Tits would give this alarm call in response to low-level threats or harmless surprises, such as a non-predatory bird or a human observer, but that the response varied according to the severity of the danger.

A very frequent vocalisation is the *contact call*. These are brief, high-pitched notes used over short distances between members of a pair or flock, and are based on A or B notes from the chick-a-dee system. The contact call has been referred to as the *sit* call for Willow Tits (Haftorn 1993a), *sip* for Marsh Tits (Morley 1953) and *tseet* for Black-capped Chickadees (Smith 1991). To my ear the contact call is a slightly harder *tsip* for Marsh Tits and a softer *sih* for Willow Tits, but the main point is that they are all variations on a simple high-pitched, chevron-shaped note. The contact call is given frequently while the birds are foraging and moving around, and, as the name implies, it seems to be aimed at keeping contact between individuals to enable them to stay together. Over longer distances, Marsh Tits also use the *pitchou* B-C notes unit as a contact call, whereas Willow Tits use the *tchay* or *tää* D note or the A-D combination of *sisi-tää* (Haftorn 1993a, Suzuki 2012).

Related to contact calls are *flight calls*, given when a bird signals to its companions that it is going to move some distance by flying off, and also when it is arriving (Figure 2.8). In Willow Tits the flight call is a thin *sie* note, and in Marsh Tits it's a harder version of the *tsip* contact call (Figure 2.7iv–v), which can be quite noticeable at short to medium range. When waiting to catch Marsh Tits at feeder traps in Monks Wood I would often be alerted to a bird arriving on the scene by these flight calls.

A call that is restricted to spring, but is heard very commonly, is the *broken dee*, used only by females to demand food from the male during the egg and nestling periods. The broken dee is acoustically similar to the calls of fledglings when demanding food from their parents,

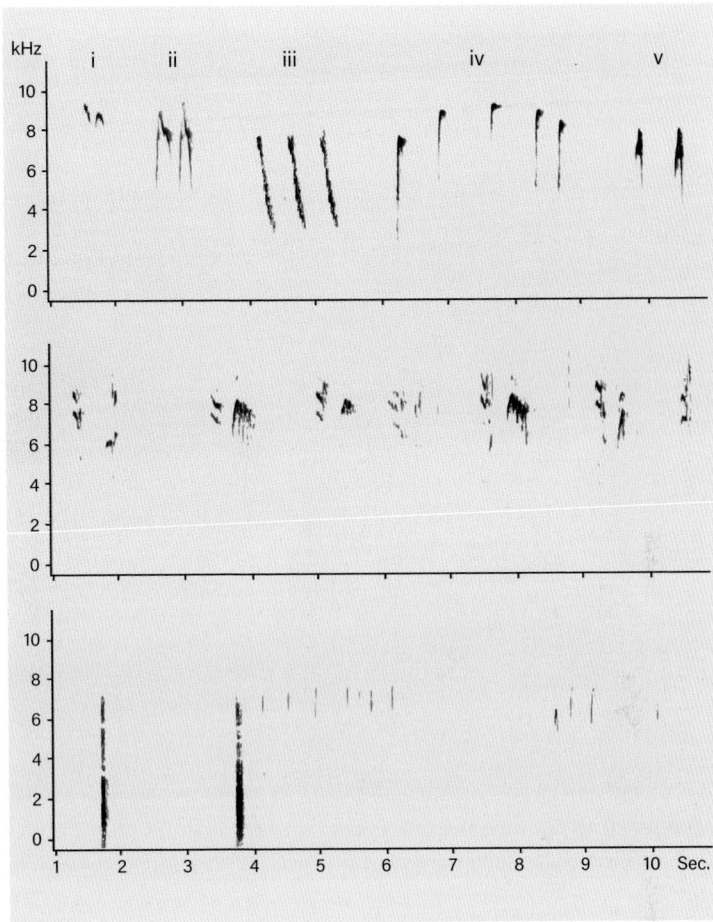

Figure 2.7. Sonograms of Marsh Tit calls from England. Top: call (i) is a 'high zee' alarm call; (ii) (*pit-pit*) and (iii) (*zrrri-zrrri-zrrri*) are agitated calls used in alarm or antagonism; (iv) and (v) are *sip* and *tsip* contact calls. Middle: extract from a series of broken dee calls from a female Marsh Tit demanding food from the male. Bottom: two squawk calls of a female Marsh Tit in the nest, followed by simple peeps (soft high calls) from six-day-old nestlings.

and is distinctive for each species. The Marsh Tit's broken dee sounds like a creaking, swinging door hinge, and is delivered as a long string of variable units that can go on for 30 minutes or more (Figure 2.7). Morley (1949) transcribed a string of Marsh Tit begging dees as *chick-erick, sik-ik, su, chick-er-ick, tik-sik-sik, su …*, and another as *see-pee-see, see, see-persee, see, see-pe …*, and these encompass the variability in the calls (Broughton 2012a).

The female Willow Tit's broken dee is a bit slower, deeper and more musical than the Marsh Tit's, with Foster & Godfrey (1950) transcribing it as a repeated *eee-da-da*. As with Marsh Tits, the Willow Tit's broken dee call is given in a long series while the pair are foraging and the female solicits food from the male. A sound recording and sonogram of a female Willow Tit's broken dee is available on the xeno-canto website (Malengreau 2020).

Figure 2.8. When setting off from one area of activity to another, Marsh Tits (shown here) and Willow Tits give flight calls to advertise their intentions and movements. (© www.garthpeacock.co.uk)

Vocal development

There is limited information on the development of vocalisations in Marsh Tit and Willow Tit nestlings, but detailed studies of Black-capped Chickadees help to fill in most of the gaps in our knowledge. The following information is largely taken from studies of Black-capped Chickadees by Clemmons & Howitz (1990), Clemmons (1995) and Baker *et al.* (2003), of Willow Tits by Haftorn (1993a), and my own study of Marsh Tits (Broughton 2019). The early development of calls seems to be very similar for all three species.

There is currently no evidence that the embryos of Marsh Tits or Willow Tits call or communicate before hatching, unlike some other birds, but this cannot be ruled out. However, nestlings are certainly vocalising on the day of hatching, producing simple *peep* calls. These calls are stimulated by the movement of the parents or other vibrations, and are given as the nestlings raise their heads and open their bills while gently swaying from side to side. Within a few days the calls are strongly associated with feeding visits, as the nestlings raise their heads and call when an adult arrives, subsiding when the parent leaves. During the first week or so after hatching, before the nestlings' eyes have opened, the parents may also stimulate their gaping with a *squawk* call, which is a brief wideband sound intended to alert the nestlings that food has arrived (Figure 2.7).

kHz
10 Marsh Tit

8

6

4

2

0

10 Willow Tit

8

6

4

2

0

1   2   3   4   5   6   7   Sec.

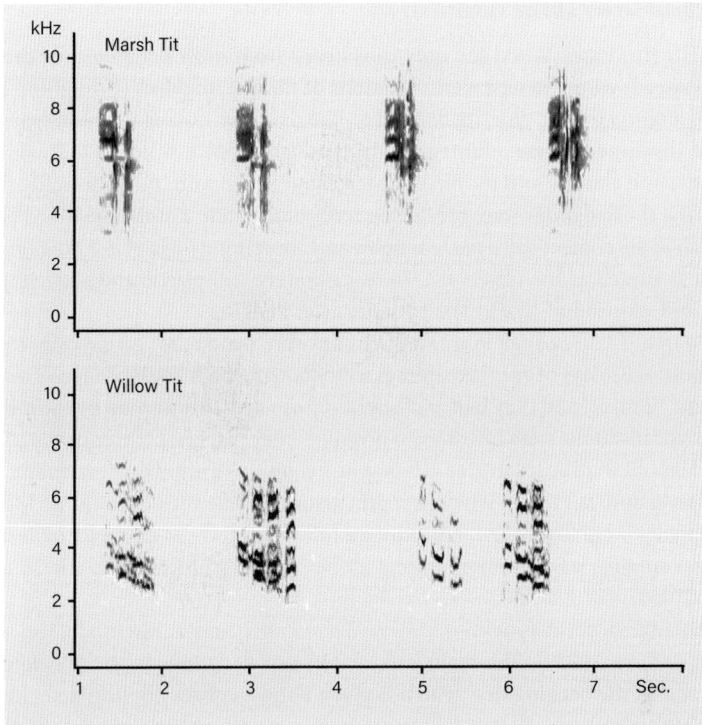

Figure 2.9. Sonograms of 'begging dee' calls of Marsh Tits and Willow Tits on the day of fledging. For Marsh Tits each begging dee is shorter, higher-pitched and more sibilant (*sur-dit*) than for Willow Tits, which are a descending series of musical notes (*dee-doo-der-der* and *dee-doo-der* in this series).

The peep calls are like proto-A notes, but over the first two weeks of life the simple peeps evolve into a more acoustically complex call known as the *begging dee*. By 13 days of age young Marsh Tits are producing begging dees with a rapid sibilant structure, sounding something like *sur-dit*, *sur-didit* or *sur-didud*, sometimes extended into a very rapid *sissississud* (Broughton 2019). The begging dees of Willow Tit nestlings are very different from Marsh Tits (Figure 2.9), and in both species these calls differ from those of all other tits and chickadees. In contrast to the Marsh Tit's short and sibilant begging dee, the Willow Tit's are slower and more musical as they descend in pitch, being transcribed as *dee-doo-der*, or *dee-doo-day*, or *pii-pää-pää*, with variations containing 2–4 notes.

Almost all of the nestling calls are involved with demanding food, the only exception being defensive or anti-predator calls. I once heard a brood of 15-day-old Marsh Tits at Monks Wood giving a growling hiss when I looked into their nest (Broughton 2005). A more common defensive call of large nestlings is the *squeal*, which is given when they are removed from the nest and handled for ringing or, presumably, when grabbed by a predator. Squeals are brief, harsh noises that have been recorded from Black-capped Chickadees and Boreal Chickadees, and apparently serve as a 'startling' device to make a predator hesitate or drop the nestling.

Vocal development after fledging

For Marsh Tits in Monks Wood the only calls heard from most fledglings for the first three or four days after leaving the nest were variations of the begging dees (Broughton 2019). One fledgling did attempt a basic chick-a-dee call on the day after leaving the nest, which sounded squeaky and disjointed. These adult-type calls gradually appear over the following days, but they are plastic and slurred, as if the birds are learning how to form the sounds. By the fifth day outside the nest the fledglings were producing recognisable attempts at various calls in specific contexts, such as *sip* contact calls, *pitchou* notes and churring strings of *dee* notes at a potential threat (such as myself as the observer). These calls were still plastic and slurred but became crystallised and more adult-like by the 10th day after fledging.

Several Marsh Tit fledglings in a family group were producing gargle calls 15 days after leaving the nest, at 35 days of age. The siblings were showing high levels of aggressive posturing, and even some fighting, and they had all dispersed from the family group by the following day. Further new vocalisations were heard from juveniles after dispersal. Three newly independent males at 31 days old were heard to give *subsong*, an extensive string of quiet warbling containing a jumble of song and call notes, which is a precursor to adult vocalisations. At Monks Wood the juvenile males were first heard to produce a full song at 31–33 days old, or 1–2 days after dispersal, and singing was common among all newly independent males by 38 days of age (Broughton 2019).

The post-fledging development of Willow Tit vocalisations is essentially the same as for Marsh Tits (Haftorn 1993a). The main call heard from fledglings in the first few days after leaving the nest is the begging dee, which is given almost constantly. The fledglings can also produce the high zee alarm call, and will freeze in response to an aerial predator. I also saw this in Marsh Tit fledglings at Monks Wood on the fourth day out of the nest, when they all froze for more than five minutes in response to a sparrowhawk passing overhead.

A week after fledging, the young Willow Tits of both sexes begin uttering subsong at around 26 days old (Haftorn 1993a). If one fledgling begins the quiet warbling this can stimulate its siblings to join in, resulting in several birds creating a soft chorus of subsong as they forage. Haftorn described how recognisable song or call elements could be heard in the subsong, and these gradually became crystallised into proper calls. The adult-type calls eventually become dominant by around two weeks after fledging, just before dispersal and independence.

Haftorn (1993a) recorded plastic gargle calls and song from Willow Tit fledglings by 30 days old, and these were more crystallised and widespread from 32 days, often accompanied by aggression between siblings. A very similar progression has been documented for Black-capped Chickadee fledglings, with initial begging calls, subsong and plastic calls crystallising into full calls, gargles and song after around 30 days of age (Clemmons & Howitz 1990, Baker *et al.* 2003).

Displays and postures

Marsh Tits and Willow Tits use an array of physical displays and postures during communication, often in combination with vocalisations. Postures and displays are stylised movements that are recognisable and repeated in specific contexts. Displays appear to be innate and develop alongside vocalisations, with recently fledged juveniles engaging in lots of posturing during the period when their vocalisations are crystallising. Susan M. Smith (1991) summarised eight displays of Black-capped Chickadees using the terminology of Susan

Figure 2.10. Two male Marsh Tits engaged in the 'ballet display' in the Monks Wood study. As the birds escalate the confrontation the one on the left uses a body-ruffling display and the one on the right adopts the bill-up posture. (© Richard K. Broughton)

T. Smith (1972), who described 14 displays for Carolina Chickadees. Most or all of these will probably have analogous displays in the Marsh Tit and Willow Tit, but the species have been relatively unstudied in this regard. Detecting and describing displays can also be difficult because of the very brief nature of many, consisting of a rapid change in posture or behaviour that may be hard to see. Recording the birds and watching back in slow motion is usually the only way to see them properly, but this also has its limitations in the field. Nevertheless, the studies by Hinde (1952) and Morley (1953), and the review by Cramp & Perrins (1993), describe displays of both species that seem to be the same as those described for chickadees by Smith (1972) and Smith (1991).

The *bill-up display* involves the bird pointing its bill directly upwards, apparently showing its black bib patch to an opponent (Figure 2.10). The bib patch is a status signal in Willow Tits (Hogstad & Kroglund 1993), with bigger patches conveying a higher dominance rank, and the bill-up display seems intended to flash this signal to intimidate rivals. The same is probably true of Marsh Tits, which also use the bill-up display. Sometimes the wings may be fluttered at the same time, or the bird may orient its entire body vertically, but it is usually silent. The bill-up display is very quick, often lasting only a moment. Great Tits use much more blatant bill-up displays to intimidate their rivals, and in this species they are easier to observe (Hinde 1952).

The *gape display* is also given as a deterrent to others, as the bird leans forward with its bill wide open, and this may be followed by a gargle call. Gape displays can often be seen at bird-feeders, where conflict is frequent among birds attracted to a small space. Hinde (1952) reported a milder form of the gape display in which Marsh Tits and Willow Tits leaned forward to intimidate an opponent without opening the bill, which was termed the 'head-forward posture'. I have also seen both species just open their bill at a rival in close proximity, without changing posture, such as when a conspecific or bird of another species arrives at a feeding station. The message of the gape display is a simple 'stay away' or 'back off'.

The *ruffled-crown display* is another very common antagonistic posture. The black crown feathers are raised for a short period, typically when the bird feels threatened or alarmed. Both species will also raise the crown feathers when being held for ringing and measuring, and will also open their bill to gape at the handler. A more extreme posture is the *body-ruffling display*, where the feathers on the back, mantle, scapulars and rump are all raised and puffed out, along with the crown, cheeks and breast feathers, to make the bird look as big as possible. During this brief display the bird might droop its wings, spread its tail and lean forward towards an opponent, or face them sideways (Figure 2.10). This may be followed by a gargle call or a direct attack. Ludescher (1973) thought the body-ruffling display was used defensively by Willow Tits, but Piaskowski *et al.* (1991) showed more convincingly for Black-capped Chickadees that it is an aggressive display.

The *sleeking display* is a submissive posture where the feathers are sleeked against the body to give the bird a slender elongated appearance, usually as it leans away from the threat of a dominant bird. The *wing flick* is used by hesitant or anxious birds when intimidated by a rival, as a momentary raising of the half-open wings. Morley (1953) mentioned how Marsh Tits make an audible noise with a rapid 'flirt or vibration' of the wings, which was perhaps the wing-flick display described for chickadees.

An intricate posturing behaviour is the *ballet display* (Smith 1991), which occurs during intense encounters. Two or more birds at close quarters rapidly change position and orientation around each other as they move between perches, with lots of calling and posturing. The dominant bird usually faces the subordinate one, which faces away, with both birds employing the bill-up, body-ruffling and other displays (Figure 2.10). In territorial disputes this ballet display is highly stylised and very noisy when two males or pairs meet at a territory boundary. The males engage and escalate through the branches as they 'dance' upwards, displacing and chasing each other from perches with lots of posturing, exaggerated movements, gargle calls and other excited calls. The females, if present, usually watch from nearby and give occasional calls. The ballet may end in a physical fight, with the two males grappling with their claws and pecking at each other, but usually it cools down into displacement or distraction activity, with one or both birds pecking at tree bark or buds. Eventually one of them leaves and the ballet is over.

Displays related to breeding or nesting include the *butterfly display* in early spring, when males float between perches in an exaggerated flight with slow wingbeats, over or around the female. The purpose is unclear, but it may be related to the timing of copulation. The *nest-site showing display* occurs during early spring when the pair are searching for potential

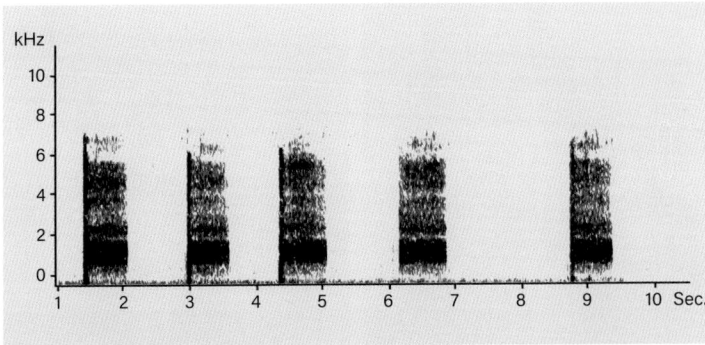

Figure 2.11. Sonogram of five hiss calls (*psshh*) of a female Marsh Tit inside the nest-cavity, given during the hissing display in response to a potential predator at the entrance. Recorded at Monks Wood.

nest-sites. On finding a possible candidate the displaying bird clings to the entrance and looks over its shoulder and inside the cavity, as if showing it to its partner.

Extended periods of *wing quivering* are given in a variety of scenarios, and it can signal a demand or anxiety. Females and fledglings quiver their wings to demand food from the male or parent while giving the food-demanding calls. Females also quiver their wings and tail while leaning forwards to solicit copulation. Both adults quiver their wings when anxious or alarmed outside the nest when a potential threat is nearby or preventing them from entering the cavity.

Other anti-predator behaviours around the nest are discussed in Chapter 6, including the *nest-site distraction display*, which is a dramatic and desperate behaviour where adults try to drive away a predator from the nest. The bird leans forward with its crown and cheeks ruffled, wings fully raised and tail spread, and swaying from side to side while giving a hissing call, sometimes followed by a diving attack. Females cornered inside the nest-cavity by a predator can also giving the *hissing display*, which Smith (1991) called the *snake display*. This is a violent defensive action by the female from within the nest, who sways from side to side with a gaping bill before repeatedly lunging forward towards the nest entrance while flicking her wings, snapping her bill closed and giving a loud hissing call (Figure 2.11). The hiss is a long (0.5–0.7 second) wideband note like a 'white noise' *psshh* sound. The hissing display is given by all tits and chickadees and has been suggested as a form of Batesian mimicry, where the birds are imitating the hissing of a dangerous snake in order to deter a predator from entering the nest (Sibley 1955).

Like Sibley (1955), however, I am not convinced that the hissing display is snake mimicry, as the birds give hissing calls in other situations, such as the nest-distraction display. I once even had a Marsh Tit hissing from within a cotton bag where it was briefly being held during ringing. Marsh Tit nestlings (and possibly Willow Tit nestlings) also hiss or growl when threatened, but without performing any kind of lunging display. The nestlings of some dome-nesting or open-nesting species, such as Wood Warblers (Maziarz *et al.* 2019), also give hissing calls to a potential threat. I suspect that the hiss of a bird or a snake is just an unpleasant blast of noise that can deter some predators, which various species have converged upon without mimicry necessarily being involved.

Another distinctive posture, not mentioned by Smith (1991), occurs when males are giving territorial song. They stand quite erect with the bill tilted upwards and the body and crown

Figure 2.12. The sitting (left) and singing posture (right) of a male Willow Tit, the latter showing the upright stance with head slightly raised, which expands and emphasises the black throat patch. Singing Marsh Tits adopt the same posture. (© Richard K. Broughton)

feathers sleeked. Meanwhile, the cheeks and especially the throat are enlarged to emphasise the black bib, probably as a plumage signal of dominance in tandem with the acoustic message of the song (Figure 2.12).

Summary

Marsh Tits and Willow Tits have a complex system of acoustic and visual communication, which is broadly shared across all *Poecile* species. The major vocalisations include the chick-a-dee complex and the gargle calls, which appear to encode information in the number, type and intensity of the notes they contain. The chick-a-dee calls of Marsh Tits and Willow Tits are also distinctive from each other, particularly the *pitchou* call of the Marsh Tit and the nasal *tchay* or *taah* notes of the Willow Tit, and these can be used for species identification.

Songs are another major component of the vocal repertoire that are used to signal territory ownership and mate availability. Songs are mostly very different between the two species, and Marsh Tits have a particularly large song repertoire among individuals and populations. Willow Tits have only four major song variants across their range, just one of which is known from Great Britain.

In both species the first vocalisations appear at or around hatching, beginning as simple sounds that increase in complexity as the nestlings grow. Early vocalisations are associated with soliciting food from the parents, but these quickly develop into a greater diversity that gradually crystallises into adult-type calls around independence. Juveniles can be singing fully formed songs when only a month old.

Songs and calls are often accompanied by visual displays and postures to emphasise areas of the plumage, such as the black throat patch. Many displays are used for intimidation or submission between conspecifics or other species in mixed flocks. Some specific displays and calls are used to repel predators attacking the nest.

The full vocal repertoires and visual displays of both species have not been documented, and there is a large amount of variation even within the better-known types. Investigating these systems of communication adds a fascinating dimension to our attempts at understanding these species and how they perceive each other and their surroundings. There is still a great deal for us to learn about how these birds communicate.

CHAPTER 3

Habitat

Marsh Tits and Willow Tits are woodland birds, like all other tits and chickadees (Figure 3.1; Snow 1954). The habitats used by Marsh Tits and Willow Tits reflect their preferences but also the local opportunities available to them, including the constraints of climate, vegetation, forest management, habitat fragmentation and interspecific competition.

The terms *woodland* and *forest* include a wide spectrum of habitats, from northern boreal forests and temperate broadleaved woodlands to dry Mediterranean forests and shrublands, as well as plantations, orchards, wood-pastures and a range of other managed or unmanaged semi-natural woodlands. The word 'forest' has a specific origin in Great Britain, referring to a medieval hunting ground of kings, but it is more widely used to mean a large expanse of woodland. Oliver Rackham (2006) explained how the terms 'woodland' and 'forest' are now used interchangeably, but with a tendency towards 'wood' for smaller patches of trees, 'woodland' for larger expanses and 'forest' for landscapes dominated by tree cover. This is generally the way I also use these terms.

More importantly for the tits than terminology, David Snow (1954) highlighted a major distinction between coniferous and deciduous woodlands, which he considered as the most important factor in habitat selection between species. Conifer-dominated woodlands and forests are favoured by some tits, such as Crested Tits, whereas deciduous or broadleaved habitats are favoured by others, like Marsh Tits and Great Tits (Table 3.1). The situation is more complex than a simple conifer–broadleaf dichotomy, however, as species can occur outside of their core habitat, although usually at a lower density.

A species' core habitat can also vary in different parts of its range. Willow Tits and Coal Tits strongly favour conifer-dominated habitats in some regions, but in others they are common in pure broadleaved woodland too (Wesołowski *et al.* 2018, Broughton *et al.* 2019). Understanding this variation in habitat use is important for interpreting the current distribution, preferences or tolerances of Marsh Tits and Willow Tits across their ranges.

Recognising this complexity, Snow (1954) summarised the primary habitat of Marsh Tits in Eurasia as temperate broadleaved woodland, but overlapping with drier Mediterranean forest to the south and colder boreal and mixed forest to the north (Table 3.1). The main habitat of the Willow Tit was described by Snow as coniferous and mixed forest, including boreal or taiga in the north and montane forest in the south, and also the birch zone north of the taiga. Snow also noted the Willow Tit's association with pure deciduous, wet or early-successional wooded habitats in western Europe.

Wooded habitats vary in age, structure and species composition within the broad climatic regions, and this determines the niches available to Marsh Tits and Willow Tits. Most wooded habitats have been fundamentally changed by human activity through deforestation, fragmentation, exploitation and management, and this is one of the overriding factors now determining the habitat quality for Marsh Tits and Willow Tits. Habitat quality is also influenced by wider landscape factors, such as the connectivity between woodland patches along hedgerows, tree lines and wooded riverbanks.

Figure 3.1. Willow Tits, along with Marsh Tits, are woodland birds that spend almost their entire lives in the cover of trees and bushes in wooded habitats. (© Philip Schofield)

All of these elements are prominent in European landscapes, which have long been heavily deforested and the remaining woodlands intensively managed. In Great Britain many of the natural Holocene habitat types that Marsh Tits and Willow Tits would have lived in, such as temperate rainforest, primary lowland forest, floodplain riverine forest and montane birch forest, have been almost completely destroyed over past millennia. What remains of the woodland habitats is heavily fragmented and degraded (Rackham 2006, Smout *et al.* 2007, Shrubsole 2022).

As a result of the huge impacts of people, the habitats now occupied by Marsh Tits and Willow Tits are a result of where populations have persisted after woodland change and fragmentation. Consequently, some caution is needed in interpreting habitat preferences based on what we can see today, especially in intensively managed landscapes like Great Britain. In many places the apparent pattern of habitat selection is a consequence of where the birds have been able to survive, despite human activity, and not necessarily the habitats they prefer.

Table 3.1. The main habitats of Marsh Tits, Willow Tits and related *Poecile* species in Eurasia, along with other sympatric tits that overlap substantially with Marsh Tits and Willow Tits in their ranges.

Species	Region	Forest ecosystem	Main woodland habitats
Marsh Tit *Poecile palustris*	Europe & east Asia	Temperate	Mature broadleaved, mixed
Willow Tit *Poecile montanus*	Eurasia	Temperate–Alpine–Boreal	Broadleaved, conifer, mixed, riverine, thicket
Siberian Tit *Poecile cinctus*	Northern Eurasia	Boreal	Northern conifer, boreal birch
Sombre Tit *Poecile lugubris*	Eastern Mediterranean	Mediterranean	Open broadleaved, wooded valleys
Caspian Tit *Poecile hyrcanus*	Caspian Sea	Hyrcanian (south Caspian)	Mixed
Black-bibbed Tit *Poecile hypermelaenus*	Southeast Asia	Temperate–Alpine	Broadleaved, conifer, mixed, montane
Sichuan Tit *Poecile weigoldicus*	Southeast Asia	Temperate–Alpine	Broadleaved, conifer, mixed
Père David's Tit *Poecile davidi*	Southeast Asia	Temperate–Alpine	Mixed, montane
White-browed Tit *Poecile superciliosus*	Southeast Asia	Alpine	High shrub zone
Blue Tit *Cyanistes caeruleus*	Europe, western Asia	Temperate	Broadleaved, mixed
Azure Tit *Cyanistes cyanus*	Eurasia	Temperate–Boreal	Mixed, open, riverine, thicket
Great Tit *Parus major*	Eurasia	Temperate	Broadleaved, mixed
Japanese Tit *Parus minor*	East Asia	Temperate	Broadleaved, mixed
Coal Tit *Periparus ater*	Eurasia	Mediterranean–Temperate–Alpine–Boreal	Mature broadleaved, conifer, mixed
Crested Tit *Lophophanes cristatus*	Europe	Mediterranean–Temperate–Alpine–Boreal	Mature conifer, mixed
Varied Tit *Sittiparus varius*	East Asia	Temperate	Broadleaved, conifer, mixed

Sources: Snow 1954, Harrap & Quinn 1995.

What are 'natural' forests?

The natural habitats of Marsh Tits and Willow Tits are the forests that recolonised Europe and northern Eurasia when the climate warmed after the last ice age, around 16,000 years ago (Kirby & Watkins 2015). As the ice retreated the forest expanded from refugia around the Mediterranean and Black Seas, and many woodland birds followed as it crept northwards (Pavlova *et al.* 2006). However, Willow Tits apparently recolonised Europe from populations that expanded from east Asian forests that persisted throughout the ice age (Kvist *et al.* 2001), and also from ice-free river valleys in northern Eurasia (Pavlova *et al.* 2006). Marsh Tits probably also spent the last glaciation in Asian refugia and recolonised Europe from the east (Tritsch *et al.* 2017).

Europe's forests reached their greatest extent around 7,000 years ago as tree-covered landscapes that were interspersed with open habitats (Kirby & Watkins 2015). Old-growth forest is a climax vegetation typified by abundant large trees, high species diversity and huge amounts of deadwood, which Oliver Rackham (1976) popularised as the 'wildwood', unaffected by human civilisation. Primeval forests in the lowlands and river valleys would also have been much wetter than modern woodlands in drained landscapes, with lots of seasonal flooding and standing water (Figure 3.2). Tansley (1939) suggested that Great Britain was largely covered by this primary forest before clearance for farming, which would have meant vast areas of habitat for woodland birds. However, Vera (2000) argued that abundant large herbivores across Europe would have maintained a mosaic of open grassland, shrubland and mature forest in a dynamic equilibrium, analogous to modern-day wood-pasture. A good account of this debate is given by Kirby & Watkins (2015) and Peterken (2023).

Figure 3.2. Undisturbed near-primeval forest at Białowieża in eastern Poland, showing high species and structural diversity among the trees, lots of standing and fallen deadwood, and lots of water. (© Richard K. Broughton)

Rather than being a purely academic discussion, it is essential to consider how natural forests existed in the landscape to better understand the ecological needs of woodland species like Marsh Tits and Willow Tits. It seems probable that a spectrum of wooded landscapes existed in Europe and Asia, from extensive old-growth forest to open wood-pasture and shrublands. The original forest was unlikely to have been completely untouched by people, who would have influenced woodland from the early Holocene through fire, hunting and other resource use (Kirby & Watkins 2015, Peterken 2023). This early influence means that the concept of 'naturalness' is subjective, but may be considered as a gradient where the most natural woodlands have not been directly influenced by people on a large scale (Rackham 2006, Peterken 2019). At one extreme would be a primeval primary forest, and at the other a wholly artificial plantation, with varying degrees of managed and semi-natural woodlands in between. On this spectrum, Peterken (2019) argues that managed or planted woodlands can increase in their naturalness as time passes from the last human intervention, such as felling or planting. This increased naturalness is likely to benefit Marsh Tits and Willow Tits as degraded or planted woodlands become rewilded, or wilder, and converge with their original habitats.

When considering the range of wooded habitat types that Marsh Tits and Willow Tits occupy across their large ranges, and in which habitats they do best, it is useful to consider the broad forest biomes. In Eurasia the greatest relevance to Marsh Tits and Willow Tits are the boreal, temperate deciduous and Mediterranean forest biomes (Figure 3.3), and these contain all of the woodland habitat variants that both species currently inhabit.

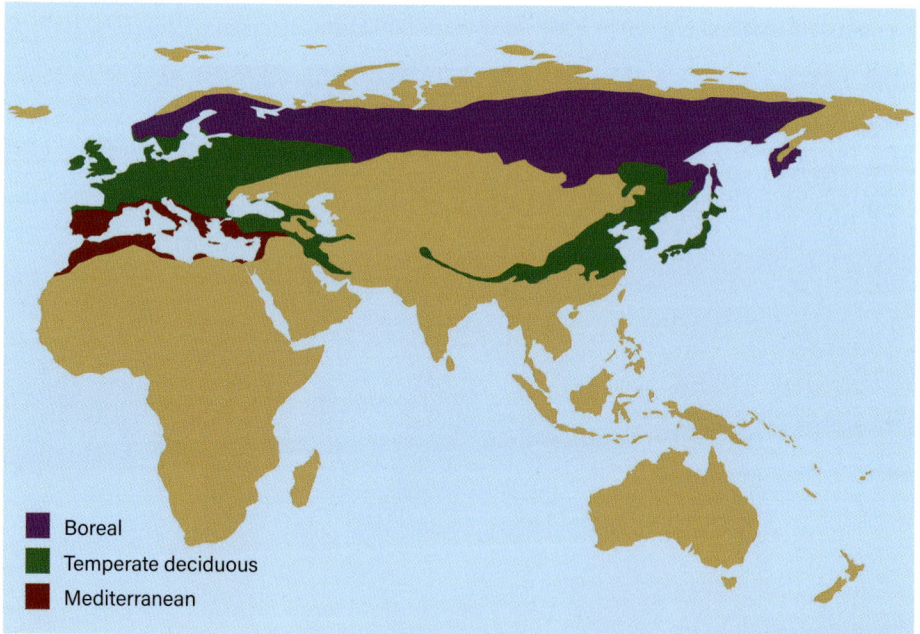

Figure 3.3. The major forest biomes of Eurasia where Marsh Tits and Willow Tits live. The temperate deciduous forests in Europe and east Asia are the primary domain of Marsh Tits, which also extend into cooler regions of the Mediterranean biome. Willow Tits occupy most of the temperate and boreal forest biomes.

The temperate deciduous (broadleaved) forest biome

The temperate deciduous forest biome stretches across most of Europe in the mid latitudes, from northern Iberia, Ireland and Great Britain, through southern Scandinavia and central Europe, and eastwards to the Urals. A further expanse occurs in east Asia, including northern China, Korea and Japan, and through the Himalayan chain. The distribution of temperate deciduous forest across Europe and east Asia closely matches the Marsh Tit's global distribution (Figure 1.9). Underlying this is the fact that Marsh Tits are habitat specialists of deciduous woodland and forest, and this is their primary habitat.

The Marsh Tit is increasingly recognised as an indicator species for mature or ancient broadleaved woodland, and this makes them an important sentinel for habitat quality in this biome. Willow Tits also occur widely in temperate broadleaved forest, but have a more northerly distribution that extends into the boreal forest biome (Figure 1.8). Whereas Marsh Tits are habitat specialists, Willow Tits are more generalist and occupy a variety of broadleaved, mixed and coniferous habitats in the temperate and boreal zones, from shrublands to mature forest.

The natural vegetation of the temperate deciduous forest biome encompasses a range of broadleaved trees, including oaks, limes, beeches, ashes, maples and hornbeams, with aspens, alders and willows in wetter areas. Understorey shrubs include hazels, hawthorns, Blackthorn and hollies, with bamboo thickets (as an honorary shrub) in east Asia. There is usually a rich ground layer of herbs and grasses. The tree canopy can be continuous or relatively open among a woodland-grassland mosaic, as in wood-pasture. On the Atlantic fringe the moist temperate rainforests have stunted canopies of oak, birch, Common Ash and Common Hazel and an abundance of epiphytic mosses, lichens and ferns. Natural forest is punctuated by open bogs, lakes, grasslands, montane or lowland heaths, fringed by transitional pioneers like birches and willows. Around wetlands and along rivers the woodland can be swampy on waterlogged soils with seasonal flooding. Marsh Tits and Willow Tits can occur in all of these woodland types at varying densities.

Europe's original temperate deciduous forest has been almost completely lost to clearance for agriculture and transformation by forestry management. Only in a few places can we get an idea of the extensive forests that Marsh Tits and Willow Tits likely inhabited before humans transformed the continent. Some wood-pasture survives in Transylvania, where domestic cattle and sheep replace wild herbivores in maintaining open woodland and grassland (Hartel *et al.* 2015). Meanwhile, the Białowieża National Park in eastern Poland is perhaps the best-preserved fragment of closed-canopy temperate forest in Europe, widely used as a benchmark for understanding what primeval forest would have looked like (Tomiałojć & Wesołowski 2005, Jaroszewicz *et al.* 2019). The Białowieża National Park is also the location of one of the most important Marsh Tit studies (see Chapter 1).

Białowieża National Park

The Białowieża Forest is a 1,500km² complex of tree cover, meadows, rivers and bogs that straddles the Polish–Belarusian border. The forest has a variable history of management and logging across its area (Latałowa *et al.* 2018). On the Polish side, however, the strictly protected Białowieża National Park is a 47km² block within the wider forest that has never been logged. The national park retains a primeval character of a high diversity of trees and

Figure 3.4. European Bison in the Białowieża National Park, Poland, and damage on a young Common Hornbeam tree, caused by bison or Red Deer rubbing their horns/antlers and stripping the bark. The bark damage may decay to form tree cavities that are used by nesting Marsh Tits. (© Richard K. Broughton)

shrubs, a complex multi-layered structure, huge amounts of standing and fallen deadwood and a rich ground flora.

Almost all of the native mammals and birds of the late Holocene ecosystem are apparently still present in the forest, including Marsh Tits, Willow Tits and their competitors and predators, such as Great Tits, Blue Tits, Crested Tits, Eurasian Sparrowhawks, Great Spotted Woodpeckers, Pygmy Owls, Common Weasels, Forest Dormice and Pine Martens. Large native herbivores include Red Deer, Eurasian Elk and European Bison, which may interact indirectly with Marsh Tits and Willow Tits by damaging trees and creating potential nest-cavities (Figure 3.4; Broughton *et al.* 2022a). The top predators include Eurasian Wolves, Eurasian Lynxes and the occasional Brown Bear, meaning that the Białowieża National Park is an almost intact forest ecosystem, with the only missing large mammal being the globally extinct Aurochs.

The forest of the Białowieża National Park is a patchwork of deciduous, coniferous and mixed stands. There are open glades and bogs within the forest where Common Cranes breed and Black Storks forage. Rivers create wide openings fringed by reeds, willow shrubs and wet woodland, occupied by Eurasian Beavers. Dominant canopy trees include Pedunculate Oak, Common Hornbeam, Small-leaved Lime, Norway Maple, Scots Pine and Norway Spruce. In the wetter areas are Black Alder, Common Ash and birches. Common Hazel, Rowan and Bird Cherry occur in the understorey. Deadwood of all sizes accounts for a quarter of the woody biomass, comprising standing snags (tall dead stumps/trunks), fallen logs and branches (Bobiec 2002). The canopy trees reach huge sizes, with the tallest forest giants in excess of 55m tall.

The Białowieża National Park is a magnificent place, and the massive trees reminded me of the Iron Age logboats from Great Britain that were carved out of huge straight trunks to navigate rivers and estuaries (Millett *et al.* 1987). I had never seen trees big enough to produce such logboats until I went to Białowieża (Figure 3.5), but the archaeology shows that similar primeval forest with massive trees must once have been present in Great Britain. This was a vivid realisation of just how much western European forests have been altered by people, and how woodland species like Marsh Tits and Willow Tits have had to adapt.

Other small pockets of primary deciduous forest are rare in Europe and generally not well preserved. The Białowieża National Park is probably the closest we can get to observing Marsh Tits, Willow Tits and other European woodland birds in a 'natural' habitat of lowland closed-canopy forest. It is fortunate that in the 1970s Polish ornithologists established long-running research of the bird communities in the national park (Wesołowski *et al.* 2022, Anon. 2022). These studies have been described as 'a window into the past', showing the species densities, nest-site selection, breeding success and predation rates in natural forest (Tomiałojć & Wesołowski 2005, Wesołowski 2007a).

Marsh Tits and Willow Tits in the Białowieża National Park

Marsh Tits are common in the Białowieża National Park, with an average of 15 pairs per km^2 and a slight increase in abundance from 1975 to 2019 (Wesołowski *et al.* 2022). They are most abundant in deciduous stands, including swampy riverine forest dominated by Black Alder, Common Ash and birches with a tangled understorey of shrubs, fallen trees, nettles and ferns (Figure 3.6), and also in drier stands dominated by massive Small-leaved Limes, Pedunculate Oaks and Common Hornbeams (Figure 3.7). Fewer Marsh Tits are found in the coniferous areas of the forest dominated by Norway Spruce and Scots Pine with young birch, Common Hornbeam and European Aspen.

Willow Tits are also resident in the Białowieża National Park, but they are not abundant. In contrast to the Marsh Tits, Willow Tits have not yet been studied in detail in the forest, and little is known about them. Breeding Willow Tits mostly occur within the conifer-dominated areas, but even here they can be hard to find, at densities of 3–4 pairs per km^2 (Wesołowski

Figure 3.5. An immense Pedunculate Oak in the Białowieża National Park, with Professor Tomasz Wesołowski for scale (attending a frass collector to monitor caterpillar abundance). The tree is approximately 40m tall. (© Richard K. Broughton)

Figure 3.6. Marsh Tit breeding habitat in swampy riverine forest in the Białowieża National Park. (© Richard K. Broughton)

Figure 3.7. Marsh Tit breeding habitat in primeval oak–lime–hornbeam forest in the Białowieża National Park. (© Richard K. Broughton)

et al. 2022). I have found Willow Tits in the national park only in areas dominated by very large Scots Pines and Norway Spruces, with lots of standing and fallen dead trees and dense patches of regenerating birch, hornbeam, young pines and spruces (Figure 3.8). Willow Tits forage in these thickets and understorey as well as in the canopy trees, and the stumps and taller snags of dead pines and spruces provide lots of potential nest-sites. Crested Tits and Coal Tits are also prominent in these areas, but there are few Marsh Tits, Great Tits or Blue Tits.

Figure 3.8. Willow Tit breeding habitat in primeval coniferous forest stands of the Białowieża National Park. (© Richard K. Broughton)

It is interesting that in the Białowieża National Park the Willow Tits do not occur in the swampy riverine forest of Black Alder, Common Ash and birch along the Narewka River. This would superficially seem to be good habitat for them, with abundant deadwood. Similarly, in the alder-swamp floodplain forest in the Polesia region extending into neighbouring Belarus, Marsh Tits are present but Willow Tits are not (Sahvon 2007). Elsewhere in eastern Poland Willow Tits can be found in many wet or riverine woodlands, such as in the Morgi Forest (Sikora 2021), and I have seen them in the Biebrza Marshes and in broadleaved riverine woodland and shrubland along the Wisłoka River at Mielec. In this wider context, the absence of Willow Tits from swampy riverine forest in the Białowieża National Park is puzzling. The habitat may have too many competitors, including Marsh Tits, Blue Tits and Great Tits, and this could be restricting the Willow Tits to coniferous stands through interspecific competition. Similar competition was suggested in Sweden, where Marsh Tits occurred in deciduous forest and Willow Tits were limited to the coniferous, but were able to expand into the deciduous habitat in places where Marsh Tits were absent (Alatalo *et al.* 1985). Nevertheless, Willow Tits seem able to live alongside Marsh Tits in a range of other woodlands (Ludescher 1973, Skórka & Wójcik 2003, Sikora 2021), so the apparent habitat separation in the Białowieża National Park remains unexplained for now.

SEMI-NATURAL TEMPERATE WOODLANDS

Across Europe and east Asia the original temperate deciduous forest is now heavily fragmented by deforestation for agriculture, industry and urbanisation. Many of the remaining woodland patches are not very natural but have been shaped by management and exploitation. Many Marsh Tits and Willow Tits therefore live in semi-natural woodlands that have been modified by human activity, or secondary woodlands that have regrown on sites that were previously cleared of the original primary forest. This is the general situation across

Figure 3.9. Common Beech and Black Alder forest in the Söderåsen National Park, southern Sweden, one of the largest remaining expanses of semi-natural native forest in northern Europe. Marsh Tits live here. (© Richard K. Broughton)

Great Britain and most of the rest of Europe, where the larger expanses of native forest are restricted to national parks (Figure 3.9).

People have long exploited and managed woodlands for timber, firewood and other products, and also for hunting, shooting and foraging by domestic animals. These activities transform the nature of woodlands for Marsh Tits, Willow Tits and other forest species. For example, the logging of old-growth forests reduces the age profile of trees and favours commercially valuable species like oaks and spruces, while removing or suppressing others, such as birches and aspens. Artificially high numbers of domestic or wild herbivores, bolstered by eradication of predators such as Eurasian Wolves and Lynxes, can then inhibit the regeneration of young trees and shrubs. This all results in a simplified habitat structure and an impoverished species composition. The response of birds to the habitat changes resulting from forestry and woodland management is a major topic in ornithology (Avery & Leslie 1990, Mikusiński *et al.* 2018).

For Marsh Tits and Willow Tits the habitat quality of semi-natural woodlands differs from that of natural forests, but it is not always obvious if they are inherently worse. If we use the Białowieża National Park as a benchmark of naturalness, then the abundance of Willow Tits in semi-natural or altered woodlands can be far higher than the 3–4 pairs per km^2 seen in Białowieża (see Chapter 5). For example, various types of semi-natural woodlands in England, Germany, Finland and Japan had Willow Tit densities around 3–5 times higher than in the Białowieża National Park (Ludescher 1973, Nakamura 1975, Orell & Ojanen 1983, Broughton *et al.* 2020). However, lower densities of Willow Tits than in Białowieża have been recorded in Swedish and Norwegian woodlands that have been subject to management (von Brömssen & Jansson 1980, Hogstad 1987).

For Marsh Tits, an abundance of around 15 pairs per km^2 in the benchmark Białowieża National Park is very similar to what we find in many semi-natural broadleaved woodlands across Europe (Ludescher 1973, Amann 1997, 2003, Broughton *et al.* 2006, Carpenter 2008). Some of these woodlands could reflect a bias of selecting study sites precisely because they have particularly good populations of Marsh Tits, but densities were also similar in rather generic ancient woodlands in the landscape surrounding Monks Wood (Broughton *et al.* 2018a). It appears that Marsh Tits widely reach an optimum or maximum density in semi-natural woodlands that is similar to that found in primeval forest.

BRITISH ANCIENT WOODLAND

Broadleaved ancient woodland is a culturally and ecologically important semi-natural habitat in Great Britain, defined specifically as woodland dating to at least the year 1600 (1750 in Scotland). Many ancient woodlands are much older than this, with parts of Monks Wood and Wytham Woods in England probably being continuously wooded since prehistory. Although Great Britain's total woodland cover is currently around 13 per cent, ancient woodland accounts for only 2.5 per cent of the land area, with the great majority occurring in England and Wales. The remaining British woodland consists of younger natural or planted woods, particularly plantations of non-native conifers. The surviving ancient woodland is heavily fragmented across the landscape, and almost all (98 per cent) of the individual woods are smaller than 100ha in size (Spencer & Kirby 1992).

British ancient woodlands have been shaped by human management over many centuries, which is what makes them semi-natural. However, for much of the past century many woods have been only lightly managed, or essentially unmanaged, and so have increased in naturalness (Peterken 2022). These woodlands are dominated by Pedunculate or Sessile Oaks, Common Ash, Common Beech, Silver Birch and, in some Scottish woods, Scots Pine, with abundant Common Hazel, Blackthorn and hawthorns in the understorey. Ancient woodlands often have a species composition and multi-layered structure that suits Marsh Tits very well, especially when intensive management has relaxed and the woods have become wilder and more diverse.

Marsh Tits are widespread in ancient woodlands in England and Wales, and this is their core habitat in Great Britain, where they are a good indicator species for structurally diverse woods (Broughton *et al.* 2006, 2012a, Hinsley *et al.* 2007, Carpenter *et al.* 2010). Marsh Tits especially like ancient woodlands that have a dense understorey of shrubs and young trees combined with a taller and mostly closed canopy of broadleaved trees (Figures 3.10 and 3.11). The tree species seem less important than their mature structure. A high structural diversity and species composition gives Marsh Tits lots of foraging niches and a wide range of seeds and invertebrates to sustain them in their home-ranges throughout the year.

Willow Tits were also found in British ancient woodlands during much of the 20th century, but today it is easy to forget that they were ever there, as it is no longer considered to be 'Willow Tit habitat' (e.g. Pinder & Carr 2021). Until the 1990s, however, Willow Tits were a rather widespread generalist of ancient woodland and other wooded habitats on farmland, wetlands, heaths and plantations (Simms 1971, Perrins 1979). The British Willow Tit's decline during the 1970s to 1990s was far more severe in mature woodland and farmland than in the early-successional woodland around wetlands that now function as habitat refuges (Siriwardena 2004). The decline of Willow Tits in British ancient woodlands has been so complete that it is now virtually extinct in this habitat.

Figure 3.10. Marsh Tit habitat in ancient woodland at Monks Wood National Nature Reserve in Cambridgeshire, England, in May 2013. The wood has largely been managed as non-intervention for over a century and has accrued increasing naturalness, with a high species diversity and complex, multi-layered vegetation structure of ground flora, understorey shrubs and tree canopy. (© Richard K. Broughton)

Figure 3.11. Marsh Tit habitat in semi-natural ancient woodland at Wytham Woods in Oxfordshire, England, in June 2013. Limited management throughout the 20th century has promoted woodland maturity and structural diversity. (© Richard K. Broughton)

Ironically, the first detailed study of British Willow Tits was during the 1940s in Wytham Woods, a typical ancient woodland near Oxford (Foster & Godfrey 1950). Willow Tits were also present in almost every other notable ancient woodland in England from the 1970s to the 1990s (Table 3.2). Most of these populations had disappeared by the early 2000s. Perhaps the very last Willow Tit population in English ancient woodland clings on at Combe/Faccombe Woods (Figures 3.12 and 3.13), with around 20 territories in 2016 (Chalmers 2017). I found at least four pairs still present at Combe Wood during a limited search in 2023.

Table 3.2. Notable broadleaved ancient woodlands in England that held Willow Tits until the late 20th or early 21st century, and the year of last observation before local extinction.

Wood	County	Area (ha)	Last record	Sources
Wytham	Oxfordshire	415	2002	Foster & Godfrey 1950, Perrins & Gosler 2010
Monks	Cambridgeshire	160	2000	Steele & Welch 1973, Hinsley et al. 2005
Treswell	Nottinghamshire	48	2018	MacColl et al. 2014, Treswell Wood IPM Group
Lady Park	Gloucestershire	45	1980s	Peterken & Mountford 2017
Hayley	Cambridgeshire	52	1995	Madin 1979, Cambridge Bird Club 1996
Sheephouse	Buckinghamshire	58	1999	Fuller 2022
Bradfield	Suffolk	70	1990s	Fuller & Henderson 1992, R. J. Fuller, personal communication
Combe/Faccombe	Berkshire–Hampshire	450	Present 2023	Chalmers 2017, personal observation
Brampton	Cambridgeshire	132	1997	Cambridge Bird Club 1998

Dominant tree and shrub species in most woods were mature Pedunculate Oak, Common Ash, Common Hazel and hawthorns.

The ancient woodlands in Table 3.2 that previously held Willow Tits also had competing Marsh Tits, Blue Tits, Great Tits and Coal Tits, alongside predatory Great Spotted Woodpeckers. Willow Tits were thus able to coexist as part of the bird community of British ancient woodlands during the 20th century, though usually at lower densities than the other tits. This no longer seems to be the case, and Willow Tits appear unable to survive in most ancient woodlands in modern Great Britain.

The reasons for the loss of Willow Tits from ancient woodland are not fully understood. Many woods have changed over the past 70 years as they developed with little human interference, often becoming more mature and complex. These changes would seem to be superficially better for Willow Tits, such as more gaps in the tree canopy, increasing density of the understorey layer and more deadwood (Amar et al. 2010). Nevertheless, Willow Tits even

Figure 3.12. Semi-natural ancient woodland at Combe Wood on the Berkshire–Hampshire border in southern England, in March 2023. Marsh Tits and Willow Tits both breed here. The mature, unmanaged woodland sits on chalk downland at 200m above sea level, and has abundant Pedunculate Oak and Common Hazel. (© Richard K. Broughton)

Figure 3.13. A Willow Tit from the small population at Combe Wood, one of the last places in Great Britain where they can be found in ancient woodland, and one of the last locations occupied by Willow Tits in southern England. (© Peter Stronach)

disappeared from ancient woodlands that did not appear to change very much, like Bradfield and Treswell Woods, where traditional coppice management continued. As such, it is hard to see how changes in habitat structure could explain the near total loss of Willow Tits from ancient woodlands in Great Britain, especially as this species previously seemed quite flexible in a range of wooded habitats.

The wider landscape context around ancient woodlands is just as relevant as changes within the woods themselves. The widespread loss of Willow Tits from British farmland during the latter part of the 20th century would have isolated small populations in the ancient woodlands, making them more vulnerable to extinction (Siriwardena 2004). Woodland could also have become more hostile to Willow Tits due to increases of competitors and nest predators such as Blue Tits and Great Spotted Woodpeckers (Parry & Broughton 2018, Shutt & Lees 2021). More subtle changes from nutrient pollution or climate change could be associated with declines of important insect prey such as moths (Hopkins & Kirby 2007, Blumgart *et al.* 2022). Like Willow Tits before them, Marsh Tits are now also declining in ancient woodlands in parts of Great Britain, including the Monks Wood study area, and they may be suffering from some of the problems that previously affected Willow Tits.

EASTERN EUROPEAN TEMPERATE FOREST

Great Britain is at or near the western range limit of Marsh Tits, Willow Tits and the temperate deciduous forest of Eurasia, but they reach another limit in eastern Europe where forest meets the steppe grassland biome that extends throughout central Asia to China. The limit of temperate deciduous forest falls across the very large expanse of Ukraine, where Marsh Tits and Willow Tits are present in the wooded north of the country but absent from the open steppe towards the Black Sea and Crimea in the south. Marsh Tits occur in oak woodland in Ukrainian forests (Figure 3.14), as in other parts of the temperate forest biome, but are also associated with Common Beech and Common Hornbeam.

Willow Tits have a wider habitat range than Marsh Tits in this region, occurring in many woodland types but usually at low densities. In western Ukraine Marsh Tits are common in beech and hornbeam forests in Chernivtsi Oblast and into the Carpathian Mountains, whereas Willow Tits are common in oak–hornbeam woodlands and also conifer plantations (Guzy 1994a, 1994b, 1995, Bashta 1999). Both species occur in city parks in Chernivtsi in the autumn and winter (Skilsky 1998). In the Kharkiv and Sumy regions of northeast Ukraine Willow Tits are common in mixed and coniferous forests, especially in wet woodland and carr, and along river valleys among willows, birches and alders (Figure 3.15) (Belik & Moskalenko 1993, Banik *et al.* 2013, Skliar & Knysh 2016). Yehor Yatsiuk (2015) reported that Willow Tits have apparently also expanded into wet deciduous woodlands in this area in recent decades. Marsh Tits are becoming less common than Willow Tits in northeast Ukraine, but they may previously have been more widespread in deciduous woodlands (Volchanetsky 1950). The brutal full-scale invasion by Russia during the 2020s has devastated many of the forests in northern and eastern Ukraine through artillery bombardment and logging under occupation. The catastrophic ecological legacy will likely shape future distributions of Marsh Tits, Willow Tits and other forest species in this region.

Further east towards the edge of Europe, around the Volga and Kama Rivers in western Russia, the hemi-boreal and temperate deciduous forests are dominated by Norway and Siberian Spruces, Siberian Firs and Pedunculate Oaks. Marsh Tits are more common here

Figure 3.14. Oak-dominated deciduous forests where Marsh Tits occur in the Kharkiv region of Ukraine. (© Yehor Yatsiuk)

Figure 3.15. Typical Willow Tit habitat in the Kharkiv region of Ukraine: stands of birch, aspen and willow in patchy wetlands and bogs among more extensive pine forest. (© Yehor Yatsiuk)

than Blue Tits or Crested Tits despite being almost at the eastern limit of their European range. Willow Tits are even more abundant than Marsh Tits, being one of the most numerous resident woodland passerines. Both species have increased substantially in this region since the 1990s, probably due to climate change and milder winters (Askeyev *et al.* 2018). The Willow Tits are associated with coniferous or mixed forests, whereas Marsh Tits are more patchily distributed and favour the deciduous stands, especially European Aspen along rivers (Popov 1978, Askeyev *et al.* 2018).

The general picture in eastern European temperate deciduous forests is that both species seem to be doing well and are increasing in some areas. Willow Tits are widespread and abundant in a range of woodland types, including mature coniferous, mixed and riverine deciduous stands. Marsh Tits are more localised in a narrower range of deciduous habitats, especially beech and hornbeam forests, but can still be quite common. Marsh Tits peter out just beyond the fringe of Europe, where the temperate forest biome ends, but the Willow Tit's range continues across the northern boreal forests of Asia. In east Asia the temperate deciduous forest reappears, along with eastern populations of Marsh Tits.

TEMPERATE FORESTS IN EAST ASIA

Mirroring the habitat relationships of Marsh Tits and Willow Tits in Europe, similar patterns are found in the temperate broadleaved forests of China, Korea and Japan. The east Asian forests are composed of trees from many of the same genera as in European deciduous or mixed forests, including oaks, hornbeams, maples, birches, ashes, willows, elms and pines, but generally involving different species than in Europe. A notable difference from European forests is that in east Asia the understorey often contains thickets of bamboo. As in Europe, however, many east Asian forests are widely degraded and fragmented by logging and forestry, creating wide variation in habitat quality (Vandergert & Newell 2003, Tang *et al.* 2010).

Early habitat assessments for Marsh Tits and Willow Tits in Asia were complicated by different subspecies that are now treated as separate species, such as the Black-bibbed Tit and Sichuan Tit (Snow 1954, Tritsch *et al.* 2017). There are some habitat overlaps among these species, as shown in Table 3.1, but their ranges do not overlap with Marsh Tits or Willow Tits (Harrap & Quinn 1995). Overall, the habitat associations of Marsh Tits and Willow Tits in the temperate forests of east Asia are broadly similar to those in Europe.

Marsh Tits in east Asia are often found in forests dominated by oaks and conifers, such as Mongolian, Daimyo and Jolcham Oaks and Korean Pine, Japanese Red Pine and Manchurian Fir. In South Korea, Marsh Tits are common on mountainous slopes cloaked in deciduous or mixed forest (Jabłoński & Lee 2002, Rhim *et al.* 2003). They are also present but less common in more open forest in the lower valleys. Information is very scarce from North Korea, but Marsh Tits are present and possibly common in mixed forests of oaks, pines, birches and elms (Moores 2017). In northeast China, Marsh Tits also breed quite commonly in mountainous mixed forests of Mongolian and Sawtooth Oaks, Manchurian Walnut, Japanese Red Pine and Chinese Ash (Zhang *et al.* 2021).

Willow Tits in China occur in a range of secondary broadleaved, mixed and coniferous woodlands up to quite high elevations (Harrap & Quinn 1995). In the Saihanba National Forest Park, in Hebei Province, Zhang *et al.* (2020) recorded Willow Tits breeding in secondary mixed forest at altitudes of up to 1,350m. In Gansu Province Willow Tits are found in mountainous mixed forest up to 2,300m (Liu *et al.* 1989). In Japan, Willow Tits occur on all of the main islands. On Honshu they are one of the most abundant breeding birds in old-growth deciduous forest of the Ogawa Forest Reserve, alongside Japanese Tits, Coal Tits, Varied Tits and Narcissus Flycatchers (Tojo 2009). The mature tree canopy in this area is dominated by oaks and beeches, with an understorey of dwarf bamboo thickets. This multi-layered forest structure is broadly analogous to ancient woodlands in Great Britain, where Willow Tits used to occur until recent decades.

Japanese Marsh Tits occur only on the northern island of Hokkaido, but they are common in temperate deciduous forests of oaks, Painted Maple and Japanese Lime (Murakami 2002), and also in regenerating broadleaved forest dominated by Japanese White Birch, Mongolian Oak and Tree Aralia with a dense understorey of dwarf bamboos (Unno 2002). Kurosawa & Askins (2003) considered Marsh Tits to be forest generalists in Japan, but associated more with extensive tracts of woodland rather than small patches.

Across east Asia, Marsh Tits and Willow Tits both appear to be most common in the more extensive forests on mountainous slopes, which have avoided the worst of deforestation. Marsh Tits are particularly associated with the oak-rich forests in this region, while Willow Tits seem to occur widely in mixed forests at higher altitudes.

The Mediterranean forest biome

In the more arid climate of the Mediterranean biome in southern Europe and Anatolia, around the Mediterranean and Black Seas (Figure 3.3), the forests are dominated by oaks (including evergreen Holm Oak) and various pines, with deciduous oaks at middle elevations. Marsh Tits are generally restricted to belts of deciduous and mixed forest in mountainous areas or wet riverine deciduous woodlands in the lowlands. In northern Greece, for example,

Figure 3.16. A Willow Tit at Lake Kerkini in northern Greece, where the Mediterranean forest biome transitions to the temperate forest biome. (© Han Onderwater)

Figure 3.17. Marsh Tit habitat in deciduous Mediterranean woodland dominated by Common Beech, maples and willows, on the slopes of the Italian Apennine range at Barrea, in the province of L'Aquila. (© Samuele Ramellini)

where the Mediterranean biome transitions into the temperate zone, Marsh Tits are one of the commonest small passerines in the extensive forests on the southern flanks of the Rhodope mountain chain, particularly on the oak-dominated lower slopes (Xirouchakis 2005). Willow Tits are mostly restricted to the higher slopes in the cooler coniferous forest (Figure 3.16). On the hills of the Yuvacik watershed in northwest Turkey, Marsh Tits occur only in the deciduous forest of oaks, Common Hornbeam and Common Beech, but not in coniferous forest elsewhere in the catchment (Beskardes *et al.* 2018).

In the central Mediterranean, in northern Italy, a study in the Magra Valley in Liguria found that Marsh Tits were absent from pinewoods, maquis and willow shrubland in the lowlands, and were instead associated with Sweet Chestnut orchards, European Hop-hornbeam and Common Beech woodlands in the cooler hills and mountains (Farina 1983). Further south along the Apennine Mountains, Marsh Tits are more localised among areas of Common Beech (Figure 3.17), being scarce in the coniferous or mixed forests that dominate much of the region (Martini *et al.* 2021). Willow Tits were thought to be present in mixed forests in the central Apennines, but a recent assessment concluded that all records involve Marsh Tits (Brunelli & Fraticelli 2020). Sicily is perhaps the most southerly outpost of European Marsh Tits, and again they are associated with upland broadleaved forest, being absent from coniferous stands, eucalyptus plantations, Olive groves or arid lowland shrubs (Londi *et al.* 2012, La Mantia *et al.* 2014).

In northern Italy the forest changes as it rises towards the Alps. The broadleaf canopy of Pedunculate Oak, Common Hornbeam and Sweet Chestnut on the floodplains and foothills shifts to montane Common Beech at higher altitudes. In the Lombardy region Marsh Tits are an indicator species for the quality of deciduous woodlands, and a population increase during

Figure 3.18. Willow Tit habitat in the Italian Alps at Primiero San Martino di Castrozza, in the Autonomous Province of Trento. Norway Spruce and European Larch dominate. (© Samuele Ramellini)

the 1990s and early 2000s suggests that the habitat is improving (Bani *et al.* 2006, 2009). Up into the Alpine zone the woodlands become dominated by Norway Spruce and European Larch towards the treeline, and Marsh Tits peter out, being replaced by Willow Tits at these higher elevations (Figure 3.18).

The original habitats of the Mediterranean forest biome are highly transformed, however, with little or no original forest remaining. In Lombardy, despite a tree cover of around 25 per cent, almost all of the broadleaved woodlands are managed as coppice, and the coniferous stands are managed as high forest (Massimino *et al.* 2010). Plantations of poplars and non-native trees such as Black Locust are also common among the native woodland, and these are avoided by Marsh Tits (Porro *et al.* 2020).

Overall, the Mediterranean seems quite a marginal forest biome for Marsh Tits and Willow Tits, and they appear to do best at higher altitudes in the cooler forests. Climate change is likely to have a significant impact on the woodlands in this region by pushing the cooler forest types further up the mountains, which might eventually run out of space. Increased drought and forest fires are also a direct threat to local populations.

The boreal forest biome

To the north of the temperate deciduous forest in Europe and Asia is the boreal forest biome, or taiga forest, which stretches from northern Scandinavia and Finland right across northern Asia through Siberia to the Kamchatka peninsula and Japan's northern island of Hokkaido (Figure 3.3). Located mostly between the latitudes of 57° and 69° N, the boreal forests are shaped by short warm summers and very cold winters with several months of snow cover. In

Figure 3.19. Willow Tits, like this bird at Dombås in Norway, are the most common tit in the boreal forests of northern Eurasia. (© Robert Fredagsvik)

Asia the boreal forest extends further south than in Europe, through Siberia to the mountains of northern Mongolia. Compared to the temperate deciduous forest there is less diversity of trees and shrubs in the boreal biome, and these are dominated by coniferous spruces and pines mixed with deciduous birches, willows, Rowan and European Aspen.

The boreal forest, then, is a simpler and harsher environment than the milder and richer temperate deciduous forest biome. Nevertheless, old-growth boreal forest has its own structural diversity of mature trees and lots of standing deadwood and snags. The tree cover is broken by open mires, bogs, rivers and steppe grassland, and the ground layer is covered by thick mosses, lichens, heathers and dwarf shrubs, especially berry-bearing *Vaccinium* species like Cranberry.

Willow Tits breed commonly across the boreal coniferous, mixed and birch forests of northern Eurasia, where they are generally the most abundant tit species (Figures 3.19 and 3.20). Marsh Tits become rather localised where the temperate deciduous forest zone blends into the mixed and boreal forest, favouring patches of birch, willow and European Aspen (Snow 1954, Cramp & Perrins 1993). In Estonia, for example, at the transition between temperate and boreal zones in the hemi-boreal forests, Marsh Tits are restricted to deciduous woodlands or those with some admixture of Norway Spruce, including wooded swamps and birch fen (Lõhmus 2022). Willow Tits occur in all of these woodlands and also in conifer-dominated stands of Scots Pine, where the Marsh Tits are absent.

Compared to northern Europe there is much less information from the boreal forests of east Asia. The key study for Willow Tits in this region is by Pravosudov & Pravosudova (1996) in the Magadan area of northeast Siberia. The breeding habitat for these birds was mixed forest in river valleys dominated by Cajander Larch, Mongolian Poplar and Chosenia, with

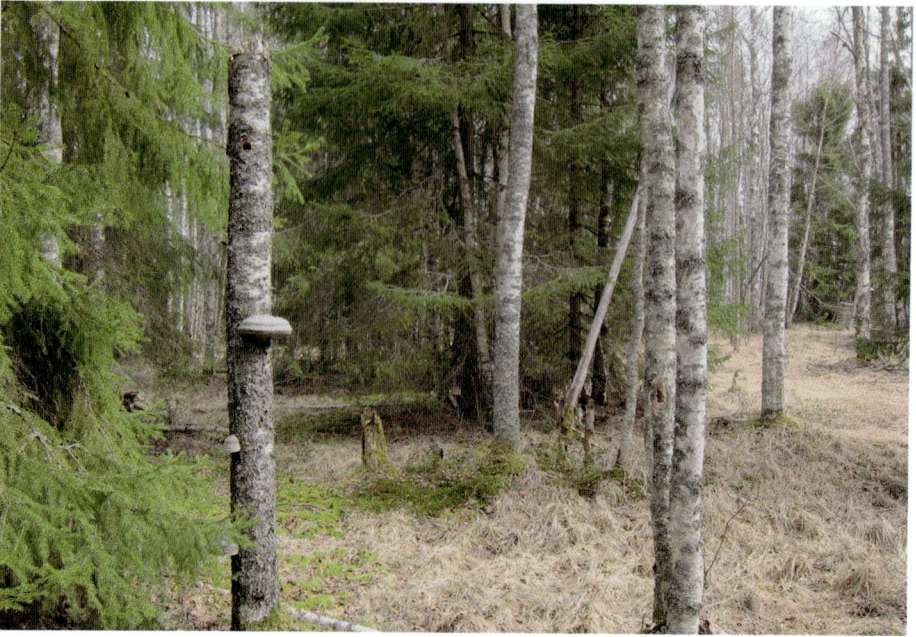

Figure 3.20. Breeding habitat of Willow Tits in the Oulu study area in northern Finland. The boreal forest is dominated by birches and spruces. The birch snag with the bracket fungus has a Willow Tit nest-hole near the top. (© Markku Orell)

Figure 3.21. Logging of Willow Tit habitat in the boreal forest in the Oulu study area, showing extensive clear-cutting and piles of felled logs encroaching on the remaining intact stands. (© Markku Orell)

78

some birch, alder and willow. Marsh Tits and Willow Tits are also common in the Mongolian boreal forests of the Khentii mountain range, with Marsh Tits favouring deciduous woodlands of birch, poplar and willow, while Willow Tits favour pine and spruce (Purevdorj *et al.* 2022). Both species are equally quite common in larch-dominated forest in the Khentii range, with Willow Tits also using pine plantations.

As with the temperate and Mediterranean forests, however, much of the boreal forest has been badly degraded by forestry management and exploitation. In northern Europe there is very little unmanaged boreal forest remaining after extensive logging and replanting with commercially managed trees grown for timber (Figure 3.21). One example in the Lycksele parish of northern Sweden documents the destruction of extensive boreal forest that was originally a relatively undisturbed mix of Scots Pine, Norway Spruce, birches, willows, Rowan and European Aspen, with lots of old trees and deadwood (Östlund *et al.* 1997). Prior to industrialisation the main disturbances were fire and Reindeer herding, but the arrival of major agriculture in the early 19th century saw increasing forest grazing by domestic animals along with drainage and extraction of deadwood and deciduous trees. Industrial forestry arrived in the 1860s and began with logging the old-growth forest, which progressed through the 20th century to modern intensive forestry. The end result is a heavily managed landscape with a simplified forest dominated by younger conifers, where deciduous trees are almost eradicated.

The example of Sweden's Lycksele parish was broadly repeated in the boreal forests of Norway and Finland, leading to a large-scale homogenisation and transformation of the forests across Fennoscandia. Further east, expanses of relatively undisturbed boreal forest still exist, particularly in Siberia, although logging and human impacts are increasingly affecting almost all of the Asian boreal forests (Gauthier *et al.* 2015).

The changes to boreal forests wrought by forestry have had long-term negative effects on their bird communities, particularly in Europe, with the combined abundance of 26 forest birds in Norway, Sweden and Finland declining by 25 per cent between 1980 and 2012 (Roberge *et al.* 2018, Lehikoinen & Virkkala 2018). Declines are most significant for forest residents and non-passerines associated with old-growth forest, such as woodpeckers, forest grouse and Willow Tits, which is highlighted in Finland as a species that has been most severely impacted by forestry (Lehikoinen & Virkkala 2018). There is also a severe long-term (30-year) decline in the abundance of Willow Tits and Marsh Tits across Sweden between 1977 and 2006, probably related to the same habitat changes, although stronger regulation since the 1990s may have stabilised these declines (Ottvall *et al.* 2009, Ram *et al.* 2017).

Plantations and managed forestry

Marsh Tits and Willow Tits generally seem best suited to their original habitats, namely large expanses of old-growth deciduous, mixed or boreal forest. The radical modification of these habitats by people has had some very negative impacts for both species, but has also led to some later attempts at beneficial interventions, such as conservation management aimed at improving or restoring habitat quality.

As mentioned earlier, a dramatic change to Europe's native forest has been the felling of the natural vegetation and replanting with near-monoculture blocks of conifer plantations. These plantations have relatively short cycles of planting, thinning, felling, then replanting, which is aimed at producing timber. This type of forestry creates a standing crop of trees in blocks of a uniform age and structure, which are harvested before reaching maturity (Figure 3.22). Broadleaved or old trees that provide habitat diversity are rarely retained, or only in small numbers. Drainage ditches and forest tracks for vehicle access help to dry out the forest and change its microclimate, and the permanent shade of high-density conifers reduces the seed-bearing ground flora. Damaged, dying and dead trees, which provide deadwood as standing or fallen logs and branches, are typically removed during sanitary or salvage logging, which further simplifies the habitat (Avery & Leslie 1990).

These fundamental changes to woodlands reduce the habitat quality for Marsh Tits and Willow Tits by severely degrading the diversity and abundance of tree species, food resources and nest-sites (Enoksson *et al.* 1995, Lehikoinen & Virkkala 2018). The catastrophic decline of Willow Tits in Fennoscandia has been driven by intensive forestry, especially the thinning and sanitary logging that removes the understorey structure and standing deadwood that is essential for foraging and nesting (Eggers & Low 2014, Kumpula *et al.* 2023). Marsh Tits are also highly sensitive to logging and are more likely to persist where larger areas of mature trees are retained in the forest matrix (Söderström 2009). This effect can even be seen on the territory scale. For example, in the Monks Wood study area the felling of a single mature oak tree, which was a significant part of the canopy in one particular territory, was followed by the resident Marsh Tit pair moving to another part of the wood. In a neighbouring wood, extensive logging and replanting in a Marsh Tit territory resulted in its total abandonment, with the birds disappearing and not being replaced over the following decade.

Large areas of British ancient woodland were felled and replanted with conifer plantations in so-called 'PAWS' (planted ancient woodland site) woodlands during the 20th century (Rackham 2006). The conifer plantations often contained spruces and Scots Pine, which are not native to England or Wales, and sometimes also Pedunculate Oak and Common Beech. Larger native woodlands were commonly targeted by foresters for this replanting, such as Bernwood Forest and Savernake Forest in southern England, and this would have involved significant degradation of huge areas of high-quality habitat for Marsh Tits (Figure 3.23). In total around 40 per cent of British ancient broadleaved woodlands were converted to PAWS between 1930 and 1980. Many PAWS woods retained small pockets of native woodland where a few Marsh Tits could persist (Figure 3.24), as well as some of the original ground flora where it was not shaded out by conifers.

In later decades some of the PAWS woods fared poorly in producing useful timber, and management was largely abandoned (Rackham 2006). In several failed PAWS woods in our Monks Wood study area, native trees and shrubs were regrowing among the plantation stands of spruces and pines, either as self-sown trees and bushes or regrowing from old stumps that had survived from the former ancient woodland (Figure 3.25). By the early 2000s these semi-abandoned PAWS woods contained a regenerated understorey thicket of native trees and shrubs in less shady places, and also some scattered older oaks retained by the foresters. Marsh Tits were present in all of these regenerating PAWS woods in the Monks Wood study area, and Willow Tits were also present until the late 1990s, with the last ones disappearing by spring 2000.

Although the Marsh Tits have lasted longer than Willow Tits in most PAWS woods, their densities are lower than in broadleaved native woodlands, reflecting the lower habitat quality (Table 3.3). Nevertheless, these PAWS woods still provide useful links in the landscape between the better woods, with birds able to move through them during dispersal and a few pairs settling to breed. Despite these consolations, it would have been

Figure 3.22. A monoculture forestry plantation of Norway Spruce in Poland, showing a typical high density of trees all of the same age, offering low structural and species diversity. (© Richard K. Broughton)

Figure 3.23. Waterperry Wood in the Bernwood Forest, in Oxfordshire, southern England. Between 1955 and 1970 almost all of the native ancient woodland was felled and replanted with Norway Spruce and Pedunculate Oak, converting it to a planted ancient woodland site (PAWS). The simplified woodland structure and species composition has resulted in Marsh Tits being virtually absent from these plantations. (© Richard K. Broughton)

Figure 3.24. A pocket of ancient woodland retained within 10 per cent of Waterperry Wood in the Bernwood Forest. The more diverse mix of native Silver Birch, Common Ash, Pedunculate Oak, Common Hazel and hawthorn provides a habitat refuge for Marsh Tits within the larger area of managed plantation (see Figure 3.23). A 2019 survey of the wood found Marsh Tit territories only within this ancient woodland area. (© Richard K. Broughton)

Figure 3.25. Upton Wood in Cambridgeshire, eastern England, in 2011. A native woodland that was converted to a planted ancient woodland site (PAWS) by foresters in the 1950s and 1960s, but largely abandoned in later decades. Falling conifers were replaced by naturally regenerating Common Ash, Common Hazel and hawthorns. The wood holds several pairs of Marsh Tits. (© Richard K. Broughton)

better for the birds if the PAWS woods had remained as native woodland habitat. When the local Marsh Tit population began declining in the Monks Wood area in the 2010s it was PAWS woods that began losing their birds first, again reflecting the lower-quality habitat and smaller populations in the plantations.

Table 3.3. Densities of Marsh Tit territories in different woods in the wider Monks Wood study area according to woodland type. Marsh Tit densities are generally lower in planted ancient woodland sites (PAWS).

Wood	Area (ha)	Territories/10ha	Woodland type
Monks	160	1.3	Native
Aversley	62	1.3	Native
Odd Quarter	13	1.5	Native
Wennington	74	1.5	Native
Holland	27	2.6	Native
Bevill's	31	0.6	PAWS
Coppingford	31	0.6	PAWS
Upton	28	1.3	PAWS

Densities are average number of territories per 10ha, during 2009–2012, derived from annual surveys. Woodland type is native ancient woodland or PAWS, where most of the native trees have been felled and replanted with non-native conifer plantations.

Plantations created on previously open sites are also common in Great Britain and elsewhere in Europe, such as non-native conifers on former moorland or heathland. These plantations are of limited value to Marsh Tits and Willow Tits, having no residual native vegetation, but can allow dispersing birds to cross an otherwise open landscape. New plantations are most useful if they link patches of native woodland and contain a deciduous component, such as a perimeter hedgerow or deciduous margins. One such plantation on lowland heath at Allerthorpe Common in East Yorkshire held Marsh Tits and Willow Tits when I surveyed it in 2010. The pine plantation had margins of birch and hawthorn, and a nearby pocket of native broadleaved woodland. Ulfstrand & Nilsson (1976) also found that Danish Marsh Tits were limited to those pine plantations next to deciduous woodland, but Willow Tits were absent. It seems that these plantations can extend the wooded habitat available to the birds, even if much of it consists of low-quality conifers.

An early study of Willow Tits in eastern England during the 1950s came from 30–40-year-old plantations of Scots Pine in Thetford Forest (Gibb 1960, Gibb & Betts 1963). The pine plantations had only a small amount of broadleaved birches and oaks, but these were heavily used by foraging Willow Tits, especially when collecting caterpillars for nestlings. They also collected moth caterpillars from Scots Pines and ate lots of pine seeds during the winter. Coal Tits were also common in these pine plantations, but Marsh Tits were absent and Blue Tits and

Great Tits were scarce. This lack of competition from dominant species may partially explain how Willow Tits could survive and breed in plantations with few deciduous trees. Eventually the Willow Tits disappeared from these plantations, and they declined to local extinction in eastern England by the 2020s.

Willow Tits still breed in some upland conifer plantations, such as at Lake Vyrnwy in north Wales. In a planted area of around 10km^2 dominated by spruces, with some oaks and willow shrubs, Willow Tits occur at a low density of only 0.9 pairs per km^2 (see Chapter 5). This sparse distribution again reflects the lower habitat quality of conifer plantations with a limited deciduous component, but this may still provide a refuge from competition from other tits in richer woodlands elsewhere.

COPPICING AND CONSERVATION MANAGEMENT

Coppicing is a forestry management technique that involves the periodic cutting of trees or shrubs at the base to produce a dense cluster of straight regrowth, which can be harvested for wood products such as poles, rods, canes, firewood or charcoal. Much has been written about the traditional craft of coppice management, its commercial aspects and the benefits for woodland species that are adapted to early-successional conditions and disturbance (Rackham 2006, Kirby 2020a, Peterken 2023). For Marsh Tits and Willow Tits, coppice management is an important factor, given the widespread influence that it has had in shaping woodlands in recent history, and also the future role it may play as a forestry or conservation tool.

Traditional coppicing is an intensive management that was widespread in Europe's deciduous temperate woodlands for thousands of years. Regular coppice species include Common Hazel, Common Ash, Sweet Chestnut, Black Alder and Small-leaved Lime, which all regrow quickly after being cut to produce prolific upright stems. Rotational cutting typically takes place every 7–25 years, removing all woody stems in a different woodland compartment during winter and early spring.

Mature or semi-mature trees ('standards') are often cultivated as a patchy canopy layer above the coppiced understorey, usually oaks and Common Ash. This *coppice-with-standards* management was widespread in European woodlands until the early 20th century, but declined with changing demand and economics. The legacy of coppicing permeates many woodlands through the dominance of oak and hazel favoured by the foresters, and also a general lack of old-growth characteristics such as very old, large trees and abundant deadwood, although these are slowly increasing since management abated (Kirby *et al.* 2005, Amar *et al.* 2010). Some commercial coppicing still occurs in temperate deciduous woodlands, including in Great Britain, though at a much smaller scale than previously. In the modern era coppicing is also used as a conservation tool to create periodic open spaces within woodlands where light reaching the ground favours species adapted to those conditions, such as many plants and butterflies (Buckley 2020, Kirby 2020a).

The historical decline of coppice management in Great Britain has been mooted as a potential factor in the decline of Marsh Tits and Willow Tits via subsequent maturation of woodlands (Fuller *et al.* 2005, Lewis *et al.* 2007). The timing does not quite fit, however, as the proportion of British woodland under coppice management was only 15 per cent by the late 1940s and just 1 per cent by the 1960s (Mason 2007). Coppicing was long gone from most British woods by the time that population trends of Marsh Tits and Willow Tits were being monitored. Where coppicing still occurs, the major impact for both species is the effect on woodland structure and composition. This depends on the length of the rotation between cutting, the density and maturity of any standard trees, and the size of the area being cut. Traditionally a coppice compartment would be at least 0.25ha, but it can be much larger. Immediately after a cut there is no understorey in the compartment, as the shrubs have been removed and the woody debris cleared away. That leaves a simplified woodland structure of a bare floor below a fairly open tree canopy, although standard trees may be absent too. The regrowth of shrubs is rapid, with several metres of dense growth appearing within a couple of years. Regrowth provides a thick understorey for a number of years until it reaches the desired height and size and is cropped again, when the process restarts.

For both Marsh Tits and Willow Tits, the area being coppiced can represent a significant portion of a pair's territory, with the periodic removal of the understorey layer that they rely on. Although the vegetation loss is temporary, those birds resident within the affected territory will be displaced or have to cope until it regrows. This can be compounded by the occasional felling of standards for timber, such as large oaks, meaning a substantial loss of habitat volume for the birds in that territory. If the removal of vegetation is great enough in relation to the size of the woodland, then the territory may become unviable for several years, leaving the resident birds with nowhere else to go. It is therefore critical to consider the issue of scale when using coppicing for conservation.

For Marsh Tits, which are specialists of mature forest, coppicing would superficially seem to have a negative impact. For Willow Tits, which are often associated with early-successional woodlands in Great Britain, it may seem like a more positive intervention despite the limited

Figure 3.26. Coppicing under way at Bradfield Woods in Suffolk, eastern England. Almost all woody vegetation is cut within the compartment and stacked for removal. A fresh crop of dense stems will regrow from the stumps and will be cut again in 25 years' time, but Marsh Tits in this territory will initially have to cope with the temporary loss of habitat. (© Richard K. Broughton)

Figure 3.27. A Marsh Tit blending into the regrowth of coppiced Common Hazel at Bradfield Woods during a survey in April 2015. (© Richard K. Broughton)

deadwood. In fact, studies from ancient woodlands show a mixed response of Marsh Tits and Willow Tits to the presence or absence of coppicing, and some results show little effect at all, so the costs and benefits are not straightforward. Bradfield Woods in Suffolk is a rare example of where coppice-with-standards management has continued since the 13th century. The 70ha woodland has numerous compartments of coppiced Common Hazel, Common Ash and Black Alder, with Common Ash and Pedunculate Oak standards (Figure 3.26). Marsh Tits and Willow Tits were both present in the late 1980s (Fuller & Henderson 1992). Both species avoided the coppiced compartments immediately after cutting, but increased their usage as the understorey grew back after 3–4 years. In another study in the 48ha Treswell Wood in Nottinghamshire, Willow Tit activity was greater in un-coppiced parts of the wood, but in the coppiced areas they were more often found in younger regrowth, possibly foraging on a proliferation of seeding plants (MacColl *et al.* 2014). Nevertheless, both Treswell and Bradfield eventually lost their Willow Tits despite continued coppice management.

Unlike Willow Tits, populations of Marsh Tits have persisted in Treswell Wood and Bradfield Woods, where they are mostly associated with older vegetation rather than the early stages of coppice regrowth (Hinsley *et al.* 2009, MacColl *et al.* 2014). Our colour-ringing surveys of Marsh Tits at Bradfield Woods during 2014–2017 found a good number of territories (Figure 3.27), similar to the density in other ancient woodlands where coppicing had long since ended (Table 3.4). These comparisons suggest that the Marsh Tits managed quite well alongside traditional coppice-with-standards at Bradfield, although this was not universal, as Table 3.4 shows that densities in some other coppiced woods were lower than where coppicing had ceased.

Table 3.4. Marsh Tit densities in English ancient woodlands that have retained extensive coppice-with-standards management, compared to woods where extensive coppicing ceased at least 70 years earlier.

Wood	Area (ha)	Management	Territories/10ha	Survey period
Treswell	49	Coppiced	0.6	2005
Ravensroost	48	Coppiced	0.6	2019
Brasenose	56	Coppiced	1.1	2019
Bradfield	70	Coppiced	1.6	2014–2017
Aversley	62	Unmanaged	1.0	2008–2010
Monks	160	Unmanaged	1.4	2004–2011
Wennington	74	Unmanaged	1.5	2009–2017
Combe	128	Unmanaged	1.9	2010
Sheephouse	58	High forest	1.9	1984–2015

Densities are average number of spring territories per 10ha. The survey period for Monks Wood was limited to when the Marsh Tit population was stable. All woods were in regions with good Marsh Tit populations at the time of surveying.

Sources: Hinsley *et al.* 2007, Fuller 2022, author's own data.

At Monks Wood the breeding Marsh Tits showed a strong avoidance of a small (7.5ha) block of active or recent coppice management within the larger 155ha ancient woodland. Over 21 years of study none of the Marsh Tit pairs nested within this coppice block (Figure 3.28). Instead, the birds placed their territories among the older, unmanaged areas of ancient woodland that had been maturing for a century (Broughton *et al.* 2006, 2012a). The coppice block had less structural complexity and a limited tree canopy, with only a few decades of regrowth.

Since the end of widespread coppice management in Great Britain by the middle of the 20th century, semi-natural woodlands like Monks Wood are either increasing in naturalness or requiring restoration, depending on your point of view (Crane 2022, Peterken 2022). A lack of coppicing can be seen as a conservation problem because of the loss of the periodic open habitats, early-successional regrowth and the species that depend on it (Kirby *et al.* 2016, Peterken 2023). Meanwhile, deadwood and mature forest specialists should benefit

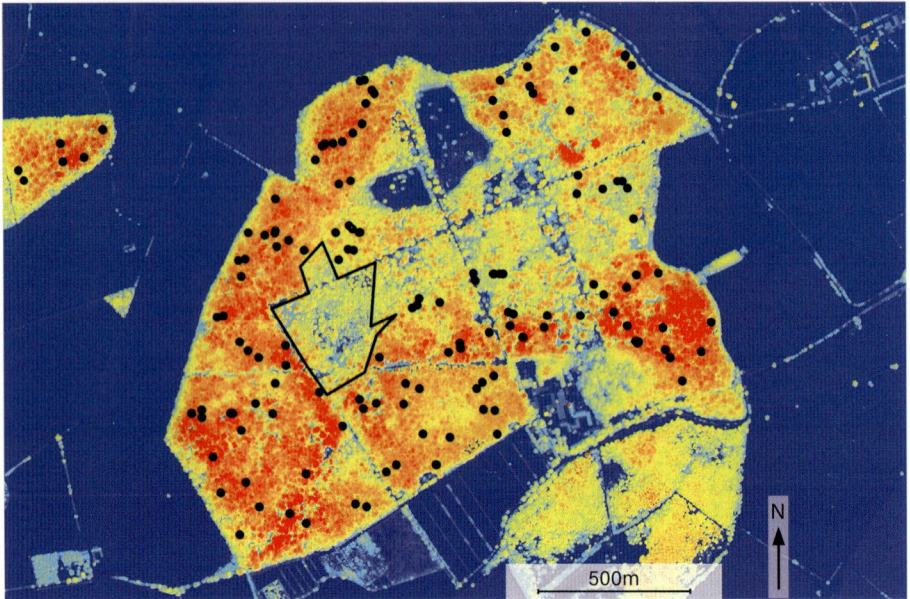

Figure 3.28. Distribution of 221 Marsh Tit nests (black dots) in Monks Wood over 21 years, showing avoidance of a 7.5ha block of active/recent coppice (outlined black) within the wood. Vegetation heights are indicated from lidar data, where cooler (blue) colours are the ground surface and low vegetation, and warmer (orange/red) colours are taller trees and shrubs up to a maximum of 27m. Lidar data were captured by the NERC Airborne Remote Sensing Facility, via the CEDA repository (CC BY 3.0).

from reduced management, and over time an increasing dominance of natural processes should lead to gaps opening in the canopy as trees and shrubs die, collapse and are blown down in storms.

The similar or lower densities of Marsh Tits in coppiced woods compared to unmanaged or mature ancient woodlands underlines that traditional coppicing is unlikely to be a significant factor in the species' decline in Great Britain (Table 3.4). For British Willow Tits, too, the presence or absence of coppice management has not affected local declines. Willow Tits were lost from Treswell and Bradfield Woods despite continued coppicing, yet are hanging on at the unmanaged Combe Wood. Elsewhere in England, another study showed that there was no difference in the amount of coppicing between woods where Willow Tits were still present and those where they had been lost (Lewis *et al.* 2007, 2009a). Furthermore, the targeted creation and maintenance of early-successional woodlands by planting, natural colonisation and coppicing was unsuccessful in maintaining or expanding Willow Tit populations in the Dearne Valley in South Yorkshire (Pinder 2021).

In summary, these various results and observations point to the widespread ending of woodland coppicing not being a major cause of national declines of Marsh Tits or Willow Tits, nor its reintroduction being a panacea for recovery. The use of coppicing or other woodland management as a conservation tool has so far had mixed or modest results at best (Pinder & Carr 2021, Bellamy *et al.* 2022). The creation of much more native woodland of all kinds may be a better way of maintaining populations, by increasing the buffering and connectivity of existing habitat (Broughton *et al.* 2013).

New woodlands

Marsh Tits and Willow Tits readily use new woodlands if they are within reach of existing populations. In the temperate forest zone new native woodlands are planted by governments, charities, companies and individuals. In Great Britain there have been national policies over recent years to significantly expand the cover of native woodland for multiple benefits, including biodiversity, and this could significantly aid both species. More native woodland would help to expand and link up existing habitat, although the intrinsic habitat quality of new plantings depends on species composition, tree density, establishment success, and whether or how it is subsequently managed (Herbert *et al.* 2022). Dense plantations of limited tree species will be of lower habitat quality for Marsh Tits and Willow Tits than more naturalistic plantings of mixed trees and shrubs at varying densities, which can develop into structurally diverse native woodlands.

In a study of small (up to 13ha) native woodlands planted on English farmland as part a national scheme, Marsh Tits and occasionally Willow Tits were using them after just 30 years (Dadam *et al.* 2020). Marsh Tits were more likely to use those new woods that were more mature, had a shrubby understorey and were near to other woodlands. This shows how the design of new woodland plantings, in terms of structure and location, can assist woodland specialists like Marsh Tits.

REWILDING AND NATURAL COLONISATION

In recent decades the conservation movement in Europe has become more ambitious in trying to restore ecosystems with a greater degree of naturalness. The process of *rewilding* can be defined as the development of self-sustaining and resilient ecosystems that develop in a self-willed manner, with little or no human intervention and no defined outcome (Pettorelli & Bullock 2023). Flagship rewilding projects often involve the restoration of landscapes driven by the reintroduction of large herbivores, such as at Oostvaardersplassen in the Netherlands, where an extensive mosaic of wetlands and grassland has also involved habitat creation for Willow Tits in peripheral wet shrubby woodlands (Figure 3.29).

Another common variant is *passive rewilding*, which is widely resulting in new woodlands appearing due to spontaneous natural colonisation of open land by trees and shrubs. Passive rewilding can be planned by ceasing management and allowing natural processes to return, such as the natural colonisation or regeneration of woodland onto open land. In many cases, however, passive rewilding involves an expansion of wooded habitats onto unintentionally abandoned farmland or post-industrial landscapes, and in much of Europe this is one of the biggest land-use shifts of recent decades (Ustaoglu & Collier 2018). In central and southern Europe this development has been associated with some declines of farmland birds but increases in forest birds (Orłowski & Ławniczak 2009, Zakkak *et al.* 2015). In the Italian Alps, for example, woodland birds including Marsh Tits and Willow Tits were using regenerating Common Beech woodlands on abandoned meadows after 30–50 years of tree growth (Laiolo *et al.* 2004).

In Great Britain new woodlands and shrublands are also developing spontaneously by passive rewilding, which is providing new opportunities for woodland birds. Wild oakwoods have been recolonising abandoned commons around Greater London for a century or more (Farjon 2022). At Knepp Wildland in Sussex, another flagship rewilding project, new areas of natural woodland and shrubland have been developing via natural colonisation after intensive farming was

Figure 3.29. A Willow Tit in a willow shrub at the Oostvaardersplassen rewilding site in the Netherlands. (© Han Onderwater)

withdrawn (Kirby 2020b). At Monks Wood, planned passive rewilding on abandoned fields saw rapid colonisation of new woodland due to abundant seed sources in the surrounding woods, and the presence of thrushes and Eurasian Jays to help disperse them. Two open fields next to the ancient woodland became completely wooded within only a few decades, mostly by Pedunculate Oaks and hawthorns (Broughton *et al.* 2021a). Marsh Tits from adjacent territories regularly foraged in the younger mosaics of shrub thickets, saplings and flower-rich grassland in the initial 15–35 years, the first territorial birds were recorded after 36 years, and after around 50 years the woodland was mature enough for Marsh Tits to breed (Figure 3.30).

Sites that are far from woodland seed sources are colonised much more slowly by trees (Hughes *et al.* 2023). Thorny shrubs often appear quickly, however, transported as seeds by berry-eating thrushes. Willows and birches can also arrive as seeds transported by the wind. The result of slow regeneration and succession is a patchwork of grassland, shrubland and young woodland that can be increasingly important in Great Britain for birds like Willow Tits and Willow Warblers (Bellamy *et al.* 2009, Broughton *et al.* 2020, 2022b). However, where new habitats are far from remaining populations (maybe more than 2–5km) then they are unlikely to be colonised by Willow Tits or Marsh Tits, as they will be beyond the typical dispersal range of enough birds that could find them.

Elsewhere in Great Britain, Willow Tits have been able to exploit new wooded habitats that have developed on post-industrial brownfield sites across the Midlands and north of England, particularly around 'flashes' of water after gravel extraction or the end of coal mining (Figure 3.31). Post-industrial sites around Wigan, Cheshire and north Nottinghamshire have been restored or rewilded as mosaics of woodland, shrubland and wetland that have provided important habitats for Willow Tits (Lewis *et al.* 2009a, Carr & Lunn 2017, Broughton *et al.* 2020). The abundance of young Black Alder, birches and willows on these sites provides suitable

Figure 3.30. Natural colonisation of woodland on previously open fields at Monks Wood. On the left, a mosaic of shrubland and flower-rich grassland has developed after 25 years and is used for foraging by Marsh Tits from the adjacent mature woodland. On the right, woodland development after 61 years, with Marsh Tits resident and breeding. (© Richard K. Broughton)

Figure 3.31. Willow Tit habitat in early-successional birch and alder woodland developing around a wetland 'flash' at Amberswood Common in the Wigan study area, northwest England. (© John Gibson)

breeding habitat, but there are high nest losses to Blue Tits and Great Spotted Woodpeckers (Lewis *et al.* 2009a, Rustell 2015, Parry & Broughton 2018).

Although Willow Tits will occupy early-successional woodlands on wetlands on abandoned land (Stewart 2010, Broughton *et al.* 2020), Marsh Tits can only colonise later when the tree canopy begins to mature above the understorey shrubs (Hinsley *et al.* 2007, 2009). This woodland succession can create conflicts, as management may be considered necessary to maintain early-successional woody habitats for Willow Tits and other species, like Common Nightingales, but that would prevent mature woodland development for Marsh Tits (Lewis *et al.* 2009a). Management would also require an ongoing commitment of rotational cutting and regeneration, as outlined by Pinder & Carr (2021). Elsewhere in Europe it is the loss of mature and old-growth forest to logging that is threatening Willow Tit populations, showing how a species' conservation needs can differ across its range (Lehikoinen & Virkkala 2018, Kumpula *et al.* 2023).

The reintroduction of Eurasian Beavers to promote rewilding has been suggested as part of the solution to reversing Willow Tit declines in Great Britain, by enabling the animals to engineer new wetlands and wet woodland habitats through their stream-damming behaviour (Macdonald 2022). Although the logic for this suggestion is attractive, there are no studies showing a habitat association between Willow Tits and Eurasian Beavers. Indeed, Willow Tit populations have plummeted in Fennoscandia as beaver populations have been restored from near-extinction. Although wetter woodlands and increased deadwood would be beneficial for a wide range of forest species, reintroducing beavers is unlikely to be significant for Willow Tits. This is because beavers will not really expand the area of wooded habitat, which is the primary limiting factor for Willow Tits rather than deadwood for nesting sites. The size of beaver ponds in relation to the large area of individual Willow Tit territories also means that relatively few pairs would benefit, even if the habitat created by beavers was superior, which is unproven. Perhaps the strongest counterpoint is the ongoing decline and loss of Willow Tits at existing wetland habitats that were previously considered as strongholds, such as the Dearne Valley in Yorkshire, Far Ings in Lincolnshire and Woolston Eyes in Cheshire. This shows that the Willow Tit's problems in Great Britain go far beyond the creation of wet wooded habitats, whether that is undertaken by people or by beavers.

Urban environments

Towns and cities are generally hostile places for Marsh Tits and Willow Tits. The low tree cover and poor connectivity between urban green spaces means that both species are unlikely to maintain populations in densely urban environments. Their poor dispersal abilities outside woodland also mean that urbanised landscapes are significant barriers to movement (Shimazaki *et al.* 2016). Marsh Tits and Willow Tits are therefore often absent from urban parks and gardens, especially in Great Britain. This contrasts with Great Tits, Japanese Tits or Blue Tits, which regularly breed and winter in urban environments across Eurasia.

In very large cities, like London, Marsh Tits occur only in some peripheral woodlands on the very edges of the sprawling conurbation, but they are declining even here. For example, on the northern edge of London the large expanse of ancient woodland in Epping Forest lost its Marsh Tits in the late 20th century. Occasional dispersing birds may appear in some of the outer London parks, but they don't make it to the inner-city green spaces, such as Hyde Park and Regent's Park. Even in the large and well-wooded Richmond Park in southwest London, Marsh Tits stopped breeding in the 1930s and haven't been seen since 1972 (London Natural History Society 2014). Willow Tits have only been recorded in Richmond Park on a handful of occasions, and not since 1987.

In light of the unsuitability of the London conurbation today, it is ironic that the first known specimens of British Willow Tits were collected in 1900 at Coldfall Wood, between Finchley and Muswell Hill in what is now north London. Coldfall was then a rural woodland but is now 8.5km inside the M25 orbital motorway, deep within the urban sprawl. Needless to say, Willow Tits have long since disappeared from Coldfall Wood, and no longer occur anywhere near London.

Even in cities with a large amount of woodland cover, Marsh Tits and Willow Tits are limited to the larger habitat patches. Sweden's capital of Stockholm has patchy tree cover across

40 per cent of the city and its suburbs, but Marsh Tits and Willow Tits rarely occur in pockets of woodland under 30ha, and have a high likelihood of being present only in woodlands of at least 200ha in the outer suburbs (Mörtberg 2001). Meanwhile, in the South Korean capital of Seoul, Marsh Tits have maintained populations in the large forests on the hills and mountains that have been engulfed by the metropolis (Hong & Kwak 2011).

In smaller cities, the urban centre is often not physically far from peripheral populations of Marsh Tits or Willow Tits, and so birds can penetrate deeper into the city via green corridors. In the Polish city of Wrocław the urban rivers are flanked by swathes of trees and shrubs, giving Marsh Tits enough habitat to breed deep into the city (Kopij 2019). More Marsh Tits breed in the outer suburbs than in the inner city, however, and populations in the urban parks and riversides have declined over time (Tomiałojć 2011).

In the small city of Lahti in the Finnish boreal zone, Blue Tits and Great Tits are the most abundant species wintering within the urban area, being far more common than in rural forests nearby (MacGregor-Fors 2022). Willow Tits are also resident and common in the rural forests but they are not present within the city. This difference in urban use between the tits is seemingly related to extensive bird-feeding by people in Finnish cities, which is readily exploited by mobile Blue Tits and Great Tits but not by the Willow Tits, which remain within their forest territories.

Marsh Tits and Willow Tits are more likely to occur in small towns and villages than in larger urban areas. In Białowieża village in Poland, a few kilometres from the Białowieża Forest, Marsh Tits regularly breed in wooded parks and managed green spaces close to the village centre, and they forage on old fruit trees in gardens (Figure 3.32).

Figure 3.32. Suburban breeding habitat of Marsh Tits on Park Dyrekcyjny Street in Białowieża village, Poland. A Marsh Tit nest is located 10m high in the tree on the left (circled), within a small park surrounded by houses, office buildings and roads. The pair did much of their foraging in nearby gardens. (© Richard K. Broughton)

Marsh Tits and Willow Tits both incorporate suburban gardens and parks into their foraging ranges if they are close to their home-ranges in nearby woodland, especially where they find bird-feeders (Plummer *et al.* 2019). Willow Tits in the Wigan study area regularly use gardens and suburban green spaces near to their main shrubland and wet woodland habitats. However, radio-tracking in the Dearne Valley in late winter and early spring showed that the local Willow Tits used gardens only occasionally (Pinder & Carr 2021). In the Monks Wood study some of our colour-ringed Marsh Tits occasionally left their home-ranges in the woodlands to visit bird-feeders in nearby village gardens during winter, sometimes travelling up to 500m to do so. The birds must have been exploring outside the woods in order to find these food sources, using hedgerows and trees as corridors, and perhaps following other tits in a foraging flock. These sightings in village gardens were quite rare during more than 20 years of observations around Monks Wood, probably because most gardens were so far away.

Overall, the use of urban or suburban gardens by Marsh Tits and Willow Tits seems to depend on their proximity to existing habitat, maybe because leaving the cover of woodlands to visit gardens increases exposure to predators such as Eurasian Sparrowhawks. Bird-feeders can be a significant attraction, but heavy competition from dominant Blue Tits and Great Tits can limit the ability of Marsh Tits and Willow Tits to access the food, so the risks are not always worth the reward (Broughton *et al.* 2022c, Maziarz *et al.* 2023).

Farmland and hedgerows

Open farmland and fields are actively avoided by Marsh Tits and Willow Tits. Despite this, farmland is the dominant landscape surrounding most of the woodlands where the birds live, and so they have to cope with the agricultural environment to some extent. The degree of agricultural intensification and woodland fragmentation is the key to how well the birds can cope in the farmed landscape. The progression of intensive farming throughout Europe in the 20th century has resulted in well-documented declines of many farmland bird populations (Newton 2017). Even as woodland birds, Marsh Tits and Willow Tits are also affected by agricultural intensification, because of the widespread loss of woody habitats like hedgerows, trees, small groves and even entire woods that the birds use. Intensification has also involved the increased use of fertilisers and pesticides that reduce food availability on farmland and can drift into woodlands (Gove *et al.* 2004).

Hedgerows are a major feature of farmland in Great Britain and elsewhere in western Europe, where they are an important semi-natural woody habitat. Hedgerows come in various types, from regularly trimmed lines of dense shrubs to unmanaged rows of uneven height and width. Many hedgerows also incorporate some mature trees, such as Pedunculate Oak or Common Ash (Figure 3.33). Other linear woody features are essentially lines of trees that may have developed from unmanaged former hedgerows. Narrow strips of woodland also occur along streams, riverbanks and field boundaries. These woody linear features are essential for Marsh Tits and Willow Tits in connecting pockets of woodland and allowing birds to move between them through farmland landscapes.

Relatively few Marsh Tits actually breed within hedgerows or other linear woody features on farmland, as maintaining a sufficient territory area along a narrow line of trees and shrubs is impractical. Marsh Tits are more likely to use taller hedgerows and tree lines to move between patches of woodland (Broughton *et al.* 2021b), and I occasionally found them

Figure 3.33. A typical hedgerow on English lowland farmland at Hillesden in Buckinghamshire, dominated by hawthorns with some mature Pedunculate Oaks. These linear woody corridors allow Marsh Tits and Willow Tits to disperse across open landscapes. (© Richard K. Broughton)

using hedgerows to link together small woods several hundred metres apart, to form a single wooded territory. Without the hedgerows connecting the woods the individual patches would not have been viable as territories.

Willow Tits seem more able than Marsh Tits to use hedgerows and other linear woody features as core elements of their territory. Radio-tracking in northern England found Willow Tits living in linear woody habitats along railway lines and waterways, which also connected other blocks of habitat (Pinder & Carr 2021). Wooded old railways, canal sides and riverbanks are important in both rural and post-industrial areas, such as Yorkshire, Greater Manchester and County Durham, creating interconnected breeding habitat (Carr & Lunn 2017). On mixed farmland at one of my monitoring sites at Hull, Willow Tits based their territories around large, unmanaged hedgerows along canals and drainage channels, mostly old hawthorns with some Common Elder, willows, Common Ash and Sycamore, but this population went extinct in 2000 (Broughton 2002). In rural Ceredigion in west Wales, 18 Willow Tit territories were found in bushes and trees along a 23km linear stretch of a decommissioned railway line (Morris 2021). I have also found Willow Tit territories on mixed farmland in western England along hedgerows and tree lines linking small wooded copses (Figure 3.34). Some of the Willow Tit pairs on the Berkshire–Hampshire border around Combe Wood base their territories along hedgerows on ancient 'green lanes' or 'hollow ways', which are old wooded tracks beneath a canopy of Common Hazels, hawthorns, Pedunculate Oaks and Common Ashes (Figure 3.35).

An analysis of the Willow Tit's decline in Great Britain during 1965–2000 showed that farmland originally accounted for a third of the plots on which they were present in the national Common Birds Census (Siriwardena 2004). Although Willow Tits were more abundant in

Figure 3.34. A Willow Tit territory along a linear woody field boundary on mixed farmland at Weston Jones in Staffordshire, England, where a pair were present and a nest-site found in 2015. A mixed-species hedgerow with a section of poplar trees joins a small (1ha) wooded plot containing Black Alder, Pedunculate Oak and a small pond. (© Richard K. Broughton)

Figure 3.35. A 'green lane' at Combe in Berkshire, southern England, occupied by resident Willow Tits. The ancient trackway through farmland is flanked on each side by linear strips of unmanaged mature shrubs and trees, forming a canopy overhead. (© Richard K. Broughton)

wetland and woodland plots, those birds in the farmland matrix would have been important in linking together the populations in the patchy woodlands and shrublands. The population decline was steepest on farmland, and Willow Tits have now virtually disappeared from this habitat. This left the remaining populations isolated in the woodland refuges, meaning that Willow Tits had to disperse much further to try and find other birds, which would be less successful (see Chapter 7).

Farmland is therefore important to Marsh Tits and Willow Tits because of its dominance as the landscape fabric in many areas. Policies that affect farmland also govern the semi-natural habitats within it, such as hedgerows and woodland strips, which the birds need to move across agricultural landscapes. Taller and wider hedgerows are clearly of more use to both species than short and heavily managed ones, as they provide more cover in which to move through the landscape safely, and more food in the form of seeds and invertebrates (Staley *et al.* 2012, Facey *et al.* 2014). As a breeding habitat, farmland has been important more as a low-density 'sea' of birds that link stronger populations in the 'islands' of better habitat, although for Willow Tits in Britain this is largely a thing of the past.

Key habitat features for Marsh Tits and Willow Tits

Spanning across three forest biomes in Europe and Asia, the ranges of Marsh Tits and Willow Tits encompass a lot of variation in the habitats in which the birds are found. Trying to tease out broad generalisations of habitat preferences from the many studies and situations is a challenge. There are some general themes, however, and these are outlined below.

MARSH TIT HABITAT STRUCTURE AND COMPOSITION

In the temperate forest biome the woodland structure seems to be of critical importance for Marsh Tits. A high-quality habitat has a complex structure of multiple layers of vegetation, including a ground flora of herbs and grasses, an understorey layer of shrubs and saplings, and an overarching canopy of mature trees. Within these layers, more complexity is provided by the structural architecture of large branching trees, climbing plants, thickets of Brambles or bamboos, snags and fallen deadwood, tree-fall gaps and tree cavities. This dynamic structure of mature woodlands provides myriad niches and microhabitats that Marsh Tits can exploit throughout the year (Hinsley *et al.* 2007, Broughton *et al.* 2012a).

In Monks Wood the structural diversity of the ancient woodland comprises a diverse tree canopy 25–30m tall with natural tree-fall gaps, and a varied understorey of patchy thickets and tangled branches up to 8m tall (Hill & Broughton 2009). Using lidar, a laser range-finding method (Figure 3.28), we mapped the height of the entire tree canopy and understorey in great detail and compared these between areas where the Marsh Tits were placing their territories and where they were not (Broughton *et al.* 2006, 2012a, Hill & Broughton 2009). This showed that Marsh Tits were favouring the areas where trees averaged at least 15m tall, canopy closure was 80 per cent or more, the understorey layer covered at least 45 per cent of the area, and understorey shrubs averaged around 3–4m tall. These results might be quite site-specific in their detail, but they point to a general preference for a more mature, multi-layered and complex woodland structure (Figure 3.36).

The Monks Wood results were supported by a broader study showing that the coverage and height of the understorey shrub layer was critically important to Marsh Tits (Hinsley

Figure 3.36. A complex, species-rich, multi-layered woodland structure at Monks Wood, preferred by the local Marsh Tits. A ground layer of Brambles and herbs merges into an understorey of shrubs up to 8m tall, below a mature tree canopy reaching up to 25m or more. (© Richard K. Broughton)

et al. 2007). Examining breeding territories in five ancient woods across England, this study found that the tree canopy varied between sites but the understorey shrub layer was far more consistent. This suggested that the size and composition of canopy trees was less important than having a dense and diverse understorey, especially the layer of vegetation at heights of 2–4m. Another study of 180 woods across southern Great Britain also showed that Marsh Tits were associated with greater understorey diversity and density in the 2–4m layer (Carpenter *et al.* 2010). This slice of the understorey vegetation might reflect prime foraging locations for Marsh Tits, or could just be an indicator of structurally diverse woodland. Regardless, these studies all underline that Marsh Tit habitat quality is strongly linked to woodland structure. The multiple layers of vegetation within ancient, old-growth and unmanaged forest generally provide the greatest amount of the complexity that Marsh Tits seem to need.

Marsh Tit studies can give the impression of some affinity for forests rich in mature oak trees. However, detailed studies in Europe have shown that there is not an especially strong relationship between Marsh Tits and oaks, or any particular tree species. Instead, mature oaks are probably more important as an indicator of a species-rich and structurally diverse woodland. Although oak trees are famed for hosting abundant caterpillars and other invertebrate prey (Southwood 1961, Smith *et al.* 2011), Marsh Tits do not select oak-rich areas of woodland in which to nest, nor do they nest in oak trees themselves very often. Oaks are not even the most popular trees for foraging when feeding nestlings. Instead, Marsh Tits more often nest in and among Common Ashes, Field Maples, Common Hornbeams or Small-leaved Limes (Wesołowski 1996, Broughton *et al.* 2011, 2012b). Given a choice in a diverse woodland, Marsh Tits feeding their nestlings tend to forage on Common Ash, Field Maple, Common Hazel and hawthorn ahead of Pedunculate Oak (Carpenter 2008).

During the winter, however, Marsh Tits did show some preference for foraging on oak trees at Monks Wood and Wytham Woods (Gibb 1954a, Broughton *et al.* 2014). The birds often foraged on the bare twigs, deeply creviced bark and moss-covered limbs of large oaks, which provide lots of different substrates and niches. Generally, however, throughout their range Marsh Tits will forage on just about any native broadleaved trees that are available, showing a preference for diversity rather than any particular species (Figure 3.37). This diversity provides a succession of resources that Marsh Tits can use throughout the year, such as nest-sites and roosting sites, along with seeds or invertebrates on the twigs, leaves, flowers, buds or bark (Gibb 1954a, Amann 2007, Carpenter 2008). Common Beech, maples, pines and spruces are all valuable as seed-producing trees, and in early spring Marsh Tits will also spend lots of time foraging in the canopy of elms, aspens, birches and willows, attracted by invertebrates or nectar in the flowers and catkins.

As with trees, Marsh Tits use a wide variety of understorey shrubs, and diversity is the important element, rather than the presence of any given species. Nevertheless, Marsh Tits spend a lot of time foraging in the understorey, and in Great Britain some well-used species at various times of the year include Common Hazels, hawthorns, Blackthorn, Sallows, Common Elder, Crab Apple and Spindle (Gibb 1954a, Carpenter 2008). Creepers can be important to Marsh Tits, especially big tangles of Honeysuckle reaching into the canopy trees, and thickets of Bramble nearer the ground or rambling through other shrubs. The understorey shrubs and young trees provide invertebrates on their foliage and stems, but some also produce abundant flowers in spring (Blackthorn, Sallow) and berries in summer and autumn (Spindle, Honeysuckle, Crab Apple), which Marsh Tits can exploit for nectar and seeds. A few shrubs can also provide good nest-cavities, especially Common Elder and Crab Apple.

Figure 3.37. A Marsh Tit foraging on Common Beech. Tree diversity is more important than any particular species of tree. (© David Tipling)

Figure 3.38. Flowery and herb-rich rides, clearings and woodland edges are important seed sources in late summer and autumn. This uncut ride edge in Monks Wood has lots of knapweed, whose seeds are a favoured food of Marsh Tits in late summer. (© Richard K. Broughton)

The ground-layer vegetation is an essential component of Marsh Tit habitat, particularly those tall plants that produce relatively large, oily or abundant seeds, like hemp-nettles, thistles, burdocks and knapweeds. Most herbaceous plants need good light or half shade, and the structural diversity of woodlands is important in providing suitable clearings such as glades and tree-fall gaps, along woodland edges or around ponds, bogs and seasonal wetlands. In managed woodlands this herb layer can be encouraged along rides (Figure 3.38) and buffered edges, and after felling or coppicing. Marsh Tits also forage on bare ground and in leaf litter on the woodland floor, picking up fallen seeds and fruits such as beechmast and Crab Apples. Again, diversity of habitat structure and composition is key to providing Marsh Tits with a range of niches and resources throughout the year. The more complexity the better.

WILLOW TIT HABITAT STRUCTURE AND COMPOSITION

Young woodlands or shrublands are suboptimal habitat for most cavity-nesting birds, like Marsh Tits, Blue Tits or Great Tits, but they can provide niche opportunities for Willow Tits in landscapes where competitors are otherwise abundant, such as in Great Britain. In contrast to the multi-layered structure of mature ancient woodland and old-growth forest, immature woodlands with a young age profile are more limited in their three-dimensional structure. The shorter shrubs and trees create only a single major layer of woody vegetation, usually under 10–15m tall. Natural tree cavities for nesting or roosting are scarce in these young woodlands, as the trees have not had enough time to develop them. This lack of a mature tree canopy and nest-cavities means that the Willow Tits have an advantage in a relatively low-competition environment. Not only are Willow Tits less reliant on tree-dwelling caterpillars than other tits (Cholewa & Wesołowski 2011), but they are able to excavate their own nest-cavity in young trees as slender as 6.5cm in diameter (Parry & Broughton 2018).

At the Wigan study site in northern England most of the tree and shrub canopy within Willow Tit territories was less than 10m tall (Broughton *et al.* 2020). This contrasts with most of the tree canopy being over 15m tall in Marsh Tit habitat at Monks Wood (Figure 3.39). A typical Willow Tit territory in the Wigan study contained around 7ha of wooded habitat with a height profile heavily skewed towards young trees and bushes up to 7m tall, which were only around 25–30 years old. Almost half of the average territory also consisted of open rough grassland and tall herbs, or waterside reeds and sedges, in a patchy habitat mosaic.

Other studies in Great Britain have also found Willow Tits occupying relatively young woodland. In an analysis of 241 woods across England and southern Scotland, Willow Tits

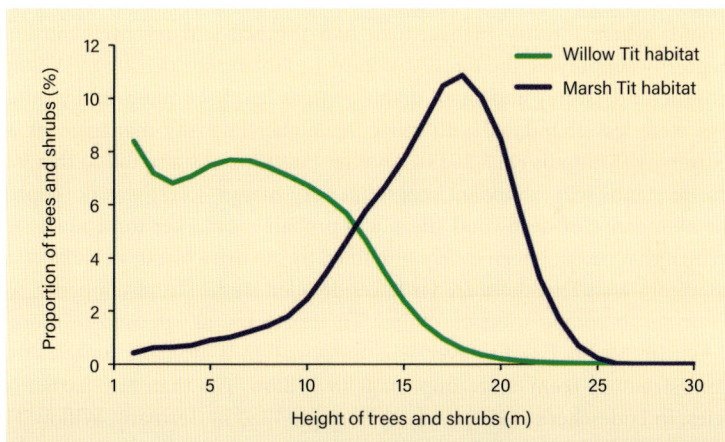

Figure 3.39. Woody vegetation height profiles for Willow Tit habitat at Wigan and Marsh Tit habitat at Monks Wood. Willow Tit habitat comprises 287ha of young woodland where two-thirds of the vegetation is below 10m tall. Marsh Tit habitat is 160ha of mature woodland where two-thirds of the vegetation is above 15m tall. Height profiles are calculated from airborne lidar from 2022 (Wigan) and 2014 (Monks Wood). Vegetation below 1m was excluded. Lidar data for Wigan contain Environment Agency public sector information licensed under the Open Government Licence v3.0. Lidar data for Monks Wood were captured by the NERC Airborne Remote Sensing Facility, accessed via the CEDA repository (CC BY 3.0).

were more likely to occupy sites with a shrubby structure and shorter tree canopy, rather than more mature woodland (Stewart 2010). In a related study of 65 woods in Nottinghamshire, in central England, the likelihood of Willow Tits being present was 60 per cent for sites with trees of 20–25 years old but fell to 15–30 per cent for sites over 80 years old (Lewis *et al.* 2009a). The canopy height and tree size were greater in the unoccupied woods, indicating a decline in occupation as the woodlands aged.

These findings from British studies point to Willow Tits being more likely to occur in younger, early-successional woodlands with dense shrubby vegetation. Another feature that was significant in two of the English studies was that soil moisture was higher in woodlands where Willow Tits were still present, compared to drier soil conditions in woods that did not have Willow Tits (Lewis *et al.* 2007, 2009a). A similar result was found between regions with declining and stable populations, where the latter tended to have wetter soils (Stewart 2010). This difference could be related to woodland structure if younger woods are wetter, or it may reflect woodland size if larger woods are buffered against the drying effects of wind and sun that penetrate deeper into small woods. It is hard to see how wet soil has a direct effect on an arboreal bird like a Willow Tit, so it must be an indicator for something else in the habitat, like woodland age or structure, seed availability or insect abundance, but this has not yet been explained.

Whereas British Willow Tits are increasingly associated with young woodlands and shrublands, in Fennoscandia they are associated with more structurally diverse and mature forests of conifers and birches, with lots of boggy clearings and large standing deadwood. The changes to this habitat structure wrought by commercial forestry have had hugely negative impacts on Willow Tits in this region. In 60–80-year-old forests of pine, spruce and birch in Sweden, Eggers & Low (2014) showed that Willow Tit survival and breeding success were lower in stands where forestry management had thinned and removed the understorey vegetation, greatly simplifying the forest structure. In Finland the density of Willow Tits was lower where the understorey and small-diameter trees had been removed by thinning, and where snags were less abundant, both reflecting higher-intensity management and lower structural diversity (Kumpula *et al.* 2023). Another Finnish study also found that Willow Tits favoured forest stands with abundant snags, but very mature forest was not a prerequisite; instead, the abundance of deadwood snags was probably a marker for a more structurally complex habitat that had not been overly simplified by intensive forestry (Vatka *et al.* 2014).

Compared to Marsh Tits, there are far fewer detailed studies from temperate woodlands of the tree or shrub composition of Willow Tit territories, as opposed to those used for nest-sites. As with Marsh Tits, however, what information is available suggests that habitat diversity and structure seem more important to Willow Tits than any particular tree or shrub species. In Ludescher's (1973) study at Pfrunger Ried in Germany, Willow Tits mostly foraged in birches, which happened to be the most common trees, but they also used Norway Spruce and Scots Pine extensively. In montane temperate or subalpine forest in Japan, Willow Tits used various conifers including hemlocks, pines, spruce, firs and especially larches, but spent more time foraging in deciduous trees like alders, birches, elms and willows, along with bamboo thickets (Nakamura 1970).

Broad studies of Willow Tit habitat use in Great Britain indicate some preference for deciduous trees over coniferous (Lewis *et al.* 2009a), although conifer plantations are used in some places, such as Lake Vyrnwy in north Wales. There is a preference for foraging in birches within conifer plantations and mixed ancient woodland, even where birch is relatively uncommon in those woods (Gibb 1954a, 1960, Gibb & Betts 1963). Other woody species used for foraging, more than their abundance might suggest, include Common Elder, European Aspen and Pedunculate Oak, but virtually all available trees and shrubs can be used. Stewart (2010) found that woodlands still occupied by Willow Tits tended to have abundant hawthorns, willows and Black Alder, but this was probably more a reflection of the birds occupying a damp, shrubby habitat structure rather than more mature woodland.

It is a circular question whether Willow Tits in Great Britain are selecting younger, wet woody habitats that tend to be rich in birch, alder and hawthorn, or whether the birds are favouring these tree species, which happen to have a certain structure. This question is possibly answered by studies in other regions and time periods, which show that Willow Tits are not really specialists of particular woodland communities, but are instead quite generalist in their habitat choice, though they can often be constrained in their use of woodland types. In Sweden, for example, the presence of dominant Marsh Tits limits the local Willow Tits to woodlands consisting of conifers, birch, aspen and alder, but where Marsh Tits are absent the Willow Tits are released from competition and can expand into oak and ash woodland too (Alatalo *et al.* 1985). Competition is also likely to be a major factor shaping Willow Tit habitat use in Great Britain, particularly in their increasingly concentrated use of wet woodlands rich in young birches, alders, willows and hawthorns, which do not have many natural nest-cavities for competing Marsh Tits, Blue Tits or Great Tits. That may be the strongest reason why Willow Tits are still present in these habitats.

The issue of identifying good Willow Tit habitat in Great Britain therefore requires some caution, as almost all of the research on this topic was carried out in the last couple of decades. This research period came after Willow Tits had already been in severe decline for a long time and had been virtually lost from other habitats, such as farmland and ancient woodland. The recent habitat studies therefore reflect where Willow Tits have held on for longer in Great Britain, and are maybe something of an artefact of the period of data collection. If there had been a similar analysis during the 1950s and 1960s it would have found Willow Tits thriving in a much wider range of habitats, from ancient woodland to pine plantations (Foster & Godfrey 1950, Gibb 1960).

A further complication for habitat assessments in Great Britain is that when a population is in rapid decline there will be an element of randomness to which woods remain occupied or lose their Willow Tits along the downward trajectory, due to chance events and local quirks. These stochastic factors probably influenced the results of a Willow Tit study in southern England, where the small populations involved were less than 10–20 years away from local extinction (Lewis *et al.* 2007). In this study there was no difference in any of the habitat features between occupied and unoccupied woods, such as tree canopy height and coverage, tree density and sizes, or understorey shrub density. This could be a reflection of fairly wide tolerances of Willow Tits with regard to woodland structure, or it could have been because the populations were so close to extinction that habitat quality had stopped being a major factor. Instead, the woods where the last Willow Tits were clinging on could have been mostly down to chance.

WOODLAND SIZE AND HABITAT FRAGMENTATION

The configuration of woodland patches within the landscape is critically important for both Marsh Tits and Willow Tits. Habitat fragmentation reflects the degree to which woodlands are dispersed, including the number of habitat patches, their size and proximity to each other, and how well they are connected by other habitats like hedgerows (Figure 3.40). There could be an

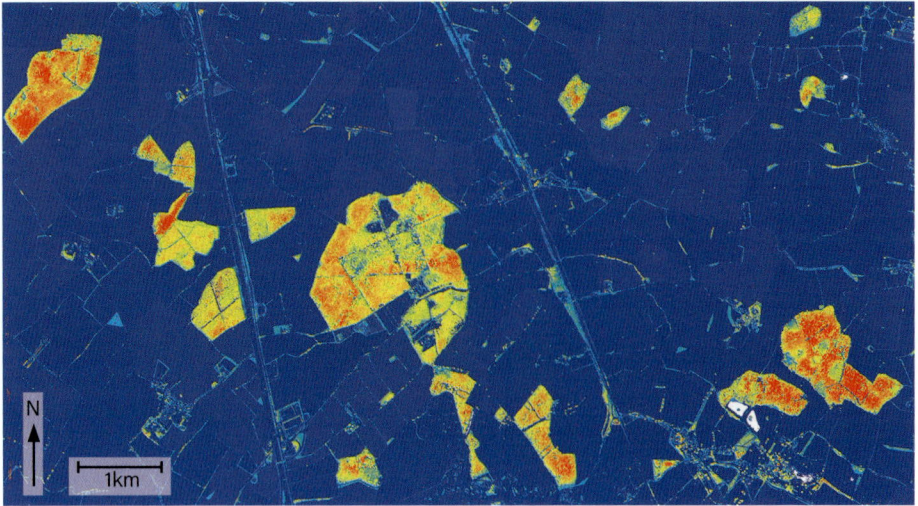

Figure 3.40. Habitat fragmentation of woodland patches in the Monks Wood study area, shown in a lidar canopy-height model from June 2014. Cooler, blue colours represent open fields and low vegetation, and warmer, redder colours represent taller shrubs and trees up to 30m tall. The woods have sharp boundaries in the agricultural landscape, with few connecting hedgerows between them, inhibiting the movements of Marsh Tits between woods. Willow Tits were locally extinct by 2000. Lidar data were captured by the NERC Airborne Remote Sensing Facility, accessed via the CEDA repository (CC BY 3.0).

excellent patch of wooded habitat, but if it is too small to hold a single territory or too far away for dispersing birds to find it, then it will not be occupied no matter how good the vegetation structure and variety of trees and shrubs.

Studies in the Netherlands, England and Japan show that larger woods in well-wooded landscapes are more likely to support and retain populations of woodland specialists like Marsh Tits (van Dorp & Opdam 1987, Hinsley *et al.* 1995a, Kurosawa & Askins 2003). In smaller and more isolated woods, Marsh Tits are more likely to disappear from one year to the next and have more years of absence, because when resident birds die they are not easily replaced by immigration (Hinsley *et al.* 1995b). As outlined in Chapter 7, Marsh Tits are not very good at dispersing between woodland patches, as the open fields in between act as barriers, reducing their ability to reach more distant woods (Broughton *et al.* 2010). Hedgerows and trees leading from the home wood can act as corridors across the open fields that channel dispersing birds in certain directions, but unless they soon lead to other good woodlands then the birds are likely to perish in the open landscape rather than settle into the local population (Alderman *et al.* 2011). In eastern England some classic studies showed that Marsh Tits never settled in woods smaller than 1ha and their likelihood of occupation only reached 90 per cent for woods of around 15ha or more (Hinsley *et al.* 1995a, 1996). Based on these studies, the average-sized wood in England, which is just 10ha, only has a 60 per cent likelihood of supporting resident Marsh Tits (Forestry Commission 2001).

The main reason Marsh Tits cannot make a living in isolated small woods is that their average territory size is 4–6ha (see Chapter 5), so a wood of 10ha only has enough room for one or two pairs. Smaller woods also contain more woodland edge relative to interior habitat, and at Monks Wood we found that Marsh Tits were sensitive to edge effects extending up to

100m into the wood, and especially within 50m of the perimeter (Broughton *et al.* 2012a). Marsh Tits are less likely to use this edge zone, where the tree canopy is shorter and more exposed to wind and rain, although they sometimes forage among dense shrubs on sunny woodland edges (Melin *et al.* 2018). Hewson & Fuller (2006) classified Marsh Tits and Willow Tits as woodland interior species, and showed that their activity within 50m of the woodland edge was only around half of their activity in the deeper parts of woodlands.

In Great Britain we found that Marsh Tits and Willow Tits are both more likely to persist in regions with more woodland coverage (Broughton *et al.* 2013). The amount of woodland was a good indicator of areas where both species had gone extinct during the 1980s and 1990s, which had happened where the median coverage was below 4 per cent of the landscape. Areas with woodland cover under 3 per cent were particularly unsuitable. This analysis showed that having more woodland habitat in the landscape seems to buffer the birds from the causes of decline, probably because there are more birds locally that can easily move around woodlands to form new pairs or replace losses. The smaller and more isolated populations in poorly wooded areas were the first to disappear.

Even where there is superficially a heavily wooded landscape, the better broadleaved habitats can be fragmented within a matrix of poor-quality plantations. In one such area of central Sweden the remaining patches of deciduous trees were limited to semi-natural refuges around lakes and in abandoned pastures (Enoksson *et al.* 1995). The surrounding conifer woodland was managed as forestry monocultures of Norway Spruce and Scots Pine that offered little habitat for Marsh Tits. The patches of deciduous trees were essentially habitat islands, almost like fragmented English woods in a farmed landscape, and Marsh Tits were only likely to occur where several broadleaved patches occurred close together to form a meaningful area of good habitat. Although Marsh Tits can disperse more freely through conifer plantations than in agricultural fields or urban streets, the study demonstrated how suitable habitat can be fragmented even in apparently forested landscapes.

In the boreal forests of Finland, studies by Siffczyk *et al.* (2003) and Kumpula *et al.* (2023) neatly showed the impact on Willow Tits of habitat fragmentation caused by industrial forestry. Wintering groups of Willow Tits avoided areas of recent clear-cut logging and replanting, instead preferring intact mature forest and wooded bogs. The fragmented areas of mature forest were left as islands among clear-cuts and replantings, so the Willow Tits had to disperse further between areas of good habitat and then expand their home-ranges to compensate for the habitat loss. The birds were forced to work much harder to cover twice as much area as in the intact forest, which would have had energetic consequences for them.

These studies highlight the serious and insidious effect of habitat fragmentation on Marsh Tits and Willow Tits. Smaller habitat patches support fewer pairs, and individual birds are constrained in their movements during foraging and dispersal. These constraints squeeze populations, making them less resilient to other pressures or random events, and can eventually lead to local extinction and a low likelihood of recolonisation.

COMPETITORS AND PREDATORS

Throughout this chapter, the subject of interactions between competition and habitat quality has arisen from time to time. The example of Swedish Willow Tits increasing their use of deciduous woodland in the absence of dominant Marsh Tits implies that competition can reduce the habitat quality for Willow Tits (Alatalo *et al.* 1985). A more subtle effect has been seen where dominant Crested Tits exclude Willow Tits from prime foraging locations, reducing the intrinsic habitat quality for birds unable to access the best feeding sites (Hogstad 1978). In both examples the habitat would be of higher quality if the competitor species were absent, or far less common.

Abundant Blue Tits and Great Tits can also reduce the habitat quality for Willow Tits by stealing lots of nest-cavities and causing breeding failures, as discussed in Chapter 6. Dominant Blue Tits and Great Tits also affect the foraging opportunities for subordinate Marsh Tits by displacing them from good feeding sites via interference competition (Maziarz *et al.* 2023). These British studies suggest that the increased abundance of Blue Tits and Great Tits since the 1960s (Harris *et al.* 2022) has likely reduced the general habitat quality of woodlands for Marsh Tits and Willow Tits through increased competition pressure.

Competition is likely to be more acute where bird-feeders have become abundant in gardens and nature reserves since the 1970s, increasing the survival, productivity and abundance of local Blue Tits and Great Tits (Gosler 1993, Robb *et al.* 2008). A study in Scotland showed that Peanuts from garden bird-feeders were the most frequent food in the diet of Blue Tits and suggested that virtually the whole British population had access to such feeders (Shutt *et al.* 2021). In much of Great Britain it's hard to see how Marsh Tits and Willow Tits can avoid these subsidised competitors, and this could have important implications for them, degrading the habitat quality of woodlands by artificially inflating the populations of dominant species (Broughton *et al.* 2022c, Maziarz *et al.* 2023).

Similar suspicions have been raised in Fennoscandia, where increased bird-feeding and climate change create better conditions for Blue Tits and Great Tits (Orell 1989, Broggi *et al.* 2021, 2022). This comes at the potential expense of Willow Tits, which lose their advantage of being better adapted to cold winter conditions because the other species are increasingly released from those limitations by supplementary food and warmer nights (Pakanen *et al.* 2018, Kumpula *et al.* 2023). As a consequence, the Willow Tits will lose any competitive edge that they have for surviving in boreal conditions.

Nest predation is also a serious issue for Willow Tits that could influence habitat quality, at least in Great Britain and other parts of the temperate zone. Great Spotted Woodpeckers are the major predator of Willow Tit nests (Ludescher 1973, Parry & Broughton 2018). The big increase in Great Spotted Woodpecker abundance across Europe, including a 139 per cent increase in Great Britain since the 1990s (Harris *et al.* 2022), will evidently make woodlands more hostile for Willow Tits and reduce the habitat quality for them. This will be compounded where nest predation is combined with nest-site competition from increased numbers of Blue Tits and Great Tits, leading to high rates of nest loss (see Chapter 6).

Summary

Marsh Tits are habitat specialists of broadleaved temperate forests and woodlands, being most abundant in mature, structurally complex habitats. Marsh Tits can be considered as an indicator species for high-quality ancient woodland and old-growth temperate forest.

Willow Tits are more generalist than Marsh Tits, occurring in a wider range of wooded habitats in the temperate and boreal forest zones, including at higher altitudes into alpine forest. Willow Tits are most abundant in boggy boreal forest, old woodland and lowland riverine forest, but also occur in birch woods, conifer plantations and early-successional woodlands where other tits are less abundant.

Woodland structure and species diversity are more important than the presence of any particular tree or shrub species. Marsh Tits prefer a complex, multi-layered woodland with a diverse understorey shrub layer below a tall canopy of mature trees. Willow Tit habitat associations vary over space and time, with a preference for old-growth forests in Fennoscandia and northern Asia, but in western Europe they are increasingly associated with (or restricted to) young woodlands and shrublands, particularly around wetlands. Willow Tits were formerly widespread in a variety of farmland and woodland in Great Britain, including ancient woodland, but are now rare or absent in such habitats.

Intensive forestry management and conversion to plantations simplifies the habitat structure and composition by degrading the understorey layer, removing deadwood and creating an even-aged, low-diversity tree canopy. Forestry can also fragment native woodland habitat within a well-wooded landscape of poor-quality plantations. Forestry and intensive woodland management are major drivers of catastrophic Willow Tit declines in Fennoscandia, and likely also in Asian boreal forests.

Woodland conservation management has so far had little success in supporting populations of Marsh Tits or Willow Tits in Great Britain. Small-scale and long-rotation coppicing in small parts of a woodland may help diversify the understorey structure in some sites, but the presence or absence of coppicing does not appear to be a major factor for either species. Unmanaged ancient woodlands have been increasing in structural diversity and naturalness over time, which appears to suit Marsh Tits in particular.

Increased competition from other dominant tits, or nest predation pressure from Great Spotted Woodpeckers, may reduce habitat quality for Marsh Tits or Willow Tits, but this requires further research. In the meantime, a precautionary approach may be advisable to avoid assisting dominant competitors via bird-feeding and nest-box provision that favours Blue Tits and Great Tits.

Marsh Tits are sensitive to habitat fragmentation and are more likely to occur in large woodlands and forests in well-wooded landscapes. Isolated and small woods of 5ha or less are unlikely to be regularly occupied. Willow Tits are also sensitive to habitat fragmentation but are better able to use marginal woody habitats outside woodlands for foraging and breeding. Hedgerows and tree lines are also vital as dispersal corridors for both species, but it is important that they connect good habitat rather than lead away from woodlands into empty farmland.

Ultimately, Marsh Tits and Willow Tits do best in extensive, varied forests and woodlands that are well connected and able to support many territories in a low-competition environment. There are considerable ongoing threats to habitat quality, including deforestation, intensive management, habitat fragmentation and climate change, which will continue to negatively affect populations of Marsh Tits and Willow Tits.

CHAPTER 4

Food and foraging

Marsh Tits and Willow Tits are omnivores that eat a wide range of plant and animal material. They find or catch their food in a variety of resourceful ways, and they also instinctively plan ahead by caching excess food items to retrieve later. Foraging takes up most of their daylight hours, and both species make good use of their agility and adaptability. They will methodically search foliage, branches, deadwood and the ground, and can hang upside down or briefly hover to reach food items. Much of their food is arboreal or terrestrial, but they also catch aerial insects and even some aquatic prey.

Like other tits and chickadees, Marsh Tits and Willow Tits have dextrous claws and a stout bill with which they can hold, manipulate and process food items by pecking, tearing and hammering open tough objects, and removing unpalatable parts (Figure 4.1). Using their bills, they are capable of digging into decaying wood or large fruit, such as apples, to find invertebrates or seeds inside. Both species can also carry large or multiple food items for some distance to consume it in cover or to cache it.

Information on the diets of Marsh Tits and Willow Tits comes from various sources, including ad hoc or semi-structured observations of birds eating various things in the wild, as well as direct sampling of the adults' digestive tracts or the nestlings' crops. Taken together, these various data give a reasonable picture of the diet and foraging behaviour of both species, but there are still knowledge gaps.

Diet of adults and juveniles

Systematic data for the proportions of different foods in the diets of full-grown Marsh Tits and Willow Tits are not easy to come by, as it is often unclear what birds are eating when observed in the wild. In the past birds were shot to quantify food intake by analysing their stomach or gizzard contents (Betts 1955). Lethal methods are now out of the question, although the data from such studies are still some of the best we have.

Non-lethal methods of analysing dietary intake can use microscopic or genetic analyses of droppings collected from birds caught for ringing. One problem is that many Marsh Tits and Willow Tits are caught using bait, such as Sunflower seeds, and so their droppings will contain lots of bait and skew the natural food intake. Catching and sampling enough of these birds without using bait is not straightforward, due to their low densities.

SEEDS

Plant materials are a fundamental part of the diet of Marsh Tits and Willow Tits, and the bulk of this is the seeds of trees, shrubs and herbaceous plants, including forbs, grasses and sedges. The variety of seeds consumed by Marsh Tits and Willow Tits across their ranges might run into hundreds of plant species (Cramp & Perrins 1993). Analysis of the proportions of different seeds in the diet is very difficult from autopsy analyses, as the birds often break up the seeds into fragments while eating them, so they cannot always be identified. Consequently,

Figure 4.1. A Marsh Tit in Monks Wood processing a Common Hazel bud by dextrously holding it down and pecking at it, to peel away the outer layers and extract a food item from inside. (© Richard K. Broughton)

much of the information on which seeds are eaten also comes from direct observation of the birds foraging on the plants.

Foraging Marsh Tits in Europe will eat most seeds that they can carry and break open, from the very tiny violet and birch seeds up to the size of Sweet Chestnuts (Cramp & Perrins 1993, Amann 2007). Marsh Tits also eat seeds of coniferous spruces, pines, firs, cedars and larches, which are taken from opening cones or from the ground. Deciduous tree seeds that are eaten include Common Ash, Field Maple and even the hard seeds of Common Hornbeam. Beechmast (from Common Beech) is a favourite where it is available, and Marsh Tits will collect it from seed cases or from the ground during winter (Gibb 1954a). Seeds from various berries and fruits are also consumed, as discussed below. There are no records of acorns or hazelnuts being eaten by Marsh Tits or Willow Tits.

Marsh Tits eat many seeds from plants in the herb layer, which they can collect directly from the seedheads. At Monks Wood particular favourites include knapweeds, thistles, scabious and burdocks, which grow in sunny glades, woodland edges and clearings (Figure 4.2). Marsh Tits will leave cover to fly out into surrounding fields, hedges and meadows to reach these plants in late summer and autumn, hanging from the plants to pull out seeds or removing the whole seedhead, which they take back into cover to dismantle.

Betts (1955) found the seeds of violets and Wood Sorrel in the gizzards of most Marsh Tits sampled in southwest England during late summer and autumn, with violet seeds again in late winter. Across Europe the seeds of Common Hemp-nettle and Cabbage Thistle are especially

Figure 4.2. A Marsh Tit in England foraging on burdock burrs, which contain clusters of oil-rich seeds. (© www.garthpeacock.co.uk)

Figure 4.3. A Marsh Tit collecting seeds from Common Hemp-nettle in August at Söderåsen National Park, southern Sweden. (© Richard K. Broughton)

favoured by Marsh Tits during late summer and autumn (Figure 4.3), with birds flying up to 80m into the open to reach them (Ludescher 1973, Amann 2007).

There are fewer studies of seeds in the diet of Willow Tits. In Germany, several sources reported that Willow Tits ate much the same seeds as Marsh Tits, again favouring thistles and Common Hemp-nettle, and also tree seeds like pine and spruce (Ludescher 1973, Cramp & Perrins 1993, Glutz von Blotzheim & Bauer 1993). Japanese Willow Tits were recorded eating seeds of thistles, Japanese Hops and Japanese Sumac (Nakamura & Wako 1988). It is not clear whether Willow Tits eat beechmast, as Gibb (1954a) did not see any taken at Wytham Woods. Beechmast is not a tough seed to open, with Marsh Tits and Coal Tits both handling it without a problem, so it may be that the Wytham Willow Tits were kept away from beechmast by competition from the more numerous Great Tits, Blue Tits, Marsh Tits and Coal Tits.

BIRD-FEEDERS AND SUPPLEMENTARY FOOD

Marsh Tits and Willow Tits will both eat a wide range of foods provided by people, either deliberately at feeding stations or inadvertently as crop plants or food provided for other species. Bird-feeders and bird-tables are readily used by both species where they are in or near their home-ranges. Where local tits have previous experience of bird-feeders, they can remember or quickly learn from others how to exploit them. At Monks Wood, individual Marsh Tits remember how to use baited cage-traps despite almost a year passing since they last saw one. I watched a new bird learning from others how to use a trap within a few minutes of first seeing it. Where artificial feeders are new to an area, however, Marsh Tits or Willow Tits can take several weeks or even months to start using them (Urhan *et al.* 2017).

Figure 4.4. A Marsh Tit taking a Sunflower seed from a bird-feeder in Sweden. (© Peter Stronach)

Figure 4.5. Garden bird-feeders disproportionately benefit common and dominant Blue Tits and Great Tits, which may then compete with Marsh Tits and Willow Tits in the wider area. (© David Tipling)

Both species really like oily Sunflower, Nyjer and Hemp seeds (Figure 4.4), but will also eat cereals such as Wheat, Barley, Oats, Millet and Maize if nothing else is available. Pumpkin seeds and Peanuts are also eaten at feeding stations, as well as animal fats like suet or lard. In Great Britain many of the cereals and Sunflowers are common as crops in the wider countryside, or as conservation plantings to feed farmland birds over the winter. Marsh Tits and Willow Tits will happily exploit these sources too, if they are nearby. Sunflower seedheads are especially popular with both species. Marsh Tits and Willow Tits can even become habituated to take Sunflower seeds or Peanuts from the hand, particularly in boreal regions.

I have seen Marsh Tits visiting the seed hoppers that provide Wheat seeds to Common Pheasants in woodlands used for hunting. One Marsh Tit that we radio-tracked at Monks Wood travelled over 1.5km each day to visit a pheasant feeder that was stocked with grain dispensed at the base. The bird took one seed at a time and ate it in nearby cover.

Although garden bird-feeders are used by local Marsh Tits or Willow Tits, they are visited by many more of the competitor species, especially dominant Blue Tits and Great Tits (Figure 4.5). In a study at Wytham Woods a high abundance of dominant tits limited the Marsh Tits' access to feeding stations, which had a strong impact on their foraging activity (Maziarz *et al.* 2023). Any benefits for some Marsh Tits or Willow Tits that access the bird-feeders are likely to be outweighed by inflating the abundance of the more dominant tits, which could then impact them more widely through interference competition for nest-sites and foraging spaces (Broughton *et al.* 2022c).

Yet another risk of feeding stations is that of increased predation. In contrast to natural food sources, which are short-lived and dispersed, a fixed and regular artificial food source can be visited very often by Marsh Tits or Willow Tits so long as seeds are available for caching (Figure 4.6). This can make them a predictable target for ambush predators such as Eurasian Sparrowhawks or domestic cats. If a Blue Tit or Great Tit is killed by a predator at a feeding station then there will be a large pool of others in the area, so the loss can easily be absorbed or replaced. For low-density Marsh Tits and Willow Tits, however, only a few resident birds from adjacent home-ranges will be visiting a fixed feeding station. With a small or negligible floating population, particularly in Great Britain, the loss of even a single bird during winter or early spring is far less likely to be replaced, and so this could mean the failure of the breeding territory for that year (Broughton *et al.* 2011).

For these reasons, artificial feeding stations in or around Marsh Tit and Willow Tit habitats are potentially detrimental to small and vulnerable populations. This is especially so in Great Britain, where their low density and disparity in numbers with the dominant species is most acute. Overall, Marsh Tits and Willow Tits can usually manage just fine without bird-feeders, due to their in-depth knowledge of their home-range's resources and their ability to store food, which gives them a natural advantage over Blue Tits and Great Tits. Providing supplementary food eliminates this advantage, as the dominant Blue Tits and Great Tits are then never subjected to a tough period of low food availability that would otherwise limit their abundance (Broughton *et al.* 2022c).

Figure 4.6. A Willow Tit collecting multiple Sunflower seeds at a feeding platform at Pennington Flash in Greater Manchester, northwest England. (© Philip Schofield)

CACHING

Marsh Tits and Willow Tits cache or hoard food, as do all *Poecile* species and also Coal Tits, Crested Tits, Varied Tits, nuthatches and corvids (Sherry 1989, Brodin 2005). Caching involves hiding food items in various places around the home-range, which the birds return to hours, days, weeks or even months later. Caching enables the birds to store excess food, and seems to be related to their year-round residency in territories or home-ranges. By spreading food resources over time, caching enables the birds to remain resident in one place throughout the year, even in the boreal and alpine winters. However, in order to benefit from their caches the birds are tied to their home-range and so cannot wander off on exploratory foraging trips like Blue Tits or Great Tits. Food caching is therefore a behaviour that is intertwined with the social organisation of Marsh Tits and Willow Tits (Ekman 1989, Matthysen 1990, Dhondt 2007).

Both species predominantly cache seeds, and they do it most actively from late summer and throughout the winter until early spring, with a distinct peak in autumn. Caching can occur throughout the year, however, and invertebrates or berries can also be stored away. During autumn, Swedish Willow Tits cached a quarter of their invertebrate prey (mostly larvae/caterpillars) and three-quarters of the seeds that they found (Brodin 1994a). There are also observations of Marsh Tits caching breadcrumbs and small pieces of fat (Almond & Almond 1950, Robinson 1950).

Caching occurs when the birds have an excess of food in front of them, such as masting trees, fruiting bushes, seeding herbs, opening cones on conifers, or artificial gluts like a bird-feeder. This stimulates them to collect the food and hide it, and they generally do this in concentrated periods of activity until the food is depleted. Caching can also occur during more general foraging, when the birds will eat items as they are discovered but then occasionally cache something.

When faced with a concentrated seed source at a bird-feeder, Marsh Tits and Willow Tits are selective in which seeds they take. They will pick up and drop several seeds in turn before selecting the ones they want, presumably testing them somehow in their bill, maybe for size or firmness. Both species can carry at least three large seeds (e.g. Sunflower) at a time, holding one or two in the bill, one deeper in the mouth and one in the throat. When dissecting a Marsh Tit that had flown into a window near a garden bird-feeder, I was surprised to find it still had intact Sunflower kernels held in its mouth and throat.

Marsh Tits and Willow Tits can cache hundreds of food items per day and up to 150,000 in a year (Brodin 2005), taking seeds from bird-feeders at a rate of one per minute (Cowie *et al.* 1981). They are *scatter hoarders*, caching each food item individually in a different location to create a spread of hidden food around their home-range (Figure 4.7). Food can be cached within a few metres of where it was found or carried for distances of over 100m, but most caching occurs within 50m of the food source. Items collected within a short period are often cached in loose clusters within 10m of each other, and seeds are hidden at lower densities as the distance increases from a food source.

Food items are cached in a variety of nooks and crannies, usually above the ground, and are simply pushed into the crevice or substrate. Popular cache sites for both species include cracks or fissures in tree bark, under loose bark on deadwood, among the mosses or lichens on tree trunks and boughs, in crevices in open conifer cones, and at the ends of broken twigs or dead, hollow plant stems, such as Cow Parsley or Common Nettle. In wet habitats the birds will hide food inside the hollow stems of Common Reed.

Figure 4.7. Willow Tits are scatter hoarders, placing seeds individually in nooks and crevices around their home-range. (© Robert Fredagsvik)

Cowie *et al.* (1981) reported that most Sunflower seeds hoarded by Marsh Tits at Wytham Woods were pushed into the soil, leaf litter or moss on or near the ground, although this might have been biased by detection methods limited to 3m in height. At Monks Wood I have most often observed Marsh Tits caching Sunflower seeds at 5–10m above the ground, especially on mature oak trees. Some individuals may have a preference for certain types of hoarding place, hiding more items in bark crevices or dead stems rather than under moss, for example, and this can be consistent for a while and then change to another type of caching site.

Marsh Tits and Willow Tits retrieve the seeds by remembering their location, and they have a physiological adaptation in their brains that gives them enhanced memory abilities. The hippocampus region of the brain, which is associated with spatial memory, is significantly larger in Marsh Tits and Willow Tits than in tits that do not cache food, such as Blue Tits and Great Tits (Brodin & Lundborg 2003). Caching behaviour is therefore innate, and it develops almost as soon as the juveniles are independent of the family group just a few weeks after fledging (Clayton 1992).

Retrieval times for caches placed around a highly concentrated food source, like a bird-feeder, are generally within a few hours or days. Where food hotspots and caches are more naturally dispersed, such as seeding plants or trees spread throughout the home-range, then retrieval times can average several weeks or more (Nakamura & Wako 1988, Brodin 1992). Willow Tits appear to store their food caches over longer periods than Marsh Tits, with the longest recorded gap between caching and retrieval in the wild being 14 weeks for Willow Tits and four weeks for Marsh Tits (Brodin 2005). However, experiments have shown that Willow Tits do not have better memories than Marsh Tits, as both species performed just as well in recovering cached food after an imposed interval of 17 days, retrieving the items within minutes (Healy & Suhonen 1996). Retrieval success does decline over time, though, with Willow Tits and Black-capped Chickadees successfully recovering almost all stored items within two weeks of caching, but after four weeks their success rate was similar to random searching, showing that their memories were fading (Brodin 2005).

Cached food items may be lost to other foraging Marsh Tits or Willow Tits that find them by chance, but conspecifics within the same home-range minimise these losses by caching in different foraging locations (Brodin 1994b). In a Swedish coniferous forest there was a very low rate of loss for experimental food caches placed by researchers, at just 1.3 per cent daily, showing that hoarded food rarely disappeared and that caching was thus a reliable strategy for local Willow Tits (Brodin 1993). In the deciduous Wytham Woods in England, however, experimental caches mimicking those of Marsh Tits lasted less than a day (Cowie *et al.* 1981). This more rapid disappearance was probably due to higher rates of pilfering by a greater abundance of other tits. Put simply, the food was more likely to be stolen by competitors, and so the local Marsh Tits could not afford to leave their caches hidden for very long.

Great Tits and Blue Tits will regularly steal cached food items from Marsh Tits and Willow Tits, either by coming across them while foraging, or by actively following a caching bird and taking the seed as soon as it leaves. I have observed this behaviour at Monks Wood, where it was clearly profitable for a Great Tit or Blue Tit to follow a Marsh Tit from a bait station as it cached Sunflower seeds in surrounding trees. There seemed to be no reaction from the Marsh Tit that had lost its cache, as if it wasn't paying attention. Great Tits can even watch from a distance and remember the locations where Marsh Tits or Willow Tits are caching food items, returning up to a day later to pilfer them (Brodin & Urhan 2014). Oddly, Marsh Tits themselves do not appear to watch and steal from others (Urhan *et al.* 2017).

This *kleptoparasitism* from dominant Blue Tits and Great Tits is yet another aspect of how competition is likely to be more acute where these species are particularly common, such as in the deciduous woods of Great Britain.

BERRIES AND OTHER FRUITS

In Monks Wood the Marsh Tits seemed to be obsessed with Honeysuckle berries during August and early September (Figure 4.8). They would spend hours pulling off the berries one at a time and taking them to a perch to process, pulling out the seeds and dropping the berry skins. The birds looked quite comical as they carried away the bright red berry in their bill, like a clown nose. After extracting the seeds they would often cache them in the surrounding trees, only eating a few. When ripe Honeysuckle berries were available some Monks Wood Marsh Tits were so fixated that they ignored even oil-rich Sunflower seeds in my cage-traps, and so were almost impossible to catch until the berries were gone.

In late summer and autumn the toes of Marsh Tits and Willow Tits are often stained purple after the birds have fed on the seeds within the juicy dark berries of Common Elder, Bramble and Common Dogwood. As with Honeysuckle, they do not eat the berries whole but instead pick them off and hold them under a foot to extract the seed, which they hammer or peel open with their bill to eat the starchy contents. The skins are discarded, and often the pulp too.

During winter at Monks Wood I found that the Marsh Tits also exploited Wild Privet and especially hawthorn berries. Again, they would pick off a berry one at a time and take it to a favourite perch to process, pulling off the skin and dropping it, being interested only in the seeds inside. Under their perches there was a scatter of orange pulp and red skin from perhaps dozens of hawthorn berries they had processed. Hawthorn is a hard seed, and Marsh Tits

Figure 4.8. Berries of Honeysuckle (left) and Spindle (right) in Monks Wood, which are favoured food items for Marsh Tits. (© Richard K. Broughton)

open it by hammering away with their bill and peeling off layers of the woody casing until they get to the starchy interior. It seems a lot of work but must be worth it.

In the depths of winter around December and January the Monks Wood Marsh Tits also showed a strong interest in the colourful berries of Spindle (Figure 4.8), pulling out the bright orange seed from the coral-pink fruits. Spindle seemed to be almost as favoured as Honeysuckle. Another fruit they often exploited throughout the winter was the Crab Apple. These are the largest fruits that I saw Marsh Tits carrying: they picked them up from the ground in their bill and took them to a perch to work on. It was amazing to watch a Marsh Tit carrying a Crab Apple that was sometimes larger than its own head. The birds dug into the apples to reach the pips, which they ate or cached. For larger Crab Apples and cultivated apples that are too big to carry, Marsh Tits dig into them on the tree, or where they lie on the ground, to extract the pips.

Other berries and fruits exploited by Marsh Tits include Juniper, Yew, Wild Cherry, Common Buckthorn, Raspberry, Rowan, pear, Black Mulberry, Mistletoe, Snowberry, Black Bryony, White Bryony, Woody Nightshade, Wild Service, Guelder Rose and currant (Snow & Snow 1988, Cramp & Perrins 1993). It is not always clear whether the seeds or the pulp are being eaten, but in southern England Snow & Snow (1988) estimated that Marsh Tits ate the pulp in 56 per cent of 48 observations involving eight types of berry, with seeds being extracted from the rest. For all of these species, the whole berries or their seeds might be cached for later consumption.

There is little information in the literature about Willow Tits foraging on berries in Great Britain, except for Spindle and the seeds from White Bryony (Foster & Godfrey 1950, Snow & Snow 1988). I have seen Willow Tits eating seeds from Common Elder berries in England. Elsewhere there are records of Willow Tits eating the seeds or pulp of berries from many of the same species as Marsh Tits, including Honeysuckle, Bramble, Raspberry, Juniper, Rowan, Common Buckthorn, Snowberry, various *Vaccinium* species such as Bilberry, and also Cotoneaster (Cramp & Perrins 1993, Glutz von Blotzheim & Bauer 1993). Juniper berries were particularly favoured in a Swedish study, with the birds caching lots of the seeds (Brodin 1994a).

SEED DISPERSAL

As well as being major seed predators, Marsh Tits and Willow Tits might also be meaningful seed dispersers for some plants. Several observers have noted how not all of the seeds from Honeysuckle berries are actually eaten, with many being lost or dropped onto the woodland floor (Richards 1958, Snow & Snow 1988). Other seeds or berries are transported and cached on or near the ground, or end up there if they fall, where they could germinate if not retrieved by the birds. With tens of thousands of seeds cached by individual birds each autumn and winter, the scale of seed dispersal could be quite large.

Marsh Tits and Willow Tits are unlikely to be as important for seed dispersal as some of the corvids, such as Eurasian Jays, which transport larger tree seeds (especially acorns) and cache them straight into the ground. However, the tits may have a greater role in dispersing smaller seeds, such as Common Beech, Common Hemp-nettle and Honeysuckle. It is also possible that some smaller seeds that have been found in the gizzards of Marsh Tits, such as violets and Wood Sorrel, could pass through the digestive system intact, as found with Japanese Tits (Fujita & Takahashi 2009). Very little is known of the ecological role of tits as seed dispersers, and there may well be unknown relationships between Marsh Tits, Willow Tits and some plants in the forest ecosystem that would be interesting to explore.

FLOWERS, NECTAR AND SAP

Marsh Tits and Willow Tits both spend lots of time in early spring on the buds, flowers and catkins of trees and shrubs. The birds methodically inspect and probe each flower, and also tear off catkins and buds to pull them apart. Willow Tits have been reported apparently eating the anthers from European Aspen catkins and the pollen from alders, and Marsh Tits have been seen eating the flowers or catkins of Sallows, alders, birches and *Prunus* species (Cramp & Perrins 1993).

I have spent many hours during March and April watching the Marsh Tits feeding among the flowers and catkins of elms, hazels, aspens, Sallows and birches (Figure 4.9). When they are foraging on these items it is difficult to see exactly what they are eating. I suspect that they are often not eating parts of the flower or catkin itself but are instead looking for invertebrates inside them. They spend a lot of time poking into them and being very selective in pulling off certain catkins to dismantle. The bits of Common Hazel, Sallow and European Aspen catkins that they discard have brownish traces inside that clearly contained burrowing larvae, and I think this is what the birds are eating. Possible prey are the larvae of small moths or weevils. Foster & Godfrey (1950) made the same observation of Willow Tits extracting small grubs from Sallow catkins.

Both species also drink nectar from flowers. At Monks Wood the Marsh Tits were seen working through blossoming Blackthorn in March or early April and poking their bill inside each flower, rapidly pulsating their tongue to lick the sugary nectar inside. Blue Tits and Coal Tits have been recorded drinking nectar in a similar way from blossom on a Flowering Currant bush (Fitzpatrick 1994). Marsh Tits and Willow Tits also appear to take nectar from Sallow catkins in early spring, as well as searching for burrowing larvae. Kay (1985) described how Blue Tits took nectar from Sallow catkins by repeatedly poking their bills into them to reach the base of the flowers, which provided a significant source of energy. I have seen Marsh Tits and Willow Tits behaving in exactly the same way by methodically probing Sallow catkins, apparently taking nectar like Blue Tits.

Figure 4.9. Marsh Tits at Monks Wood foraging on European Aspen catkins (left) and Blackthorn buds (right). (© Richard K. Broughton)

Figure 4.10. A Willow Tit in England probing into Sallow catkins, probably drinking the nectar at the base of the tiny flowers. Note the dusting of pollen on the bird's face. (© Peter Hendry)

When the tits are probing Sallow catkins they often get their faces covered in yellow pollen (Figure 4.10). Kay (1985) suggested that Blue Tits are major pollinators of Sallows, even more so than bumblebees, as they visit many more flowers. If this is the case, then Marsh Tits and Willow Tits foraging in the same manner would also be meaningful pollinators of Sallows, and possibly also Blackthorns and other shrubs, although this is another ecological relationship that has not been studied.

In early spring, when the sugar-rich sap is flowing in the trees it can leak out where they have been damaged. Tits and chickadees have been observed drinking these droplets as they hang from broken stems, and I have seen Marsh Tits drinking sap from Field Maples at Monks Wood. Both Marsh Tits and Willow Tits have also been seen drinking sap from European Aspens and birches. Black-capped Chickadees regularly consume the frozen sap of Sugar Maples in North America by breaking off pieces of the 'icicles' (Smith 1991), and Willow Tits might do the same with sap icicles in northern Eurasia.

INVERTEBRATES

Alongside seeds and other vegetable matter, invertebrates are the other major component in the diet of Marsh Tits and Willow Tits, particularly in spring and summer. Both species will eat almost any small to medium-sized invertebrate that they can catch, and their diet is quite broad (Tables 4.1 and 4.2). Of particular importance are moths, spiders, flies and beetles, including the adults, larvae, pupae and sometimes their eggs, too.

Table 4.1. Rounded percentages of invertebrates in the diet of full-grown Marsh Tits in the Forest of Dean, southwest England. Values are derived from prey items found in the digestive tracts during autopsy.

Taxon		Winter	April	Summer	Autumn
Lepidoptera	Moths/butterflies	13	2	13	27
	Caterpillars	*3*	*0*	*4*	*22*
	Pupae	*0*	*0*	*4*	*0*
	Adults	*9*	*0*	*5*	*5*
Hymenoptera	Sawflies/ants/wasps	9	41	7	6
	Larvae	*0*	*0*	*0*	*5*
	Pupae	*0*	*0*	*3*	*0*
	Adults	*9*	*41*	*4*	*0*
Diptera larvae	Flies	2	2	2	1
Collembola	Springtails	36	2	2	1
Hemiptera	True bugs	31	4	14	51
Invertebrate eggs		2	2	2	7
Birds containing spiders (%)		70	100	73	69
Food items		599	190	205	250
No. of birds sampled		10	4	11	13

Italics show partial breakdowns of the subtotals for the respective taxon (in bold).

Winter values are pooled from December–February, summer from June–August and autumn from September–November.

Sources: Betts 1955, Cramp & Perrins 1993.

Table 4.2. Rounded percentages of invertebrates in the diet of Willow Tits from autopsy analyses of digestive tracts or (for England) the contents of droppings from live birds.

Taxon		Southern Finland	Italian Alps	Moscow Oblast	St Petersburg	Northern England
		Summer–autumn	Winter	All year	All year	All year
Lepidoptera	Moths/butterflies	57	—	14	36	16
	Caterpillars	*57*	*—*	*14*	*36*	*—*
	Pupae	*0*	*—*	*0*	*—*	*—*
	Adults	*<1*	*0*	*<1*	*—*	*—*
Hymenoptera	Sawflies/ants/wasps	4	23	6	17	—
	Larvae	*3*	*—*	*—*	*—*	*—*
	Adults	*1*	*23*	*6*	*—*	*—*
	Ants	*—*	*23*	*—*	*—*	*—*
Diptera	Flies	3	3	4	4	—
Hemiptera	True bugs	20	36	53	16	—
	Aphids	*17*	*—*	*1*	*—*	*—*
	Psyllids	*<1*	*—*	*50*	*—*	*—*
Coleoptera	Beetles	2	0	14	14	63
	Weevils	*—*	*—*	*9*	*7*	*26*
Trichoptera	Caddisflies	—	36	—	0	—
Psocids	Barklice	<1	0	0	0	—
Araneae	Spiders	7	2	4	13	16
Sample size		798	242	1,104	921	100

Italics show partial breakdowns of the subtotals for each taxon (in bold). Sample sizes are the total number of prey items in the analysis or (for England) the number of droppings analysed.

Sources: Palmgren 1932, Rolando 1983, Prokofjeva 1990, Cramp & Perrins 1993, Beilby 2019.

These invertebrates are mostly gleaned from the foliage or bark of trees and shrubs as the foraging birds search through leaves, buds, twigs and branches. Both species forage from ground level to the topmost twigs of the trees, but they both spend a lot of time foraging in the shrub layer and sub-canopy of the trees. Relatively little time is spent on the ground itself, possibly because of the danger of being ambushed by a predator. One exception is when fallen seeds are available, but even then the birds often use a low perch to first lean downwards and scan the ground (Figure 4.11). Another exception is when heavy spring rains have washed caterpillars off the trees, and they are temporarily available on the forest floor.

Figure 4.11. A Marsh Tit at Monks Wood scanning the ground for fallen seeds. (© www.garthpeacock.co.uk)

Marsh Tits and Willow Tits have some cunning tactics to find and catch invertebrate prey, and on bright days they look from the undersides of leaves against the light to see the silhouettes of well-camouflaged caterpillars. Leaf-mining caterpillars are also detected like this, and the birds then rip open the leaves to extract them. When tortrix moth caterpillars roll leaves into a cylinder to hide inside, Marsh Tits and Willow Tits simply pull off the leaves and hold them under their foot to pull out the larvae. Like catkins, buds are pulled off and dismantled to find larvae hidden inside, although some plant material is possibly being eaten too, such as embryonic flowers or leaves.

Some invertebrates are gleaned from tree trunks or boughs as the birds cling to the bark and scan for bugs, barklice or beetles. Marsh Tits and Willow Tits can cling to the bark of vertical tree trunks and hop up them while looking for prey, similar to Eurasian Treecreepers. At Monks Wood I have seen Marsh Tits doing this while picking off door-snails (possibly *Clausilia bidentata*) from the trunks of Common Ash trees before pecking them open and eating the contents. They are also capable of pulling off small pieces of loose bark, wood or bark scales to look for invertebrates, and will investigate the piles of sticks and dried leaves among old bird nests and squirrel dreys. On the high horizontal boughs of mature trees both species scan the thick growth of mosses and lichens, pulling off clumps of material and tossing

Figure 4.12. Marsh Tits in Monks Wood forage on mossy boughs of trees, where they pull out tufts of moss to search for prey underneath. (© Richard K. Broughton)

them aside to search for prey underneath (Figure 4.12). When following Marsh Tits or Willow Tits during the winter it is common to see little tufts of discarded moss drifting down from where the birds are foraging above.

Marsh Tits and Willow Tits can flycatch by rapidly flying out from a perch to catch an insect in flight, using their bill, before landing to process and eat it. The flycatching flights only involve short distances of a few metres, and are usually only for larger insects such as craneflies, moths and ichneumon wasps. The birds handle large insects by removing the wings, pecking the head and pulling out the gut and flicking it aside. I have seen Marsh Tits do this with Large Yellow Underwing moths. The legs of large spiders are also removed and discarded. At Monks Wood I watched a Marsh Tit fly up to catch a large ichneumon wasp (*Eremotylus marginatus*), and on returning to its perch the bird pulled off and ate the wasp's abdomen before discarding the rest of it, still alive. Marsh Tits and Willow Tits can also hover for a few seconds to grab a dangling food item such as a spider in a web, a caterpillar hanging on a silk thread, or a berry.

Perhaps the most creative foraging technique of these woodland birds is the capture of aquatic invertebrates. When they can catch them, Marsh Tits and Willow Tits will eat the larvae of mayflies, caddisflies and stoneflies, which live in rivers, streams and ponds. Marsh Tits have been seen venturing under ice to grab caddisfly larvae, and also taking stonefly larvae from the water surface (Cramp & Perrins 1993). In the Białowieża Forest I observed a Marsh Tit successfully catching larvae of Limnophilidae caddisflies from the surface of the Orłowska River. The large larvae were encased in capsules of duckweed and suspended just under the surface, creating small disturbances. The Marsh Tit leaned over the water from a low perch before making a short foray to hover while grabbing a larva in its bill. The bird then took its catch to a perch and pulled it out of its case to eat it.

CARRION

Tits and chickadees scavenge on carrion, such as the carcasses of forest mammals like deer, Wild Boar, Eurasian Badger, European Bison and Eurasian Elk. These carcasses represent a huge food supply for omnivorous small birds, particularly the energy-rich fat, and especially in harsh northern winters. The mammals have thick skins that are not easily opened by small birds like tits, but when frozen they can be pecked at to reach the solid fats. Animals killed or scavenged by carnivores like Eurasian Wolves, Eurasian Lynxes or eagles are also helpfully opened up.

Willow Tits are known to feed on animal carcasses in Fennoscandia and the Baltic region (Halley & Gjershaug 1998). In Latvia a carcass in winter can attract groups of Willow Tits and Crested Tits from several surrounding territories, which will defend the food resource against others (Krams *et al.* 2020). Black-capped Chickadees, Siberian Tits and Great Tits feed on animal carcasses in a similar way (Smith 1991, Cramp & Perrins 1993, Selva *et al.* 2011). There are not yet any confirmed records of Marsh Tits scavenging on carrion, such as at animal carcasses in the Białowieża Forest (Selva *et al.* 2011). A trial using animal fat (lard or suet) hung in small wire cages in the forest was also inconclusive, with no definite records of Marsh Tits visiting them (Wesołowski 1995a). Marsh Tits probably will feed on animal carcasses, however, just like other tits and chickadees, as they readily eat animal fats provided at bird-tables or as bait in other areas (e.g. Almond & Almond 1950, Haftorn 1997a).

Water and minerals

Much of the water intake of Marsh Tits and Willow Tits comes from invertebrates, and they can also sip dew and raindrops from vegetation, eat snow in winter and drink sap in the spring. When standing water is available, however, both species will visit daily to drink and bathe.

At Monks Wood the Marsh Tits make a beeline for ponds or streams in the early afternoon, where they may bathe for several minutes, always alone. The birds stand in the water and vibrate their wings to create a spray, rolling onto each side and briefly dunking their head to cover all of their plumage. They then spend about 10 minutes in nearby cover, quietly drying and preening (Figure 4.13). Black-capped Chickadees have been recorded bathing amongst dew-covered leaves and also in fresh snow (Smith 1991), and Marsh Tits and Willow Tits may do the same. Dust-bathing does not appear to have been recorded for either species.

Figure 4.13. A Marsh Tit in Monks Wood preening after drinking and bathing in a nearby pond. (© Richard K. Broughton)

Betts (1955) found small pieces grit in the gizzards of almost all Marsh Tits examined throughout the year. The grit can grind tough plant and animal material in the muscular gizzard, but there may also be some mineral intake. Female Marsh Tits were found to have eaten fragments of calcium-rich snail shell in spring, which may help in the formation of eggshells. Willow Tits have been seen feeding pieces of snail shell and grit to nestlings, possibly to provide calcium for bone development and also allow their gizzards to deal with tougher invertebrate prey (Gibb & Betts 1963).

Nestling diet

The caterpillars of butterflies and especially moths are the major part of the nestling diet of Marsh Tits and Willow Tits (Cholewa & Wesołowski 2011). Marsh Tit nestlings are fed almost exclusively on caterpillars and a minor proportion of spiders (Figure 4.14). For Marsh Tit nestlings aged 6–13 days old in Monks Wood and Wytham Woods, 94 per cent of feeds consisted of caterpillars (Carpenter 2008). For Marsh Tit nestlings aged 2–18 days old in the Białowieża National Park the proportion of caterpillars in the diet was 70–90 per cent (Wesołowski & Neubauer 2017). The caterpillars involved commonly include Winter Moth and other small Geometridae, along with tortrix moths and noctuid moths, which are all abundant on deciduous trees and shrubs in spring.

Willow Tit nestlings are fed a more varied diet than Marsh Tits (Figure 4.15, Table 4.3), with caterpillars being the primary food in only 62 per cent of studies reviewed by Cholewa & Wesołowski (2011). Other important items in the Willow Tit nestling diet are spiders, bugs

Figure 4.14. A Marsh Tit feeding its nestlings in a knothole in the Białowieża Forest. The prey item is a spider with its legs removed. Note the Marsh Tit is regrowing her tail, which was probably lost in a near miss with a predator. (© Richard K. Broughton)

Table 4.3. Rounded percentages of items in the nestling diet of Willow Tits, derived from studies analysing food delivered to chicks.

Taxon		Eastern England	Great Britain	Eastern Germany	Moscow	Siberia
Lepidoptera	Moths	62	81	19	53	46
	Caterpillars	*48*	*79*	*17*	*44*	*41*
	Pupae	*9*	*0*	*—*	*—*	*4*
	Adults	*5*	*2*	*—*	*1*	*—*
Hemiptera	True bugs	3	0	42	16	1
	Aphids	*—*	*—*	*36*	*0*	*—*
	Psyllids	*—*	*—*	*—*	*16*	*—*
Araneae	Spiders	26	0	0	19	17
Hymenoptera	Sawflies/ants/wasps	2	0	32	4	7
	Larvae	*1*	*—*	*32*	*—*	*1*
	Adults	*—*	*—*	*0*	*4*	*—*
Diptera	Flies	—	17	4	5	17
Coleoptera	Beetles	1	0	0	1	7
Other		6	1	7	2	5
No. of items		491	86	97	455	296

Italics show partial breakdowns of the subtotals for each taxon (in bold).
Sources: Gibb & Betts 1963, Prokofjeva 1990, Cramp & Perrins 1993, Beilby 2019.

and flies such as craneflies and mosquitoes (Cramp & Perrins 1993, Rytkönen *et al.* 1996). The more diverse diet fed to Willow Tit nestlings may reflect habitats with fewer mature deciduous trees on which to find abundant moth caterpillars, and perhaps the adults foraging in more diverse vegetation for spiders and flies.

Marsh Tits commonly deliver one prey item at a time to the nestlings, but a third of visits involve multiple items (Carpenter 2008, Wesołowski & Neubauer 2017). Willow Tits regularly collect multiple prey, averaging three items per prey load (Pravosudov & Pravosudova 1996). The greater frequency of multiple prey loading in Willow Tits might reflect the lower proportion of caterpillars in the nestling diet, as these larger prey are more efficiently collected and delivered singly to the brood (Rytkönen *et al.* 1996). Adults of both species process each food item when they catch it, pecking at the caterpillars' heads to kill them and remove the sharp jaws. In both species, the size of the prey load increases as the chicks grow.

Tits and chickadees have long been known to time their breeding to coincide with peak availability of caterpillar prey each spring, so that their nestlings can benefit from this peak food supply (Perrins 1965, Hinks *et al.* 2015). The advancement of spring as a result of climate change has caused some concern that tits may be unable to keep pace with the shifting of the caterpillar peak, leading to a so-called *phenological mismatch* that could reduce the birds' breeding productivity (Visser & Gienapp 2019). For Marsh Tits and Willow Tits, long-term data from the Białowieża National Park and from Oulu in Finland shows that the time of peak demand from the nestlings regularly misses the peak of caterpillar abundance by quite some way, sometimes by several weeks. This appears to be normal, however, and the birds' productivity does not seem to be adversely affected by a mismatch (Vatka *et al.* 2011, Wesołowski 2023). In fact, almost all of the phenological mismatch in Marsh Tits and Willow Tits is because the birds deliberately time their breeding so that nestlings appear ahead of the caterpillar peak. It seems that the birds breed as early as they are able to do so, because even caterpillar abundance that is well below the peak is still sufficient for them to rear a brood.

This relatively early breeding seems to give Marsh Tits and Willow Tits some potential buffering against phenological mismatch and advancement of peak food availability. The average timing of maximum caterpillar abundance could advance by one or two weeks before there was a really serious mismatch. However, if climate warming results in a shortening of the overall period of food availability, because of more rapid growth of the caterpillars, then the birds could be in trouble. This is because abundant caterpillars are important not only to nestlings but also to fledglings and inexperienced young juveniles in their first weeks of independence. The real issue with phenological mismatch is not so much peak caterpillar abundance matching peak nestling demand, but the duration of caterpillar availability matching demand across a much longer period of the life cycle, throughout the spring and summer. Unfortunately, relatively little is known about the period of early independence in Marsh Tits and Willow Tits, and how much they are being affected by any change in food availability.

Figure 4.15. A Willow Tit delivering a prey load to its nestlings in Norfolk, England. The prey consist of some small green caterpillars and other small unidentified insect larvae. (© Ashley Banwell)

Summary

Marsh Tits and Willow Tits are omnivorous and consume a wide range of invertebrates, seeds and fruits, but also more unexpected foods like carrion, nectar and tree sap. Both species have a correspondingly wide variety of foraging behaviours to exploit these resources, using their agility and intelligence. They forage from the ground layer to the tops of trees, including the shrub layer and sub-canopy, and utilise all parts of trees and shrubs.

Large amounts of food, predominantly seeds, are cached for future use. An enlarged hippocampus in their brains gives both species an enhanced spatial memory that enables them to retrieve these hidden food items up to several weeks later. Caching allows the birds to spread resources over time, helping them to remain resident in their home-range all year round, even in harsh winter climates.

Some meaningful seed dispersal may be involved with transporting seeds and fruits away from parent plants. Both species may also act as pollinators when drinking nectar from flowers, but these ecological links have not been explored.

Nestlings are fed entirely on invertebrates. Marsh Tits predominantly feed their nestlings moth caterpillars and some spiders. Willow Tit nestlings have a more varied diet, including lots of flies and bugs, although moth caterpillars still form the bulk of the prey in many areas. The period of the nestlings' peak food demand coincides poorly with peak caterpillar availability. This phenological mismatch appears to be a normal consequence of both species breeding as early as possible each spring, although it is unknown how climate change may be affecting the broader period of prey availability.

CHAPTER 5

Social organisation, territories and home-ranges

All birds must distribute themselves within the available habitat to form a social organisation. For resident and non-migratory Marsh Tits and Willow Tits this social organisation lasts throughout the year, and is based upon the *territory* and the *home-range*. Territories are exclusive patches of habitat defended against others of the same species, whereas home-ranges are larger areas that can overlap with those of neighbouring birds. A fascinating aspect of Marsh Tit and Willow Tit ecology is that both species can adopt different strategies of social organisation at different times of the year and in different parts of their ranges. Some populations appear to be territorial throughout the year while others are strongly territorial only in the spring. Some populations spend the winter in stable groups while others form casual and dynamic associations with various neighbours. What determines the different strategies of social organisation in different contexts is still not fully understood, with many questions remaining.

Figure 5.1. Each individual Willow Tit (shown) and Marsh Tit in a population must fit into the local social organisation to allow it to coexist with its neighbours, to survive and reproduce. (© Robert Fredagsvik)

Social organisation

Aside from limited periods of dispersal or irruptive movements (see Chapter 7), Marsh Tits and Willow Tits are resident and sedentary throughout the year, with settled birds rarely leaving the home-ranges that encompass all of their regular activity. Individuals living together in the same patch of woodland must therefore arrange themselves in a manner that allows them to coexist, allowing each bird to survive and hopefully breed successfully (Figure 5.1). The way that Marsh Tits and Willow Tits achieve this is by organising into pairs or groups that are distributed across the habitat, based on some simple principles. The success of their social organisation is underpinned by the birds acquiring a deep knowledge of their chosen home-range, alongside close bonds between individuals that have a strong investment in each other.

PAIR FORMATION AND THE PAIR BOND

Both species are *socially monogamous*, preferring to live in a male–female unit as the core of their social organisation. For Marsh Tits, acquiring a territory or a home-range is a prerequisite for pair formation (Southern & Morley 1950, Morley 1950). Juvenile Willow Tits in Fennoscandia may form pair bonds within temporary groups or home-ranges in the summer before establishing a more stable winter home-range, often around the nucleus of an adult pair (Haftorn 1997b). Pair formation in Willow Tits at lower latitudes is completely unstudied, but in temperate regions like Great Britain or Germany they probably settle and form pairs quite rapidly after fledging and dispersal from the family territory, similar to Marsh Tits in these areas (Ludescher 1973, Amann 1997).

The main peak of pair formation is after juvenile dispersal in the summer, or early autumn for northern Willow Tits. There is a smaller peak in late winter and early spring as some birds disperse again to find breeding opportunities after winter mortality has created vacancies. Opportunistic pair formation can occur throughout the year, however, in response to unexpected opportunities opening up through mortality or divorce (Morley 1950, Orell *et al.* 1994a, Amann 2003).

Dispersing Marsh Tits or Willow Tits have several options for forming a pair when arriving in a new area. The most effective option is to join an established territory-holder, such as a widowed or unpaired bird, and form a pair with them. This offers a proven territory and comes with the experience and status of the resident bird, which already knows the territory's resources. By pairing with this bird, the new arrival would immediately gain a survival advantage from its valuable knowledge.

The next best option is to form a new pair with another new bird that has just settled in the area. This could be in a vacant home-range, or as subordinates in a home-range that overlaps with pairs of adults or dominant juveniles. The new juveniles are inexperienced, but together a pair can discover the food sources and look out for danger, with two pairs of eyes being safer than one. A study of Black-capped Chickadees showed that being in a pair is particularly beneficial for females, as the greater vigilance for danger allows them to have higher feeding rates (Lemmon *et al.* 1997).

The risk for a pair that settle as subordinates in the home-range of dominant birds is that the latter might all survive until the end of winter, and the juveniles, now moulted first-years, will then be evicted. If one adult dies during the winter, one of the first-years can pair with the widow, breaking its pair bond with the other youngster, which is then in the precarious position of facing eviction on its own. These expelled first-years must disperse at the end of winter and hope to find a new territory vacancy elsewhere. This shows the limits of the pair bond, as it is a mechanism for a bird to improve its own chances of survival and reproduction, superseding its commitment to another individual. Nevertheless, the two birds clearly invest a lot in the pair bond, and it is not broken lightly.

Pairing can happen very quickly after two birds meet, with individuals that are definitely unpaired one day being firmly paired the next day (Morley 1950). A first meeting between two birds has not been observed in the wild, and it would be difficult to recognise, but young Marsh Tits or Willow Tits introduced in aviaries showed an initial mixture of aggression and excitement, with calling, posturing, supplanting and chasing (Nilsson 1989b, Koivula et al. 1993). In the wild, any aggression is presumably overcome quite quickly by the attraction.

The pair bonds of Marsh Tits and Willow Tits are very obvious throughout the year. The two birds generally remain close together and are in frequent contact with short calls, and they engage in territorial defence against neighbours as a united front. At dusk the male accompanies the female to her roost site (typically in a tree cavity) and also meets her there in the early morning. If the pair are temporarily separated one bird will call to locate the other, who will usually answer and come to its mate. At Monks Wood I noticed that if a male Marsh Tit cannot readily find the female then he will go looking for her and will sing to broadcast his location. I once watched a male whose mate had been killed that same day by a Eurasian Sparrowhawk: he spent an hour moving around their home-range, singing and calling, very clearly searching for her. It is impossible to know what goes on in the inner minds of these small birds, but this male was evidently anxious for the whereabouts of his partner, with whom he had built a relationship for over a year.

PAIR BOND LONGEVITY AND DIVORCE

Marsh Tits and Willow Tits generally maintain their pair bond for life, until one of them dies. There are some exceptions, such as occasional divorce, but in most cases the duration of the pair bond depends on how long the birds live. Despite many Marsh Tits and Willow Tits living quite short lives due to predation, starvation and accidents, long-term studies show that pair bonds frequently last for years.

Morley (1950) detected Marsh Tit pairs lasting up to two and a half years at Oxford, similar to those I detected at Bradfield Woods in Suffolk during the 2010s, but these studies lasted only 4–6 years, which capped the length of pair bonds that could be recorded. In Sweden, Anvén (1961) observed a single Marsh Tit territory over eleven years, and one pair of occupants survived intact for six of those years. In Norway, Haftorn (1997a) monitored the changing Marsh Tits in a single territory over a remarkable three decades, and recorded a pair remaining together for eight years. In 21 years of the Monks Wood study, two-thirds of Marsh Tit pair bonds (68 per cent) lasted for one year, and they almost always ended because one of the pair died (Figure 5.2). However, 17 per cent of pairs survived intact for two years and 13 per cent lasted for three years. These long-term studies show that any Marsh Tits that stay together for five years or more are doing very well.

Marsh Tits that live for a relatively long time usually have several pair bonds throughout their lives, as the dead partners get replaced. One male at Monks Wood that lived for nine years had a different mate in every breeding season of his life, as each female survived for only

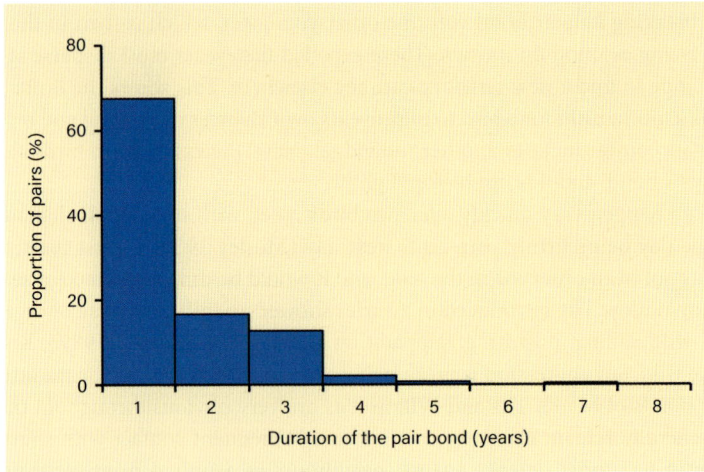

Figure 5.2. The longevity of 283 pair bonds among Marsh Tits in the Monks Wood study. The values are derived from spring monitoring of colour-ringed birds over 21 years.

a single year. In contrast, one remarkable pair at Monks Wood remained together for over seven years, having paired in their first summer and staying together for their entire lives until the female died, with the male dying just a few months later. I was particularly fond of this pair, as they appeared early in the project and I monitored them very closely throughout their long lives, including all of their nests, where they produced around 50 fledglings. I learned a great deal from those two birds, and it's always a shame when individuals with whom you have spent a lot of time come to the end of their story.

There is less information on the longevity of Willow Tit pair bonds, but they are probably similar to Marsh Tits. Pairs have been recorded lasting for at least four years in northern England and Norway (Hogstad & Slagsvold 2018). Long-term pair bonds are a general characteristic of the *Poecile* tits and chickadees, with a pair of Black-capped Chickadees lasting almost six years (Smith 1992), and Carolina Chickadees and Siberian Tits reportedly 'pairing for life' (Brewer 1961, Cramp & Perrins 1993).

Marsh Tits and Willow Tits do sometimes divorce, but it is uncommon. At Monks Wood divorce affected only 5 per cent of Marsh Tit pairs over 11 years, with four cases involving the break-up of very recent pairs that had formed just for one breeding season (Broughton *et al.* 2010). One of these females was an unpaired bird that had moved in early spring to a wood over 800m from her original territory to join a single male, thereby solving the immediate problem for both birds by being able to breed that year. Soon after the breeding season was over, however, the female moved back to her previous territory, where she then paired with a new juvenile that had just arrived after summer dispersal. In another example, an unpaired female moved 1km to breed successfully with a lone male before returning to her original territory in the summer, where she appeared to remain unpaired again for several months. These cases of divorce suggest that Marsh Tits value their original territories over their current pair bonds, or even any pair bond. The original territory may be of better quality, or the birds' deeper knowledge of its resources may be of great value to them, as it has already enabled them to survive at least one winter.

Divorce has also been recorded in Willow Tits, with one study reporting 'a few' cases among first-year pairs in Norway, but none among adults (Hogstad & Slagsvold 2018). In Finland, Orell *et al.* (1994a) found an average divorce rate of 12 per cent among Willow Tit pairs, with one female divorcing twice. First-year birds in this study were more likely to divorce than older pairs, and divorced females tended to pair again with older males, while males tended to form new pairs with younger females. The female Willow Tits appeared to be making a strategic decision to move to a better male or territory, as they bred earlier and more successfully than females who formed new pairs after being widowed, which is a forced change rather than a choice. Divorced males did not have improved breeding success in their new pairs, which suggests that females drive this behaviour and have the most to gain from it.

MULTIPLE PAIR BONDS

Although Marsh Tits and Willow Tits are socially monogamous in a male–female pair, there are some exceptions. Polyandry, when a female forms pair bonds with more than one male, has been recorded in Black-capped Chickadees (Waterman *et al.* 1989) but not in Marsh Tits or Willow Tits. Polygyny, when a male forms a pair bond with two or more females, has been recorded in all three species.

Polygyny usually occurs when a paired male forms a second pair bond with a lone female in a neighbouring territory. The two females often remain intolerant of each other and will avoid venturing deep into each other's territories, but the male is able to move across both territories and maintain a pair bond with each female. In Monks Wood polygyny was recorded on seven occasions over 21 years, with one male having polygynous pairings in two separate years, involving different females each time.

The first case of Marsh Tit polygyny that I saw at Monks Wood involved a paired male who formed another pair with a secondary female who had recently settled alone in a territory 200m away (Broughton 2006). The male moved between both territories and both females, bringing food to each of them when they began nesting. It was highly probable that the male had fertilised the eggs of both females, but when the chicks hatched in the primary female's nest the male abandoned the secondary female and was not seen to visit her again. The abandoned female was left to rear the chicks alone, which was successful.

Willow Tit polygyny was recorded in Norway for a paired male who had a secondary pair bond with a widowed female from an adjacent territory (Haftorn 1994). Unusually, both females nested about 300m apart within the male's territory. The male brought food to both females while they incubated their eggs, but only fed the nestlings of his primary female, though he did occasionally visit the secondary female after her chicks had hatched. The primary female also visited the secondary female's nest on multiple occasions, always silently and alone, sometimes clinging to the entrance but never entering or bringing food. Haftorn thought that the female was inspecting the progress of the other bird's nest, suggesting that she was well aware of the polygyny.

The examples of polygyny and divorce demonstrate that the pair bonds of Marsh Tits and Willow Tits are not rigid, and birds of either sex will take advantage of opportunities to increase their breeding prospects, should they present themselves. Nevertheless, the low frequency of these events in the long-term studies shows that the socially monogamous pair bond is a fairly robust element of Marsh Tit and Willow Tit society.

INTERSPECIFIC PAIR BONDS AND HYBRIDISATION

Very occasionally a Marsh Tit and Willow Tit appear to form a mixed pair bond with each other, or at least a very close association. This usually happens where an individual of one species finds itself among more abundant members of the other. Known as the Hubbs principle, this phenomenon applies widely to birds when two closely related species occur together but one of them is rare and has little chance of finding a mate of its own species (Hubbs 1955). Little is known of how these mixed associations work in Marsh Tits and Willow Tits, and many could just be temporary alliances to gain some of the benefits of being in a pair, such as increased predator vigilance or food discovery. Some mixed associations seem to be genuine pair bonds, however, where the birds intend to breed together.

In Great Britain, where both species are in severe decline, individuals can increasingly find themselves alone. This can happen because other birds in their population have all died and there is nobody left to pair with. Even before this happens the juveniles reared in the last few territories will disperse outwards and soon become isolated in a depopulated landscape. In these situations a lonely Willow Tit might come across a Marsh Tit, or vice versa, and begin associating with them as the 'next best thing'.

Mixed associations between Marsh Tits and Willow Tits are mostly seen during autumn and winter, with fewer records in the breeding season. At Grimsbury Reservoir in Oxfordshire, in southern England, a pair of Willow Tits and a lone Marsh Tit were present in a small wood during early 2017, but by the autumn only the Marsh Tit and a single Willow Tit remained. In the following spring it became obvious that they had formed a mixed pair and were closely associating well into the nesting season. The Willow Tit was heard singing so was probably a male, but there were no further sightings and no breeding attempt was proven, although it could have occurred out of sight. Tragically, this was the very last resident Willow Tit in Oxfordshire, and its disappearance marked the local extinction of the species.

A definite mixed pair was recorded in Belgium in 1968 and 1969, when a male Willow Tit and female Marsh Tit twice bred together in a nest-box in a park near Ghent (Dhondt & Hublé 1969). The female built a typical Marsh Tit nest and laid eight eggs each year, but only five hatched on each occasion. The chicks apparently looked like Marsh Tits and all five fledged in the first year, but none in the second year due to poor weather. This case has been accepted as confirmed hybridisation between Marsh Tits and Willow Tits (e.g. Cramp & Perrins 1993, Harrap & Quinn 1995), and it seems highly likely, but without genetic evidence it cannot be absolutely certain. Nestlings of Marsh Tits and Willow Tits are so similar that any mixed features indicating hybrids would be virtually impossible to spot.

A hybrid Marsh-Willow Tit was also suspected in Germany by Löhrl (1987), but this seems less likely. The record involved a lone Willow Tit in part of the Black Forest where the species does not breed. Two years later a female Marsh Tit in the same territory was heard giving apparently mixed calls, and Löhrl surmised from this that it was a hybrid, which seems to be very thin evidence. Hybridisation between Marsh Tits and Willow Tits has probably occurred, but has yet to be proven for sure.

Hybridisation has definitely occurred between Marsh Tits and other tits, and also between Willow Tits and other species. This is easier to determine than hybrids between Marsh Tits and Willow Tits themselves, because hybrids with other tits will have more obviously intermediate features. Nevertheless, clear cases of Marsh Tit hybridisation are rare. Duquet (1995) described an assumed hybrid between a Marsh Tit and a Great Tit in France, based on mixed features. A nest in Bosnia Herzegovina, attended by a male Great Tit and female Marsh

Tit, had five nestlings that resembled large Marsh Tit chicks, but with a black belly stripe that pointed to some Great Tit parentage (Drocić & Drocić 2013).

Willow Tits hybridise more frequently than Marsh Tits, and mixed pairs with Siberian Tits (Figure 5.3) have been reported many times in Finland (Hildén & Ketola 1985, Järvinen 1997). These mixed pairs involve females of either species that build typical nests but are attended by males of the other species. The breeding attempts often have low success because eggs do not hatch or nestlings die, which might be because of low feeding rates by the male, perhaps resulting from poor communication between the adults with their differing calls. The nestlings do not always show obvious mixed features, raising the question of whether they were true hybrids or if the male in the mixed pair was not the genetic father after all. Some broods do show mixed features, however, and these apparent hybrids are fertile and can survive to enter the breeding population (Järvinen 1997).

Other Willow Tit hybrids include a bird with obvious mixed features of a Coal Tit in southwest Finland, which had good photographic and biometric evidence (Hildén 1983). Apparent hybrids of Willow Tits and Crested Tits have also been found in Fennoscandia (Figure 5.4). In Japan, Mishima (1969) reported six cases of hybridisation between Willow Tits and Varied Tits, one of which was trapped and kept alive in captivity. A photograph showed a bird with mixed features, which reportedly called like a Varied Tit but 'finer'. What remains unclear is how many of these hybrids result from true mixed pairs rearing the chicks together, or how many might arise from casual extra-pair copulations. The causes and consequences of hybridisation and mixed pairs in *Poecile* species remain something of a mystery (Curry 2005).

Figure 5.3. A Siberian Tit at Kaamanen in Finland. Siberian Tits occasionally hybridise with Willow Tits in the boreal region. (© www.garthpeacock.co.uk)

Figure 5.4. A presumed hybrid Willow Tit × Crested Tit at the Höytiäinen Bird Observatory in Joensuu, Finland, caught during ringing operations in September 2019. Note the intermediate features of both species on the head. (© Harri Höltta)

Territories

Marsh Tits and Willow Tits in temperate regions defend their spring breeding territories most strongly between early March and mid April. Bouts of adult territorial behaviour occur on warm days from January and increasingly throughout February, and can last until late May. Cold spells in spring subdue territorial aggression, and neighbours may forage together while it lasts. In colder northern climates the pairs of adult birds might not be strongly territorial until April through to late June. Settling juveniles are also territorial in summer as they establish themselves after dispersing from the family group.

Early studies suggested that Marsh Tits were strongly territorial throughout the year in sharply defined territories, with only rare excursions outside their borders (Morley 1953, Amann 2003). Territoriality is more complex than this, however, with defence waning after the spring and territory boundaries becoming blurred. At Monks Wood we found a very rigid territory defence in the spring, and again by settling juveniles in summer, but this relaxed into overlapping home-ranges during the autumn and winter (Broughton *et al.* 2006, 2015a).

In Fennoscandia Willow Tit pairs retain their spring breeding territories into the summer, after their own fledglings have dispersed, but new unrelated juveniles may subsequently arrive to join them. The adults and juveniles then form a stable group that defends the communal territory throughout the winter (Ekman 1979a). These stable winter groups are not seen among Willow Tits in milder regions, such as Great Britain and Germany, where the breeding territories are most obviously defended in spring, similar to Marsh Tits (Ludescher 1973).

Both members of the pair, or all members of a group, take part in territorial defence at its borders by challenging intruders with calling and posturing. Ludescher (1973) mentioned how neighbouring birds of either species sometimes undertake 'provocative flights' deep into adjacent territories to call or sing, testing for a reaction. If the territory owners respond, the intruders are chased back to the boundary. Ludescher also described 'convention places' at the territory boundary where neighbours would regularly have disputes. Sometimes multiple pairs of Marsh Tits and Willow Tits would be attracted to these noisy gatherings, and Ludescher saw brief instances of mistaken identity among agitated birds chasing the wrong species.

Once nesting has begun the female's involvement in territorial behaviour becomes much reduced, although the males will still sing and defend the territory against neighbours. When the chicks hatch, both parents limit most of their activity to the area around the nest and the best foraging sites, and so territorial behaviour declines after mid April in temperate regions. Breeding is usually synchronised across a population (see Chapter 6), so most of the neighbours are also busy with feeding their chicks at the same time and few are interested in territorial activity.

MARSH TIT TERRITORY SIZES

Marsh Tit territories are well studied across Europe, giving a clear picture of how they space themselves and how much habitat is needed by a breeding pair. Territories are easily identified in early spring by the frequent singing of males and disputes between pairs at territory boundaries, especially if a researcher uses playback of recorded song to get a response (Broughton *et al.* 2018a). The range of territory sizes in Table 5.1 shows some variation between regions, which might reflect differing habitat quality, but all studies show that Marsh Tit territories are quite large. A good rule of thumb for an average spring territory is 5–6ha of wooded habitat (Figure 5.5).

Table 5.1. The size of Marsh Tit and Willow Tit spring territories in different study areas.

Location	Territory size (ha)			No. of territories	Method
	Mean	Min.	Max.		
Marsh Tits					
Bagley Wood, England	2.8	0.4	4.9	18	Mapping
Monks Wood, England	5.4	0.9	14.1	279	Mapping
Basel, Switzerland	6.9	1.7	17.0	97	Mapping
Pfrunger Ried, Germany	8.5	5.8	12.0	22	Mapping, gross
Pfrunger Ried, Germany	5.8	3.6	9.1	22	Mapping, net
Willow Tits					
Wytham Woods, England	11.0	~7	~15	2	Mapping
Wigan, England	13.7	1.1	41.3	108	Thiessen, gross
Wigan, England	7.1	0.2	29.4	105	Thiessen, net
Pfrunger Ried, Germany	7.3	4.9	9.8	21	Mapping, gross
Pfrunger Ried, Germany	5.0	3.6	8.1	21	Mapping, net
Sugadaira Hight, Japan	2.2	1.3	3.5	3	Mapping
Matsumoto, Japan	11.8	—	—	6	Mapping

Studies used the territory-mapping method or Thiessen polygons. Some studies separately calculated the gross area (including open spaces) and the net area (wooded habitat only).

Sources: Foster & Godfrey 1950, Southern & Morley 1950, Ludescher 1973, Nakamura 1975, Nakamura & Wako 1988, Amann 2003, Broughton *et al.* 2006, 2011, 2020.

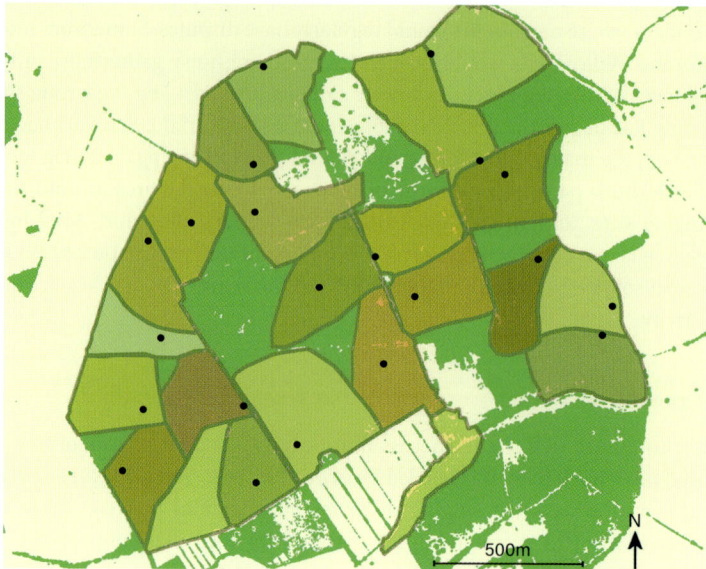

Figure 5.5. The distribution of 23 Marsh Tit spring territories at Monks Wood in 2007, with the location of nest-sites (black dots) where known. Territories averaged 5.6ha and were mapped from colour-ring observations. Lidar data for the woodland habitat were captured by the NERC Airborne Remote Sensing Facility, via the CEDA repository (CC BY 3.0).

In more open woodland the gross area of Marsh Tit territories will be larger, to account for areas of non-habitat, such as clearings. At Pfrunger Ried the gross area of Marsh Tit territories was almost 50 per cent greater than the average area of wooded habitat within them (Table 5.1). In closed-canopy woodland, however, such as at Monks Wood, almost all of the territories are fully within the woodland block (Figure 5.5), and so the territory size corresponds to the area of woody habitat defended by the birds.

The boundaries of Marsh Tit territories often seem to follow features in the landscape, such as streams, paths or clearings. This suggests that the birds use distinctive 'landmarks' to recognise their own borders and those of their neighbours. Morley (1953) described how two pairs would even divide a single tree between their territories. At Monks Wood and Bradfield Woods the Marsh Tits often used paths and rides as territory boundaries, with two males or pairs on either side of a path singing and calling at each other, sometimes flitting back and forth. As a result, many territories had some rather straight edges (Figure 5.5).

WILLOW TIT TERRITORY SIZES

Territory mapping has been used to determine the size of Willow Tit breeding territories (Table 5.1), which can extend across more inaccessible and diffuse habitats, such as dense thickets and boggy forest or wetlands (Pinder & Carr 2021). Where populations are low and declining, Willow Tits can be difficult to detect simply because the neighbours are so far apart that they have few territorial disputes. As such, the territory boundaries can be more difficult to delineate than for Marsh Tits.

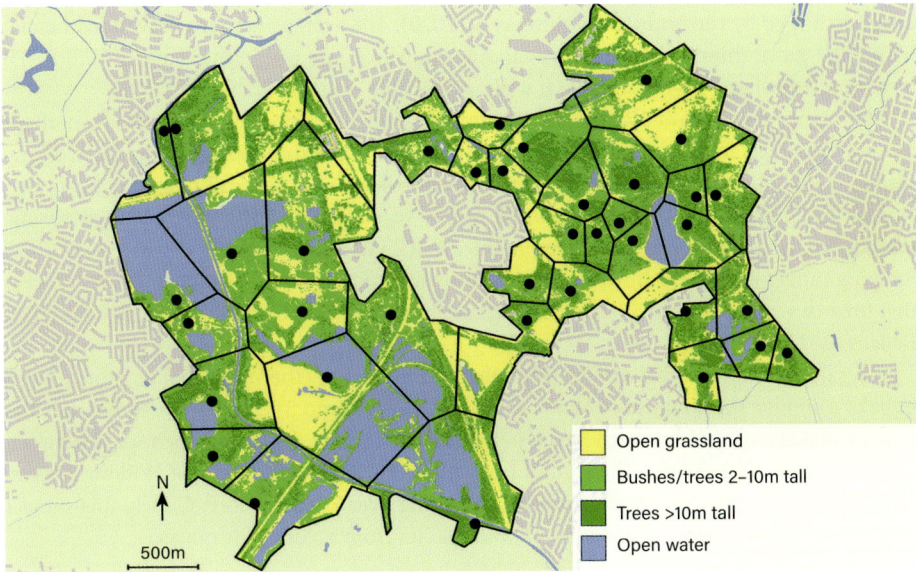

Figure 5.6. Willow Tit nests and territories from 2019 in the Wigan study area, set in an urbanised and post-industrial landscape. Each nest (black dot) is within a Thiessen polygon (black lines) that estimates territory boundaries based on nest locations. Note the smaller territories among taller trees. Nest data reproduced with permission from Broughton *et al.* (2020). Vegetation heights derived from 2019 lidar data © Environment Agency copyright and/or database right 2015. All rights reserved. Contains Ordnance Survey data © Crown copyright and database right 2023.

Willow Tit territories seem to be generally larger than those of Marsh Tits (Table 5.1). This may be related to Willow Tits often occupying more diffuse woodland or extensive shrubland, rather than a solid block of closed-canopy forest. Estimates of Willow Tit territory size can vary a lot, however, with Nakamura & Wako (1988) and Eggers & Low (2014) reporting quite small territories of only 2–3ha in Japan and Sweden. At the other extreme in Scandinavia are Haftorn's (1973) and Ekman's (1979a) estimates of 15–24ha territories. Timing may play a role in the smaller estimates, as activity becomes more limited to the habitat around the nest area once breeding has begun (Stefanski 1967, Amann 2003).

An alternative method for estimating territory size is to map the location of pairs or nests and use geometric methods to outline approximate boundaries as a best estimate. We used this approach in the Wigan study (Figure 5.6), where Willow Tits were occupying most of the available area and so the nest locations of each pair were used as the centroid (centre point) of the territory to estimate the boundaries between them (Broughton *et al.* 2020). These boundaries were created using *Thiessen polygons*, which places a border at an equal distance between each nest centroid. The result is a set of geometric polygons around each nest that divides the study area into approximate territories. The territories estimated by Thiessen polygons are unlikely to be very precise in terms of the actual boundary locations that the birds are defending, but they give a useful estimate of territory size.

The Wigan study area was a mosaic of discontinuous woody habitat among more open wetlands and grassland. We estimated the total (gross) territory area of the Thiessen polygons and also just the net area of woody habitat within them, which may be a more realistic estimate of the area used by the birds (Broughton *et al.* 2020). The average net area of wooded habitat within the gross Thiessen polygons was about 7ha, which was within the range of Willow Tit territories elsewhere, suggesting that it was a reasonable estimate (Table 5.1).

The open areas within Willow Tit territories could be important to the birds, perhaps as seed-rich vegetation around the edges of wetlands or clearings. However, radio-tracking studies from Lake Vyrnwy in Wales and the Dearne Valley in northern England indicate that Willow Tits spent most of their time within woody vegetation during winter and early spring, foraging on trees, shrubs and in Bramble thickets (Pinder & Carr 2021, Bellamy 2022). The foraging ranges of these radio-tracked birds probably reflected approximate territory sizes at that time of year, averaging 3–5ha and ranging up to 7ha in both areas (Pinder & Carr 2021, Bellamy 2022).

Overall, in temperate regions like Great Britain or Germany the Willow Tit territories seem to be of a similar size to those of Marsh Tits, averaging around 5–7ha of wooded habitat. The total defended area may be much larger, perhaps 10ha or more, including open space between patches of trees and shrubs. The studies from northern Europe suggest that Willow Tits can have significantly larger spring territories, extending over 20ha or more. Clearly this subject would benefit from more field studies of both species from across their ranges.

Breeding densities

In many studies of Marsh Tits or Willow Tits the territory boundaries cannot be adequately mapped due to the effort required, meaning that accurate territory sizes are unknown. As long as the number of territories is identified within a given area, even if their borders are not mapped, then the territory density can be calculated instead. Territory density closely reflects breeding density if most territories are occupied by pairs and not by unpaired singletons, which is a reasonable assumption in many cases.

Breeding density is an extremely useful metric for the number of pairs or territories in a given patch of habitat, and is arguably more realistic than extrapolating mean territory size to estimate a density across a wider area. Dividing a study area by the average territory size ignores any unoccupied habitat between territories, whereas dividing the habitat area by the number of territories or pairs does allow for some vacant habitat and variation in quality. Breeding density is also much quicker and easier to establish by simply finding the territorial birds, rather than mapping their territory boundaries. One limitation is that representative territory densities can only be calculated from larger patches of habitat, ideally areas of at least 30ha and preferably much larger. High-quality breeding densities rely on systematic fieldwork, and the best estimates involve colour-ringed birds to avoid double-counting of individuals seen in different places.

Within their main breeding habitat of mature deciduous woodland, the territory densities of Marsh Tits are available from only a few studies in Europe that involved colour-ringed birds, including some of those where territory sizes were estimated (Table 5.2). The densities from the Białowieża National Park provide the benchmark for primeval forest that indicates what is 'normal' for Marsh Tits under natural conditions (see Chapter 3). There are no published data for territory size in Białowieża, but there are 45 years of detailed annual census data that provide counts of spring territories in multiple large study plots, and these provide robust values for territory density (Wesołowski *et al.* 2022).

Table 5.2. Marsh Tit territory densities, from studies using colour-ringed birds and territory mapping.

Study area	Density (territories/km²)	Area (ha)	Period	Source
Monks Wood, England	13.7	160	2004–2011	Author's data
Wytham Woods, England	14.7	238	2007	Carpenter 2008
Bradfield Woods, England	16.4	70	2014–2018	Author's data
Wennington, England	17.1	101	2009–2017	Author's data
Combe Wood, England	18.7	128	2010	Author's data
Basel, Switzerland	15.1	92	1949–1954	Amann 2003
Pfrunger Ried, Germany	19.5	38	1968–1970	Ludescher 1973
Białowieża NP, Poland	15.0	138	2015–2019	Wesołowski et al. 2022

Densities are the annual number of territories for the given woodland area, averaged across the study period.

Table 5.2 shows that the Marsh Tit territory densities in the broadleaved habitat of the Białowieża National Park are similar to those from other detailed studies of colour-ringed populations in Europe, with most having 14–19 territories per km² of wooded habitat. The generally low rates of unpaired birds in these populations mean that the territory densities are close to the breeding densities (Broughton & Hinsley 2015).

Marsh Tit densities in suboptimal habitats are lower than those shown in Table 5.2. In predominantly coniferous or mixed forest plots in the Białowieża National Park there were only 2–6 territories per km². Densities are similarly low in native woodlands that have been turned into conifer plantations in England (Table 3.3).

In my own studies, the Marsh Tit territory densities have usually been estimated from surveys using a standardised playback method to map territorial birds within woodland patches. This method is reliable for estimating Marsh Tit territories and gives robust estimates for unringed populations as well as birds with colour-rings (Broughton et al. 2018a). The method involves plotting a route through suitable habitat and broadcasting playback on two visits in early spring, when Marsh Tits are most territorial and responsive. This playback method has also been used for a large-scale, county-wide survey in Sussex, southern England, in a landscape where populations were still quite strong (Black & Twydell 2022). Organised by the Sussex Ornithological Society, the survey visited woodland across 53 allocated 1km² sampling squares in 2020 and 2021. The results found Marsh Tits in 65 per cent of the survey squares at an average density of 2.8 territories per km². This relatively low density reflects the numbers of Marsh Tits that might be found in patchy woodland in modern farmed landscapes.

Several Willow Tit studies give a reasonable indication of the territory or breeding densities in different habitats across their range (Table 5.3). Densities are relatively low in the conifer-dominated areas of the Białowieża National Park (Figure 5.7), and Willow Tits are virtually absent from the mature broadleaved areas of this forest (Wesołowski et al. 2022). Densities are also low in the pine, spruce and birch forests of Norway and Sweden, but are much higher in Finland's boggy forests. Palmgren (1932) also reported Willow Tit densities of 7–25 pairs per km² in different areas of deciduous, mixed and boggy forest in the Finnish

Åland Islands, but the survey method and plot size are unclear. The overall abundance of Willow Tits in Fennoscandia has declined substantially since these studies, however, so those kinds of densities might now be hard to find.

Table 5.3. Willow Tit territory densities in various habitats, derived from systematic surveys or intensive studies.

Study area	Habitat	Density (territories/km²)	Area (ha)	Period	Source
Lake Vyrnwy, Wales	Conifer plantation & mixed woodland	0.9	3,451	2015–2019	RSPB, personal communication
Worsborough, England	Woodland–wetland mosaic	4.1	270	2015	Carr & Lunn 2017
Rabbit Ings–Carlton Marsh, England	Woodland–wetland mosaic	5.6	125	2015	Carr & Lunn 2017
Old Moor, England	Woodland–wetland mosaic	6.7	150	2015	Carr & Lunn 2017
Wigan, England	Woodland–wetland mosaic	7.3	493	2017–2019	Broughton *et al.* 2020
Wytham Woods, England	Deciduous woodland	9.1	22	1949	Foster & Godfrey 1950
Combe Wood, England	Deciduous woodland	11.8	68	2010, 2023	Author's data
Pfrunger Ried, Germany	Birch woodland	18.4	38	1968–1970	Ludescher 1973
Białowieża NP, Poland	Conifer & mixed forest	3.5	50	2015–2019	Wesołowski *et al.* 2022
Venabu, Norway	Open mixed forest	3.5	370	1990	Haftorn 1997b
Budal, Norway	Open mixed forest	3.6	5,000	1978–1984	Hogstad 1987
Gothenburg, Sweden	Conifer forest	4.4	6,000	1973–1980	Ekman 1984
Oulu, Finland	Mixed boggy forest	10.7	202	1975–1982	Orell & Ojanen 1983
Matsumoto, Japan	Conifer plantation, deciduous woods	17.5	64	1967–1970	Nakamura 1975

Densities are annual territories for the landscape area, averaged across the study period.

In Great Britain, most habitats support only low densities of Willow Tits in recent years. In the woodlands at Lake Vyrnwy in Wales, which are dominated by non-native conifer plantations in an upland landscape, Willow Tit densities are less than one territory per km². In northern England, in the mosaics of wetland, shrubland and early-successional woodlands in post-industrial landscapes, territory densities are low to moderate at around 4–7 per km².

These totals represented locally 'good' sites at the time of surveying, such as Old Moor and the Wigan study area (Table 5.3). Across the whole of the 200km² Dearne Valley landscape in South Yorkshire, which contains the Old Moor, Worsborough and Rabbit Ings–Carlton Marsh sites, the density of Willow Tits in 2015 was only 0.4 territories per km², and by 2023 this population was effectively extinct. It is interesting to note that historical and recent Willow Tit densities from mature deciduous woodland, such as Wytham and Combe, are higher than for wet shrubland and young woodland, although the survey areas are smaller.

Figure 5.7. A territorial Willow Tit among spruces in the Białowieża National Park, where they occur at relatively low densities of 3–4 territories per km². (© Richard K. Broughton)

WHY SUCH LOW DENSITIES AND LARGE TERRITORIES?

The large territories and low breeding densities of Marsh Tits and Willow Tits are characteristic of the *Poecile* genus. Siberian Tits occur at low densities in northern Europe, with around 4 pairs per km² in moderately good habitat (Virkkala 1990), but as low as 0.4 pairs per km² in sparse taiga forest near the treeline (Järvinen 1982). In mixed forests in Greece, Sombre Tits breed at densities of 12 pairs per km² (Catsadorakis & Källander 1999). Breeding densities of Black-capped Chickadees in rural Alberta in Canada were not dissimilar to Marsh Tits and Willow Tits in the richest habitats, at 15–17 pairs per km² (Desrochers *et al.* 1988). In suburban Massachusetts in the USA, however, Susan Smith (1994) found phenomenal breeding densities of 35–40 pairs of Black-capped Chickadees per km², more than double the highest densities of Marsh Tits or Willow Tits.

Figure 5.8. Great Tits breeding in Poland's Białowieża National Park occupy territories less than half the size of Marsh Tits', and so their natural breeding densities are around two or three times greater, but they are even more abundant in western European woodlands. (© Richard K. Broughton)

Even at their greatest densities, the numbers of Marsh Tits or Willow Tits are much lower than for most other tits in the same landscapes. In the natural forest of the Białowieża National Park, densities of Blue Tits and Great Tits in the broadleaved stands average two or three times higher than for Marsh Tits in the same area (Figure 5.8; Wesołowski *et al.* 2022). The territory sizes of Great Tits in these broadleaved plots are less than half the size of a typical Marsh Tit territory, with a median of just 1.9ha (Maziarz & Broughton 2015). In the coniferous areas of the Białowieża National Park the disparity is even greater, with Great Tit densities around five or six times higher than for Marsh Tits or Willow Tits.

In western Europe Blue Tits and Great Tits can reach even higher densities compared to Marsh Tits and Willow Tits. In Great Britain, the average densities of Great Tits and Blue Tits in Wytham Woods are each approximately 80 pairs per km^2 (East & Perrins 1988, Cole *et al.* 2021), which is more than five times that of Marsh Tits in the same wood, and almost nine times the density of Willow Tits (before they disappeared locally). As in the Białowieża National Park, the higher density of Great Tits is possible because their territory sizes average just 1.6ha (Wilkin *et al.* 2006), much smaller than a typical Marsh Tit territory of 5–6ha. Blue Tits can also reach extraordinary densities in Great Britain, with Martyn Stenning (2018) documenting more than three pairs per ha in the 10ha Lake Wood in Sussex. That density would mean a typical Marsh Tit territory could contain 15–18 pairs of Blue Tits. The densities of Blue Tits in studies elsewhere in western Europe often exceed the equivalent of 40 pairs per km^2, with several reporting densities equating to more than 200 pairs per km^2 (Stenning

2018). These kinds of Blue Tit densities are more than 10 times the greatest densities recorded anywhere for Marsh Tits or Willow Tits (Tables 5.2 and 5.3).

At Monks Wood, a deciduous ancient woodland, we found that even Coal Tits had densities that were 30 per cent higher than the local Marsh Tits (Broughton *et al.* 2019). Like Great Tits, the Coal Tits could achieve these higher densities by having smaller territories that averaged 3.3ha, which was only two-thirds the size of a Marsh Tit territory. These smaller territories allowed more pairs of Coal Tits to fit into the wood, despite this species generally favouring more coniferous habitats (Cramp & Perrins 1993).

The physical reason for the naturally low densities of Marsh Tits and Willow Tits is obviously their large territories, but the ecological reason for this remains unclear. After all, other tits of a similar body size can manage with much smaller territories in the same habitats. It is apparent from the way that Marsh Tits and Willow Tits settle as juveniles in the summer that they deliberately limit their own population density by quickly spacing themselves into pairs or small groups within defended areas (see Chapter 7). The limit on population density must therefore be a strategy that confers some fitness advantage of improved survival or lifetime reproductive success. The consequences of this are the quite rigid thresholds of the minimum territory size and maximum density that Marsh Tits and Willow Tits can achieve. It seems unlikely that either species could occur at more than 20–25 pairs per km².

Ian Newton (1998) reviewed the drivers of territorial behaviour and density limitation in birds, and the relevant factors for Marsh Tits and Willow Tits could be habitat quality, food resources, nest-sites and reducing the risk of local rivals having access to the breeding partner for mating opportunities. Of these factors, variable territory quality does not seem to be a major issue, as Marsh Tit territory sizes and densities are quite similar across different studies (Tables 5.1 and 5.2), although densities are much lower in suboptimal coniferous woodlands. There is also no indication that nest-sites are in short supply for either species (see Chapter 6).

Instead, the low densities and large territories of Marsh Tits and Willow Tits are probably linked to securing enough food resources in a patch of habitat to allow a pair to survive all year round. This theory is supported by the fact that even most single or widowed birds remain territorial, so they have resources to defend that are not directly related to breeding. The larger territories and lower densities in tougher environments, such as boreal forests or conifer plantations, also suggest that territories are linked to spacing of the population and securing sufficient resources.

Winter is probably the period of greatest limitation of food resources. The rapid settling and territorial spacing of juveniles in summer can therefore act as a filter to limit the number of individuals settling within an area, ensuring that the competition for winter food resources is not too severe. Spacing will also mean that Marsh Tits and Willow Tits are less likely to have their food caches discovered and pilfered by another member of the same species. In less challenging environments, such as the richer broadleaved habitats and milder climate of temperate Europe, the woodlands can support more birds in smaller territories, up to a certain point.

Aside from the brief settling phase of juveniles in summer, it is also clear that territoriality in Marsh Tits and Willow Tits is strongest just before the females become fertile in spring. As well as territorial defence, the males also show strong *mate-guarding* behaviour at this time, closely accompanying the female throughout the day (Morley 1953, Koivula *et al.* 1991). Large territories may therefore have an additional role for males, of protecting the paternity of their young by spatially isolating the female from neighbouring males.

The conundrum remains, however, that other tits can manage with smaller territories in the same habitat, regardless of whether the driver of territory size is securing food resources, caches or paternity. For example, food-caching Coal Tits are able to survive year-round in smaller territories than Marsh Tits or Willow Tits, while socially monogamous Great Tits are mate guarding within far smaller territories in spring. As such, there is no satisfactory explanation to fully explain why Marsh Tits and Willow Tits settle at such low densities compared to many other small passerines. Maybe it is the very presence of these other species that limits the densities of Marsh Tits and Willow Tits, via the extra constraint of interspecific competition that they impose.

Regardless of the reasons, the inbuilt limit to the density of Marsh Tits and Willow Tits has an important implication for conservation. The breeding population is ultimately determined by the amount of available habitat as the overriding factor governing the number of territories that an area can support. Unlike Blue Tits or Great Tits, where increasing the number of nest-sites (nest-boxes) will significantly increase the density of breeding pairs (East & Perrins 1988), providing additional nest-sites for Marsh Tits and Willow Tits cannot work in the same way. It is the overall habitat extent that is the limiting factor, and not the number of nest-sites within it.

TERRITORY STABILITY

Marsh Tits and Willow Tits are very sedentary, and the great majority of settled birds remain within the same general location throughout their lives, which can last for several years. Nevertheless, territories can vary in size and their boundaries can change over time. Ludescher (1973) recorded a Marsh Tit territory that expanded to annex a neighbouring territory after the latter became vacant. Other Marsh Tit and Willow Tit territories in that study varied in size by up to 33 per cent between years, but many changed by negligible amounts even when the occupants were replaced due to mortality.

Tomasz Wesołowski, during one of our many conversations, suspected that Marsh Tits in the Białowieża National Park increased their territory size when they were able to, such as when the population decreased and the remaining birds could make 'land grabs'. I saw conflicting evidence for territory expansion in the Monks Wood study: some males and pairs did extend their territories when a neighbour disappeared, but most territories remained quite static. For 79 Monks Wood territories of 40 males that were occupied in multiple years, the centroid of 66 per cent of them shifted by less than 100m between years. This means that the central point of the male's territory, or its 'centre of gravity', was within 100m of where it had been the previous year, showing that most birds maintained their location over time (Broughton *et al.* 2010).

If territories do get larger as populations decline, then this makes surveying more problematic. Unringed birds that are seen in different parts of an unexpectedly large territory might be wrongly assumed to be different individuals, leading to overestimation of territories or breeding density. When there are few or no neighbours then pairs can range over larger areas without any conflict, but they would not necessarily defend the entire area if they had to, so it may not all be a defended territory. This is one of the difficulties with Marsh Tits and Willow Tits in some regions, as the territories can be so large or the densities so low that identifying territory boundaries is not easy.

Home-ranges

If the boundaries of spring territories are sometimes hard to define, then identifying the extent of home-ranges at other times of year can be even more challenging. A home-range covers all of the area that a bird uses during its normal activities throughout the year. In boreal regions the home-range may be defended throughout the winter, similar to the spring territory. In temperate regions, however, neighbours will aggressively defend sharp territory boundaries during the spring, but at other times of the year the territories relax and the birds range over a wider, undefended area. These home-ranges might overlap with those of neighbouring birds, and neighbours will happily forage together outside the spring period.

Home-ranges are important when thinking of conservation and habitat management, even more so than spring territories, because it is the larger home-range that supports a bird throughout the year. It is the availability of sufficient habitat for the home-range that determines whether a pair can survive within a woodland patch.

In temperate regions the lack of strong territorial behaviour outside the spring period, and the larger areas used by the birds, means that the movements of Marsh Tits and Willow Tits can be difficult to follow. To overcome the challenges of mapping home-ranges, some researchers have used radio-tracking. Very small radio tags, weighing up to 0.5g and temporarily fitted to the birds with a small harness, or glued onto the feathers on the bird's back, can be used to follow individuals for standardised recording sessions over several weeks to plot their movements. Radio-tracking has been used to map the home-ranges of Marsh Tits in Monks Wood (Broughton *et al.* 2015a) and Willow Tits in the Dearne Valley and Lake Vyrnwy (Pinder & Carr 2021, Bellamy 2022). An alternative is to map colour-ringed birds to infer home-range sizes from field observations, but this is more approximate.

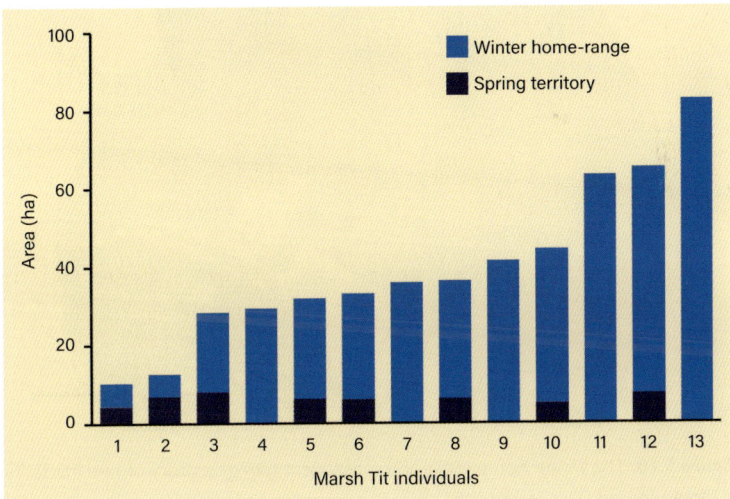

Figure 5.9. Winter home-range sizes (in hectares) of 13 Marsh Tits at Monks Wood. The birds were radio-tracked for several days or weeks during November–January and the home-ranges were derived as 100 per cent kernel estimates (see Broughton *et al.* 2015a). The areas of the defended spring territories of the same birds are shown (where this was known).

MARSH TIT HOME-RANGES

The Monks Wood radio-tracking study involved monitoring Marsh Tit movements from November to January during the winters of 2006–2008. Radio-tagged birds were each tracked on up to 11 days during a period of up to a month, before being recaptured to remove the tag. On each tracking day the bird was located using a radio receiver and antenna and then followed for up to two and a half hours as it moved around and foraged. During the tracking sessions the bird's location and activity were logged every 10 minutes, which gave us a set of data points from multiple days that could be used to model the home-range using the kernel estimation method, which delineates the areas of high and low activity based on the density of the bird's plotted locations (Broughton *et al.* 2015a).

These Marsh Tits had enormous home-ranges in winter, with the extent averaging 39ha of woodland habitat for each bird, which is seven or eight times larger than the average spring territory (Figure 5.9). The smallest home-range was 10ha and the largest was a staggering 83ha. The maximum home-range extent for each bird contained most or all of its spring territory, showing that the home-ranges were extensions of this defended area (Broughton *et al.* 2015a).

Not all of the home-range was used to the same extent, as each bird had some areas where it spent more time than in other parts. These core areas of the home-range, defined

Figure 5.10. The winter home-range and subsequent spring territory of one Marsh Tit at Monks Wood. This male's 44ha winter home-range had a 10ha core encompassing 70 per cent of the bird's activity. The 5ha spring territory falls within the home-range but only partially overlaps with the winter core activity. The home-range was derived from radio-tracking and the territory from spring mapping of colour-ring observations. Lidar data were captured by the NERC Airborne Remote Sensing Facility, via the CEDA repository (CC BY 3.0).

to encompass about 70 per cent of the Marsh Tit observations, were often of a similar size to spring territories, averaging 6ha. The core areas of the winter home-range did not necessarily coincide with the spring territory, however, as they overlapped by only 49 per cent at most, and an average of just 26 per cent across all birds. This demonstrated that the core area of the winter home-range, where the birds spent most of their time, was not in the same area as the spring territory (Figure 5.10). In other words, the spring territory was not the core of the winter home-range, and so the birds were using different parts of the home-range at varying intensity during different times of the year.

These results reiterate the important distinction between the home-range and the spring territory, as it shows that the territory alone is insufficient to provide all of the resources to support a pair of Marsh Tits throughout their annual cycle. In terms of maintaining or creating habitat for Marsh Tits, at least in Great Britain, the focus needs to be on the larger home-ranges and providing woodland habitat of 30ha or more to maximise year-round survival for a typical pair.

WILLOW TIT HOME-RANGES

There is comparatively little information differentiating the home-ranges of Willow Tits from their spring territories. In contrast to Marsh Tits, the home-ranges used by Willow Tits outside of the early spring period often do not seem to be very much larger than the breeding territories. For example, Ludescher (1973) considered that Willow Tits in Germany remained largely within their spring territory areas all year round, which averaged about 7ha (Table 5.1). During the winter, however, Ludescher noted that the Willow Tits would sometimes forage outside their territory, suggesting that the actual maximum home-range could be somewhat larger.

In Sweden, Ekman (1979a) also thought that the winter home-ranges of Willow Tits largely corresponded to their spring territories, which averaged 24ha. Other studies in Fennoscandia found similar winter home-ranges of 20–27ha (Koivula & Orell 1988, Brodin 1992), although Siffczyk et al. (2003) reported home-ranges averaging 13ha. In this latter study the home-ranges increased in size with the proportion of open habitat, such as forestry clear-cuts, which meant the Willow Tits had to compensate for fragmented woodland habitat by expanding their home-ranges. Willow Tit winter home-ranges in Japan are similar in size to those in Fennoscandia, averaging around 17–20ha for pairs or small flocks (Nakamura 1975, Nakamura & Wako 1988).

In Great Britain the only systematic investigation of Willow Tit home-ranges is the radio-tracking studies in the Dearne Valley and Lake Vyrnwy (Pinder & Carr 2021, Bellamy 2022). The core foraging areas of 3–7ha that were detected at these sites in late winter or early spring might be more representative of the defended territories at that time of year (Table 5.1). The maximum foraging range for eight birds in the Dearne Valley averaged 10ha, with one bird ranging over 17ha. These values are not very dissimilar to the other Willow Tit home-ranges, and so might be a rough approximation of the home-range sizes for these birds. There is a strong need for more studies on the habitat area used by Willow Tits throughout the year, in order to help inform conservation efforts in places such as Great Britain and Fennoscandia.

OVERLAPPING MARSH TITS AND WILLOW TITS

In many parts of their ranges Marsh Tits and Willow Tits occur together in the same habitat. In these places the territories of each species could be influenced by interspecific competition for space, as well as by intraspecific competition. However, only one study has investigated the territory dynamics or home-ranges of Marsh Tits and Willow Tits living in the same place, and that is the classic German study at Pfrunger Ried. In this study Ludescher (1973) mapped the territories of both species and found that they often coincided, sometimes very closely, with the territory of a pair of Marsh Tits overlapping almost completely with a pair of Willow Tits. Ludescher observed little aggression between the species, and these interspecific conflicts were usually very brief.

Near Uppsala, in central Sweden, Alatalo *et al.* (1985) found Marsh Tit territories coinciding with Willow Tits in mixed woodland, and they tested interspecific aggression between the two species. In an experiment, they placed a caged Marsh Tit or Willow Tit near a Marsh Tit nest in the woodland, to test how the nesting birds responded to the simulated intruder. The researchers found that the breeding Marsh Tits were most aggressive when presented with another intruding Marsh Tit, as would be expected, but they were also quite aggressive towards the simulated intrusion of a Willow Tit. Other studies have shown that Marsh Tits are dominant over Willow Tits at artificial feeding stations, and they can usurp Willow Tits from nest-cavities (Ludescher 1973, Haftorn 1993b).

Despite the dominance of Marsh Tits over Willow Tits, Ludescher (1973) showed that both species can coexist in overlapping territories and home-ranges. Birds of both species even forage together during the non-breeding period, with Ludescher finding that a third of foraging observations involved mixed groups, with up to 8–12 Marsh Tits and Willow Tits aggregating in harsh weather. There are other examples of Marsh Tits and Willow Tits cohabiting within the same woodlands throughout the year, such as Wytham Woods during the 1940s and 1950s (Gibb 1954b). More recent colour-ringing studies in southern England have found Willow Tits and Marsh Tits resident together in Combe Wood and Hen's Wood in Berkshire, Hampshire and Wiltshire (Last & Burgess 2015, my additional data). In these examples both species were clearly overlapping in their movements and habitat use within the limited woodland patches (see also Alatalo & Lundberg 1983). Another study by Alatalo *et al.* (1985) showed that where Marsh Tits were absent on Swedish islands, then the resident Willow Tits could expand their foraging into broadleaved habitats that were otherwise occupied by Marsh Tits on the mainland. This is good evidence that Marsh Tits were somehow excluding or outcompeting Willow Tits in broadleaved habitats. The degree of competition might therefore be related to habitat quality in different parts of the species' ranges, but where there is a conflict then dominant Marsh Tits will ultimately have the upper hand.

Winter flocking

During the autumn and winter, tits and chickadees of Eurasia and North America generally adopt one of two major types of social organisation, which Jan Ekman (1989) called the *basic flock* and the *discrete flock* systems (Figure 5.11). The basic flock structure involves resident birds of the same species living in partially overlapping home-ranges and associating together in casual flocks with a loose membership. In this system there is a site-based dominance, where a bird may be dominant over others in its own home-range but subordinate in a different location nearby as the flock moves around. The basic flock structure is thought to be

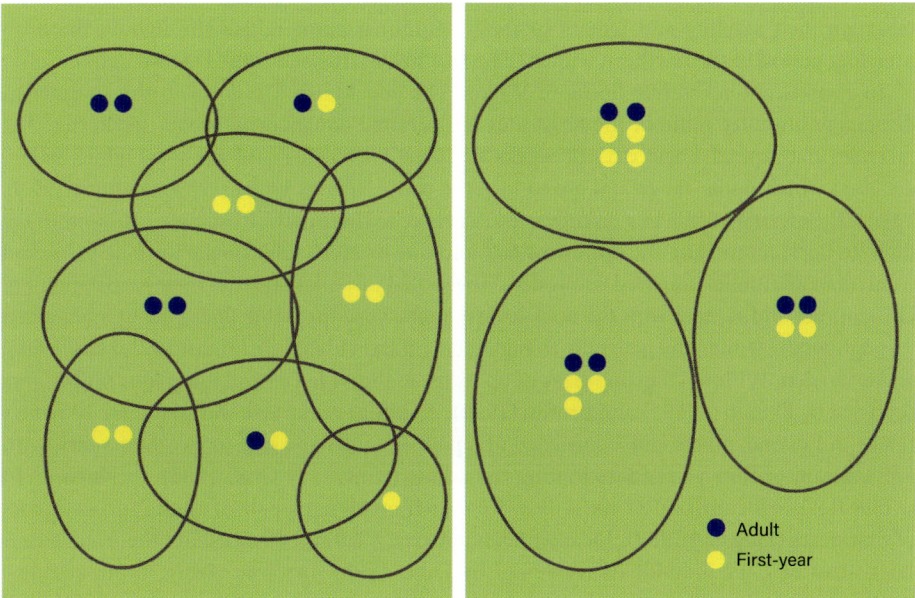

Figure 5.11. A schematic of the basic flock (left) and discrete flock (right) systems of winter social organisation, as used by Marsh Tits and Willow Tits in different parts of their ranges. Basic flocks involve pairs of birds that casually mix with neighbours in overlapping home-ranges. Discrete flocks involve a dominant pair (often adults) and one or more pairs of subordinate first-years who together defend a communal home-range against their neighbours.

the original ancestral system for tits, and species adopting this strategy include the Great Tit and Blue Tit, which do not cache food in autumn and winter (Ekman 1989, Dhondt 2007). The basic flock is flexible and birds can join when it suits them or leave to exploit food sources elsewhere during harsh weather, such as garden bird-feeders in nearby towns and villages (Gosler 1993, Hanmer *et al.* 2022). The discrete flock structure contrasts with the basic flock by involving small groups of birds in communal and defended home-ranges that do not overlap. Species in discrete flocks also cache food in winter, and the two habits of stable flocking and food caching seem to be linked. Discrete flocks are highly structured and operate with a strict social hierarchy among the birds within them. Species adopting this strategy include Black-capped Chickadees and Crested Tits. Intriguingly, Marsh Tits and Willow Tits both seem to form basic flocks in some regions and discrete flocks in others.

DISCRETE FLOCKS

Discrete flock home-ranges are each defended by an adult or dominant pair, usually accompanied by several unrelated and subordinate juveniles or first-years to form small flocks. The flocks have a stable membership of the same birds throughout the winter, which forms the discrete flock for that area, and different groups each have their own communal home-range. The discrete flock structure is very well studied in some species, especially the Willow Tit and Black-capped Chickadee (Smith 1991, Lahti 1998), and also in Scandinavian Marsh Tits (Nilsson & Smith 1988a, Nilsson 1989a, 1989b). Food caching helps these groups maintain their stability and remain in their communal home-ranges throughout the non-breeding period of the northern winter (Ekman 1989, Matthysen 1990, Dhondt 2007).

In Fennoscandia discrete flocks of Willow Tits and Marsh Tits form in the summer as dispersing juveniles settle in the territories of unrelated adults, or in vacant territories. The juveniles may spend a few days or weeks assessing each other and the local adults before deciding which home-range and group to settle with (Nilsson 1989a, 1989b, Haftorn 1997b, 1999). The juveniles settle in a broadly even sex ratio, so there may be multiple pairs within the discrete flock comprising the dominant pair and one or more subordinate juvenile pairs, and sometimes additional unpaired youngsters. The settled juveniles undergo their partial moult in late summer and remain with the flock as first-years, with the stable flocks being maintained throughout the winter and generally changing only if individuals die (Smith 1991, Lahti 1998).

In Sweden, Willow Tit group sizes of up to seven individuals (averaging four birds) were reported by Ekman (1979a) and Brodin (1992). A similar group size was found for 18 winter flocks in Finland, where most consisted of a pair of adults and one or two first-years, with a maximum of four youngsters joining the adults (Koivula & Orell 1988). In Norway, 10 winter flocks all consisted of six Willow Tits, each involving a pair of adults and two pairs of first-years (Hogstad 1987). Meanwhile, in Siberia, Vladimir Pravosudov reported winter flock sizes of 7–12 individuals (reviewed by Lahti 1998). In these northern populations, good habitat and good weather during the dispersal and settling period are associated with slightly larger group sizes in the winter flocks, presumably because of better survival of the first-year birds.

There is comparatively little information for group sizes of winter flocks of Marsh Tits in northern regions. Jan-Åke Nilsson and Henrik Smith (1988a) reported discrete flock sizes of 2–9 birds in Sweden, with most having 3–5 individuals (Nilsson 1989b). Like Willow Tits, these flocks consisted of an adult pair and several unrelated first-years, with a stable membership throughout the winter and flocks occupying non-overlapping home-ranges. Flocks containing only first-years were also present, independent of adult pairs, and these involved a dominant young pair that settled within a vacant territory and were then joined by other subordinate first-years.

The winter flocks of Willow Tits and Marsh Tits in northern Europe are mostly smaller than those of Black-capped Chickadees in North America, which can contain up to 12 birds and more than one pair of adults (Smith 1991). These large flocks of chickadees could result from a high density of garden bird-feeders in the study areas, leading to a high density of chickadees and the breakdown of flock territoriality. The rich feeding sites scattered across the suburban study area meant there was little need for the chickadee flocks to defend the resources against each other. This scenario has not yet been reported in Marsh Tits or Willow Tits, maybe due to a lower density of people and their bird-feeders across the core habitats in northern Europe.

DOMINANCE HIERARCHIES

Within the discrete flock structure a bird's rank in the social hierarchy is important for its chances of survival and reproduction, and the mechanics of this social organisation have long fascinated ornithologists. In the discrete flock system there is a linear dominance hierarchy where the bird's rank is expressed via dominance, with those individuals having the highest rank being socially dominant over the other birds in the flock. The lowest-ranked bird is then subordinate to all of the others in the flock. Dominance is shown by intimidation and physical aggression, where high-ranking birds access the best foraging spots and supplant any subordinate birds by posturing or flying at them, forcing them to move.

The dominance hierarchy in winter flocks of *Poecile* species seems to be based on a set of simple rules that have been outlined in excellent reviews by Susan Smith (1991) and Kimmo Lahti (1998). These rules are based on criteria including size, age, sex and prior occupancy, or the 'first come, first served' principle, where a bird that first joins a flock is dominant over later arrivals. Of these criteria, the effect of a bird's physical size on its social rank within the flock appears to be quite weak. Studies have generally measured size in terms of a bird's weight or mass, its wing length or the length of the tarsus. In Willow Tits, size was an important factor for rank in one study (Hogstad 1987) but not in another (Lahti *et al.* 1996), while an experimental study found that size only mattered if the birds were evenly matched in other respects (Koivula *et al.* 1993). For Marsh Tits, an aviary experiment showed that a bird's size, age or sex had a negligible effect on its dominance within a flock (Nilsson 1989b). As with the experimental study of Willow Tits by Koivula *et al.* (1993), the overriding factor was prior establishment in the flock, whereby existing flock members win every encounter with newcomers. One other factor that had some relevance for Marsh Tits was the flock size, as joining became more difficult as the number of birds increased. This suggested that the existing flock members were aware of having to balance the benefits of being in a flock with the risk of having too many competitors.

Age is closely related to prior occupancy, as the dominant pair that forms the nucleus of winter flocks is usually going to be the resident adult territory-holders. It is uncommon for adults to disperse and leave their home-ranges, so almost all of the newly arriving birds trying to join flocks will be dispersing juveniles. The prior residency rule means that the first juveniles to join the flocks gain a higher rank than later arrivals, even if this is just a matter of one day (Nilsson & Smith 1988a). A higher rank is associated with a better survival rate and a chance of inheriting the territory if one or more of the adults die over the winter (Koivula & Orell 1988, Hogstad & Slagsvold 2018). Similarly, if one of the higher-ranked first-year individuals dies then a lower-ranked bird of the same sex can be 'promoted' to take its position and rank. This would take the bird a step closer to acquiring a territory, and is better than the alternative of not joining a discrete flock, as most first-years in that situation disappear and probably die. Finding a vacancy and quickly settling in a discrete flock is a first-year bird's best chance of survival and acquiring a breeding territory at the end of the winter (Lahti 1998).

For the dominant pair of adults, the benefits of allowing several unrelated young birds to settle in their territory are not so obvious at first. The first-year birds can assist with defence of the winter home-range and its resources against neighbouring flocks, but will also consume some of the resources themselves. However, Lahti (1998) and Smith (1991) considered these questions in detail and proposed that adults and first-years alike can benefit from being in a flock because of greater vigilance against predators and greater efficiency in finding food. Forming a small flock provides safety in numbers and more pairs of eyes to spot feeding opportunities or ambushes by raptors, and also allows each bird to spend more time foraging rather than looking out for danger on its own. Dominant birds can even kleptoparasitise the food discoveries of subordinate first-years, simply stealing it from them. Another key benefit for the adults is that the youngsters can act as 'spares', or readily available replacements should an adult lose its mate before the next breeding season. This avoids the risk of an adult finding itself unable to breed if it was widowed over the winter.

BASIC FLOCKS

In the milder temperate regions of central and western Europe, Willow Tits and Marsh Tits both adopt a basic flock structure. Rather than spending the winter in stable, discrete flocks of 4–12 birds, as in Fennoscandia or Siberia, Willow Tits in the woodlands of Germany or England usually spend the winter in pairs that have only loose associations with other birds (Foster & Godfrey 1950, Ludescher 1973, my own observations). In Japan, too, Nakamura (1975) found that Willow Tits generally spent the winter in pairs whose home-ranges overlapped quite substantially with the neighbours'. In Belgium, however, Lens (1996) reported groups of 2–4 Willow Tits occupying winter territories, formed of an adult pair and up to two first-years, although it is not clear if the territories overlapped between groups. Nevertheless, it seems that in milder regions the Willow Tits relax their habit of forming discrete flocks with group territoriality.

For Marsh Tits there is also a substantial difference in their social organisation and winter flock structure between different regions. The Scandinavian studies suggest that the birds are in discrete flocks, with groups consisting of an adult pair and several unrelated first-years that maintain communal, non-overlapping home-ranges (Nilsson & Smith 1988a, Nilsson 1989b). Further south in Germany, however, the birds have a pair-based social organisation in winter, rather than forming discrete flocks (Löhrl 1950, Ludescher 1973). In Switzerland, too, juvenile Marsh Tits settle in home-ranges that partially overlap several adult territories, rather than forming a stable group (Amann 2003, 2007). During the winter the adults and first-years associate in casual small flocks that have a loose, changing membership, with no sign of group territoriality. This is a typical basic flock, and these studies from Switzerland and Germany indicate a clear absence of a discrete flock structure among Marsh Tits in central Europe. Even further south in Greece, Panayotopoulou (2005) found that Marsh Tits in mixed flocks typically numbered 1–4 birds, with a rounded average of two, indicating pairs and occasional lone individuals.

In the Monks Wood study we also found an obvious basic flock structure among the Marsh Tits (Broughton *et al.* 2015a). By radio-tracking birds during the winter period we could record their ranging and also which other birds they were associating with over hours, days and weeks. We also recorded associations between birds at temporary feeders, which showed the stability of the relationships on different days and at different locations, giving a measure of how often birds were seen together. The results showed that the Monks Wood Marsh Tits spent the winter in individual home-ranges that partially overlapped with several others. The birds were usually in groups of two or three, but occasionally up to nine Marsh Tits were together in a mixed-species flock. The only strong relationships were between members of a male–female pair, with mostly weak associations between other birds. There was no sign of stable groups, flock territoriality or a communal home-range. The large, individual home-range of each Marsh Tit (averaging over 30ha) brought it into contact with almost half of the other birds in the study group, although most of these associations only involved occasional encounters.

The reasons for the different structures of social organisation between populations are not well understood. Species that cache food, like Marsh Tits, Willow Tits, Coal Tits and Black-capped Chickadees, are also the ones that adopt a discrete flock structure in northern regions. Ekman (1989) suggested that discrete flocking was primarily linked to food caching and habitat quality, where a tough environment would result in limited resources and a greater reliance on cached items. A small discrete flock of birds defending a large group territory would make sense for protecting their food caches by limiting the competition from other birds wandering in and finding them.

The hypothesis of habitat quality and resource richness does seem to match the general pattern of changing social structure with latitude. Marsh Tits, Willow Tits and Coal Tits form discrete flocks in Scandinavia, where the winters are relatively harsh and food is harder to find. In the milder winters of England, Spain or central Japan, however, these species adopt a more basic flock structure in a more benevolent environment (Nakamura 1975, Brotons 2000, Broughton *et al.* 2015a). Nevertheless, even in mild regions these three species still obsessively cache food during autumn and winter, undermining any direct link between caching and the discrete flock structure. The drivers of winter flock structures therefore remain elusive.

FLOATING INDIVIDUALS

Not all Marsh Tits and Willow Tits have a typical home-range in winter, whether in a basic or a discrete flock. A small contingent of the winter population seem to have extremely large or vague home-ranges, which can cover extensive areas of the landscape and cross the more usual home-ranges of many other birds. These mobile birds have been called *winter floaters* by Jan Ekman (1989), Susan Smith (1991) and Olav Hogstad (2014). In the early study of Marsh Tits at Oxford, Averil Morley (1953) referred to floating first-years as 'landless' birds, which did not seem to have an identifiable home-range at all, probably because it extended far beyond the limits of the study area.

During the radio-tracking study at Monks Wood, we discovered that floating Marsh Tits do indeed have regular home-ranges, but they are just very large and difficult to define unless tracking the birds over a wide area (Broughton *et al.* 2015a). The floating young male that we tracked over 11 days at Monks Wood had a home-range of 63ha, which was more than double the average size for other birds that we followed. I remember tracking this individual one day as it quickly moved 1.6km in a steady direction over a matter of minutes, as well as moving 800m in another direction on another day. This young male clearly foraged across a very large area (Figure 5.12), but it was familiar enough with its local environment to travel long distances to food sources that it knew about.

In Norway, Willow Tit floaters would also travel over similarly large distances in relatively short periods, with two birds covering 2km over a couple of hours (Hogstad 2014). The typical winter home-ranges of the discrete flocks in this population were around 25ha, and floaters would regularly range over between two and eight of them, so they were thought to have been ranging over about 100ha in total. This is a huge foraging range for a small resident songbird.

Their large home-ranges mean that floaters can be difficult to study, and there is not much information on how frequent they are among winter populations, especially for Marsh Tits. The available studies suggest that floaters are an uncommon but a regular part of the autumn and winter populations of Marsh Tits and Willow Tits, representing about 5–10 per cent of first-year birds (Nilsson & Smith 1988a, Hogstad 2014). Floaters can be males or females, but they are low-ranking and unpaired, moving around by themselves and casually associating with other birds but with no strong social bonds. Because they are often alone, floaters have to spend more of their time watching out for predators and less time foraging. As such, they tend to be in poorer condition than the regular settled birds, and they also have a lower likelihood of survival, with almost all of them disappearing before the breeding season (Nilsson & Smith 1988a, Smith 1991, Hogstad 2014).

Figure 5.12. Observations (black dots) of a first-year male Marsh Tit during autumn and winter at Monks Wood, showing it to be a floater with a very large home-range. The blue square marks the nest from which the bird fledged a few months earlier, before dispersing. Woodland and hedgerows above 2m tall are shown in green. Lidar data were captured by the NERC Airborne Remote Sensing Facility, via the CEDA repository (CC BY 3.0).

In light of the poor chances of joining the local breeding population, what are the advantages of adopting the strategy of being a winter floater? This is a question that several researchers have asked, particularly Susan Smith (1984, 1991) for Black-capped Chickadees, Olav Hogstad (1990a, 2014) for Willow Tits, and Jan Ekman (1989) for parids in general. The consensus is that being a floater is better than nothing, and these birds are making the best of a bad situation. Floaters have been unsuccessful in establishing themselves in a flock, pair or home-range after dispersal, probably because all of the available vacancies have been taken by birds that arrived before them. Their only options are to leave the area for an unknown destination, or to settle as a low-ranking floater living outside the main social organisation. Adopting a floater strategy gives them a chance to wait over the winter and hope that a settled bird dies, freeing up a vacancy in one of the flocks or home-ranges. This would allow the floater to pair with the widowed bird and slot into a social position.

The very large home-ranges used by floaters brings them into contact with many other conspecifics in other home-ranges. Consequently, floaters are in a better position to detect a new vacancy than a low-ranking first-year individual in a discrete flock, which may only know about the situation in its immediate area. This is one tangible benefit of being a floater, having the ability to monitor a large area for any vacancies arising. It is a high-risk strategy, however, as Marsh Tits, Willow Tits and other chickadees have quite high survival rates over the winter period. For floaters, this means that their chances of eventually finding a vacancy are low, and that is why so few of them join the local breeding population. Ekman (1989) also estimated that being a floater is only worth the gamble if there are few other floaters around, so that an individual stands a stronger chance of slotting into a scarce vacancy if it appears. If no vacancies appear before the spring, when the settled birds establish their exclusive breeding territories, then floaters will eventually have to leave the area and join the small wave of spring dispersal, and most of them are likely to perish in the process.

MIXED-SPECIES FLOCKING

Whether as individuals, pairs or small groups, Marsh Tits and Willow Tits will commonly associate with other woodland birds in mixed flocks, particularly in the autumn and winter. In Europe these mixed flocks are dominated by other tits, and in Great Britain these include Blue Tits, Great Tits and Coal Tits. Other species joining these mixed flocks may be Long-tailed Tits, Eurasian Nuthatches, Eurasian Treecreepers, Goldcrests, Great Spotted Woodpeckers and occasionally Lesser Spotted Woodpeckers (Morse 1978). In Greece the mixed flocks joined by Marsh Tits can include many of the same species but also another *Poecile*, the Sombre Tit (Panayotopoulou 2005). In Japan and Korea the mixed flocks can include Willow Tits, Marsh Tits, Varied Tits, Japanese Tits, Coal Tits, Eurasian Nuthatches, Long-tailed Tits, Goldcrests, Great Spotted Woodpeckers and Japanese Pygmy Woodpeckers (Jabłoński & Lee 2002, Krams *et al.* 2020).

The mixed flocks joined by Marsh Tits or Willow Tits can be very small, consisting of just one or two other birds, or they can be very large indeed. At Monks Wood I have seen mixed flocks of over 200 birds in early autumn, dominated by many juvenile Long-tailed Tits, Blue Tits and Great Tits, and containing up to 10 Marsh Tits. For radio-tracked Marsh Tits at Monks Wood during the winters of 2006–2008 we used instantaneous sampling to record every 10 minutes during tracking sessions whether the bird was in a mixed flock or foraging alone (or in a pair). Some birds were sampled for only one tracking session on a single day, while others were sampled on up to seven days in sessions totalling 12 hours, so the data were a bit variable. However, the results showed that the wintering Marsh Tits were mostly in mixed flocks with other species, which accounted for 85 per cent of their time on average (Figure 5.13). These mixed flocks usually contained at least two Marsh Tits, but two-thirds of them contained fewer than 10 birds of any species.

The flocking data from Monks Wood were similar to observations at Wytham Woods, when PIT-tagged birds were logged in mixed flocks at feeding stations (Maziarz *et al.* 2023). In this study Marsh Tits were in mixed flocks for 90 per cent of observations, spending the remaining 10 per cent alone or in pairs with another Marsh Tit. Two of the Monks Wood Marsh Tits spent at least half of their time outside of mixed flocks (Figure 5.13), one of which was the first-year male floater mentioned above. The other bird was a first-year female, and the similar amount of time that she spent foraging alone suggests that she may have also been a floater, with her home-range being calculated at 27ha.

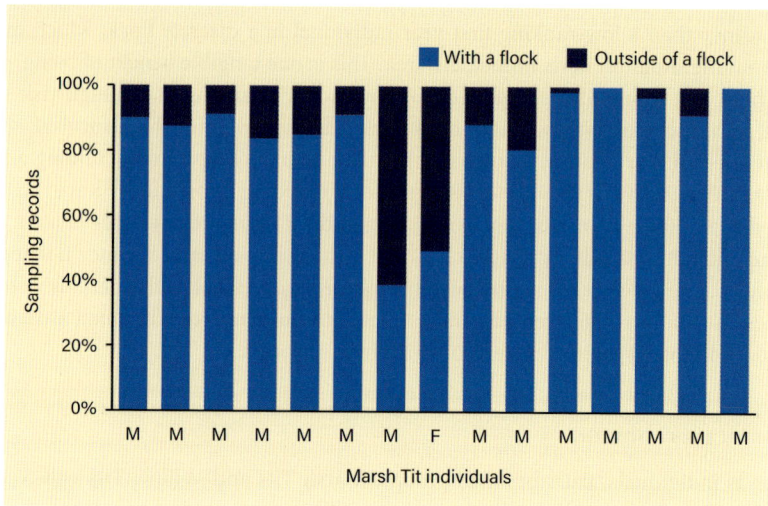

Figure 5.13. Marsh Tits joining mixed-species flocks at Monks Wood during winter. The columns show the proportion of sampled time that 15 individuals spent with a mixed-species flock of other woodland birds, or outside of the flock (alone or with another Marsh Tit). The birds were sampled at 10-minute intervals during radio-tracking sessions totalling at least 1.5 hours over 1–7 days. Most birds were males (M), with one female (F).

DOMINANCE IN MIXED-SPECIES FLOCKS

When joining mixed flocks, Marsh Tits and Willow Tits are usually subordinate to larger species, such as Great Tits, Japanese Tits and Eurasian Nuthatches, but they are also subordinate to the similar-sized Blue Tits. In Fennoscandia and the Baltic region, and also in Belgium, flocks of Willow Tits have been reported associating with groups of Crested Tits in mixed compound flocks (Figure 5.14), but the Crested Tits are always dominant (Ekman 1979a, Lens 1996, Krams *et al.* 2020). In Norway, Haftorn (1993b) suggested that Willow Tits were even subordinate to the smaller Coal Tit, although Hogstad (1978) thought that Coal Tits were probably dominated by Willow Tits. The dominance relationship between Willow Tits and Coal Tits is still unclear, but Marsh Tits are certainly dominant to Coal Tits wherever both species coincide (Figure 5.15). As with discrete flocks, the social dominance in mixed flocks is demonstrated by species gaining better access to food or foraging sites by using intimidation or force on other species, so that the subordinates have to give way to the dominants.

The dominance hierarchies within mixed-species flocks have important implications for where birds forage within the woodland habitat. In two classic studies from Wytham Woods, John Gibb (1954a) and Douglass Morse (1978) outlined how the different tits appeared to be segregated in their foraging niches. Each species was noticed to feed at different heights within the woodland profile, on different parts of trees and bushes, or exploiting different foods in different ways. For example, Marsh Tits foraged disproportionately on the lower branches of trees, and more on oaks compared to dominant Blue Tits or Great Tits. Meanwhile, Willow Tits spent much more time foraging on Silver Birch and Common Elder compared to other tits. Willow Tits were also the only tit not to feed on beechmast, which was a favoured winter food of Great Tits, Blue Tits, Marsh Tits and Coal Tits.

Figure 5.14. A colour-ringed Crested Tit in the Oulu study area of northern Finland. Crested Tits are dominant over Willow Tits where they share woodland habitats. (© Anne Laine)

The degree to which these foraging patterns are a preference, or how much they are due to avoidance or competitive exclusion by dominant species, is a question that has occupied behavioural ecologists for decades. The studies by Gibb (1954a) and Morse (1978) helped to shape the hypotheses of David Lack (1971) and Christopher Perrins (1979) at Oxford, which centred on the idea that the different tits have evolved different niches in order to avoid interspecific competition. The early evidence from Wytham Woods suggested that there was enough stable segregation of foraging places to allow the different species to coexist in the same habitat without suffering too much competition.

Later thinking suggested that the foraging behaviour of tits in mixed flocks is more strongly influenced by interspecific competition than was previously supposed, and this competition is ongoing and dynamic (Alatalo 1982, Dhondt 2012). If this is true, then Marsh Tits and Willow Tits are in constant competition with other tits for space and resources and do not have rigid niches. The different foraging patterns observed by Gibb (1954a) and Morse (1978) would then be an indicator of competition, otherwise Marsh Tits and Willow Tits would forage in consistent places even when dominant tits were absent. Field studies and experiments show that this is not the case, and the birds' foraging patterns do change when dominant species are not present. For example, where Willow Tits and dominant Crested Tits foraged together the former were largely excluded from feeding in the best parts of the trees; when Crested Tits were absent, Willow Tits expanded their foraging into these favoured areas, from where they were previously excluded (Hogstad 1978, Lens 1996). Marsh Tits in mixed flocks also foraged in different places when Great Tits were absent, being more likely to feed on the ground (Jabłoński & Lee 2002), suggesting that they were being prevented

Figure 5.15. Marsh Tits (right) are dominant over Coal Tits (left) wherever both species occur together. (© Loic Poidevin/Nature Picture Library)

from foraging there by the larger dominant tits. Experimental removal of dominant species has demonstrated that these competition effects are real (Dhondt 2012), which raises the question of why subordinate Marsh Tits and Willow Tits join mixed flocks in the first place.

The benefits for both species in joining a mixed flock probably centre on safety. As with same-species flocks, more eyes scanning for predators means that individuals can spend more time feeding rather than looking out for danger. In very cold weather Willow Tits are more likely to join together in larger mixed flocks, as this allows each bird more time to forage and feed while the others provide extra vigilance (Hogstad 1988).

The costs and benefits of joining mixed flocks can be finely balanced. In Wytham Woods a study by Farine *et al.* (2015) analysed the transmission of information within a mixed flock after the finding of a novel food source, in this case feeding stations baited with Sunflower seeds. The tits were fitted with PIT tags, allowing each bird and its associates to be logged by recorders at the network of feeding stations. Powerful analyses of the millions of interactions and movements plotted the social networks between the birds, showing that Marsh Tits were usually the first to discover the feeding stations. The dominant Great Tits and Blue Tits would then acquire this knowledge from watching the Marsh Tits, and exploit the food source themselves. It was unclear whether the dominant tits were 'parasitising' the Marsh Tits, effectively stealing their knowledge (and food), or whether the Marsh Tits were recruiting the flock of other tits to the feeding site, possibly to provide extra vigilance against attacks from predators such as Eurasian Sparrowhawks. In Japan, Suzuki (2012) showed that when

160

Willow Tits discover a new food source they call to attract other Willow Tits and other tit species, which would support the recruitment hypothesis.

Where the dominant tits are very abundant then the costs of flocking may outweigh the benefits, and the subordinate species may begin avoiding the larger mixed flocks. In another study from Wytham Woods, the foraging activity of Marsh Tits was investigated at feeding stations when the mixed flocks contained varying numbers of dominant Blue Tits and Great Tits (Maziarz *et al.* 2023). The results showed that Marsh Tits became less likely to join the flocks as the number of Great Tits and Blue Tits increased. In larger flocks the Marsh Tits also foraged at the feeding stations less frequently, suggesting that they were avoiding or being excluded by the dominant birds. This apparent competition had a big impact, shifting the foraging pattern of Marsh Tits so that they avoided the feeding stations later in the day and later in the winter, which is when the feeders were visited by the largest flocks of Great Tits and Blue Tits.

This evidence from Wytham Woods and elsewhere shows the difficulty that Marsh Tits and Willow Tits can face when joining mixed-species flocks, where they are subordinate to Great Tits, Blue Tits or Crested Tits. In some circumstances the benefits of extra vigilance can outweigh the costs of competition from dominant birds, but where dominant birds become abundant then Marsh Tits or Willow Tits may find that the increasing costs of competition make it no longer worth joining these flocks. The ability to avoid the flocks may not always be available, however. The study by Maziarz *et al.* (2023) showed that dominant Great Tits and Blue Tits were so common and ubiquitous in Wytham Woods that there was nowhere that the Marsh Tits could go to completely avoid them. If this scenario applies more widely to woodlands in Great Britain or elsewhere, then Marsh Tits or Willow Tits would be constantly exposed to significant competition and the negative consequences that go with it.

Summary

The social organisation of Marsh Tits and Willow Tits is built around the pair bond, with the birds investing a great deal in one another. Divorce is uncommon, so pair bonds usually last for life, with some pairs surviving together for up to eight years.

In spring the pair defend a large exclusive territory, and they generally remain in the vicinity for the rest of their lives. Some birds switch territory locations between years after being widowed, forming a new pair with a bird nearby. Large territories mean that breeding densities are naturally low, and the populations of both species are limited by habitat extent.

The territory expands into a larger home-range in autumn and winter, averaging 20ha or more for both species. In temperate regions with milder winters, neighbouring Marsh Tits can have overlapping home-ranges and form small loose groups in basic flocks, having strong bonds only with their mate. Information for Willow Tits is scarce in temperate regions, but they may winter as pairs or in loose groups like Marsh Tits.

In colder climates, such as Fennoscandia, Marsh Tits and Willow Tits form discrete flocks made up of a small, stable group of a dominant (usually adult) pair and several unrelated first-year birds. The discrete flock defends a communal home-range against neighbouring flocks throughout the winter. It is still largely unknown what determines whether Marsh Tits and Willow Tits spend the winter in discrete flocks with a stable membership, or in basic flocks with a casual membership. Likely factors include the richness of the habitat and food resources, and the harshness of the winter climate.

A small proportion of juveniles do not settle in typical home-ranges, but as lone 'floaters' that range across a very large area. Floating is a high-risk strategy, as the low-ranking youngster must wait for a settled resident to die before it can get a chance to replace it. This is the floaters' best option for acquiring a breeding territory at the end of the winter. Their extensive home-ranges allow them to monitor many local birds to detect when a vacancy arises. Unsuccessful floaters join the small wave of dispersal at the end of winter, and most probably perish.

Marsh Tits and Willow Tits will join mixed flocks of other woodland birds in winter, but they may experience interspecific competition from more dominant species, such as Great Tits, Blue Tits and Crested Tits. This competition can influence when and where the subordinate Marsh Tits and Willow Tits can forage, and even their broader habitat selection.

Many aspects of the social organisation of Marsh Tits and Willow Tits remain poorly understood, due to limited studies. Even basic information is vague or lacking from many parts of both species' ranges, such as territory or home-range sizes and winter flocking structure. These studies can be labour-intensive, but further investigation into the impacts of interspecific competition would be particularly valuable.

CHAPTER 6

Breeding

For Marsh Tits and Willow Tits, breeding is a challenging and risky business. The birds are pinned down to a predictable location around the nest for about two months, and the female must also spend lots of time sitting inside a nest-cavity, where she can be discovered by a predator. If the birds make a poor choice of nest-site, or are just unlucky, then they can lose their lives. Marsh Tits and Willow Tits are extremely committed in their drive to breed, but on rare occasions they will play safe to prioritise their own survival by abandoning the breeding attempt. Some pairs may even decide not to breed if conditions are poor.

There is a predictable sequence to the different stages of breeding, from finding a nest-site and building the nest to laying and incubating the eggs, feeding and protecting the nestlings, and then guiding the fledglings until they become independent. Some of these roles are undertaken by only one member of the pair, but are almost always supported by the other.

Timing of breeding

Both species breed around the same time in the northern spring. In temperate regions, such as Great Britain, most nesting activity falls between April and June, but this shifts to May–July for Willow Tits in northern boreal regions. Both species are overwhelmingly single-brooded, nesting only once per year unless their first attempt fails, in which case they might try again. Re-nesting mostly happens when first nests are lost at the building, egg-laying or early incubation stages. If a nest is lost late into incubation or after hatching, pairs will only very rarely try again, so their breeding season is generally over. Genuine second broods have been reported sporadically for Willow Tits in Fennoscandia (von Brömssen & Jansson 1980), but never for Marsh Tits.

Older females begin egg-laying around two days earlier than first-year birds, possibly due to greater experience, better territories or being in better condition than youngsters (Smith 1993a). Birds also breed earlier if they are provided with supplementary food, hinting at the role of body condition (von Brömssen & Jansson 1980), but this is not necessarily a good thing if it decouples their timing from the phenology of natural food.

Breeding begins earlier in warmer regions and in warmer springs, reflecting the earlier development of shrubs, trees and the tits' invertebrate prey, especially moth caterpillars (see Chapter 4). In Sweden the weather window from mid March to late April is a major driver for the timing of breeding in Marsh Tits, with warmer temperatures driving earlier budburst on the trees and shrubs (Andreasson et al. 2023). This is similar to the Białowieża Forest, where warmer springs and earlier budburst are associated with earlier breeding (Wesołowski 2023). Individual Marsh Tits in this population can shift their egg-laying by up to three weeks in different years to match the forest phenology (Wesołowski et al. 2016).

There is a long-term trend of breeding getting earlier as a result of climate change. Marsh Tits in Sweden have advanced the average timing of their breeding season by nine days over the past 40 years, and by 11 days in Great Britain over the last 50 years, while British Willow Tits have advanced by 14 days over the same period, but Finnish Willow Tits only by two days since the 1970s (Vatka *et al.* 2011, Massimino *et al.* 2023, Andreasson *et al.* 2023). Ultimately, Marsh Tits and Willow Tits breed as soon as conditions allow each spring, and most local pairs start breeding within a week of each other (Pakanen *et al.* 2016b, Wesołowski 2023).

Nest-site availability

Marsh Tits and Willow Tits are cavity-nesters that breed inside holes in the trunks, stems, roots or branches of trees and shrubs. The Willow Tit is a primary cavity-nester that excavates its own nest-hole, whereas the Marsh Tit is a secondary cavity-nester that uses pre-existing holes. Marsh Tits do often enlarge existing holes through further excavation, so they are also partial excavators. Willow Tits very occasionally use an existing cavity, but will generally deepen it. Both species also roost inside tree holes outside of the breeding season, but sometimes roost in the open in trees or shrubs.

Within their large territories, both species usually have many potential nest-sites. In natural or semi-natural woodlands there is no sign that either species is prevented from breeding due to a lack of options (Figure 6.1). Marsh Tits typically nest in small cavities in living trees or shrubs, and these are superabundant in the near-primeval forest of Poland's Białowieża National Park (Wesołowski 2007b). In central European birch forests, Marsh Tits in one study had up to 57 usable tree cavities per territory, and at a second site they used only 10 per cent of the available cavities (Ludescher 1973, Markovets & Visotsky 1993). In the deciduous woodlands of Monks Wood and Wytham Woods we found no shortage of suitable nest-cavities for Marsh Tits (Carpenter 2008, Broughton *et al.* 2011).

Figure 6.1. Combe Wood in Berkshire–Hampshire, southern England, in early April. Undisturbed or unmanaged native woodland such as this contains abundant nest-sites for Marsh Tits and Willow Tits among the complex variety of living trees and standing deadwood. (© Richard K. Broughton)

Willow Tits need decaying deadwood in which to excavate their nest-cavity, and this is also widely available. A survey of British woods in the 1980s and 2004 showed that standing deadwood was common and increasing in most regions (Amar *et al.* 2010). As with Marsh Tits, the large territories of Willow Tits often have enough standing deadwood to provide multiple nest-sites. In our Wigan study 81 per cent of Willow Tit nests were in natural deadwood, despite up to five artificial nest-sites being provided in each territory (Parry & Broughton 2018). This suggests that the birds had plenty of natural sites even in woodland that was only 30 years old. A similar picture is seen in studies ranging from birch forest in Germany to boreal birch–conifer forest in Finland or Mongolia (Ludescher 1973, Orell & Ojanen 1983, Bai *et al.* 2005).

In heavily managed woodlands and plantations, where deadwood and cavity-bearing trees are typically removed by foresters, potential nest-sites for Marsh Tits and Willow Tits can be unnaturally scarce (Sells 1998, Kumpula *et al.* 2023). Over-management also reduces the habitat quality, potentially making the territory unviable regardless of nest-site availability. Retaining standing deadwood of all sizes, and a variety of living trees of various ages and conditions, is important for retaining Marsh Tits and Willow Tits in managed forests.

Choosing a nest-site

Little is known about how the birds choose where to nest within their territory. The female makes this decision, and at Monks Wood we found that first-year Marsh Tits breeding for the first time placed their nests randomly within their territory, or at least we couldn't detect any pattern. However, the older females nested in the central part of their territory (Broughton *et al.* 2012b). All females nested among a more mature tree canopy, and so it may be that older females simply centred their territory on good habitat that could provide plentiful nest-sites and food.

Marsh Tits begin inspecting tree holes in the winter, but intensive prospecting becomes conspicuous in March and early April (Morley 1953, Ludescher 1973). The female leads the pair through the territory as they flit between trees and shrubs, moving up and down the trunks and limbs to examine any cavity or crevice. At a potential hole the finder hangs from the entrance rim, calling and looking over its shoulder to attract its partner in a *nest-site showing display* (Smith 1991; see Chapter 2).

Willow Tits show the same exploratory behaviour on deadwood in early spring (Foster & Godfrey 1950, Ludescher 1973). In western Europe they begin trial excavations during March, spending around 10 days pecking at stumps, snags or dead branches. These trial excavations create only shallow depressions or small holes up to 5cm deep, perhaps testing the integrity of the deadwood before settling on a final location.

When Marsh Tits have selected a cavity the female spends some time pecking around the entrance to remove any living tree tissue and prevent narrowing while the hole is in use (Wesołowski 1995b, 2013). Without this, the tree's callus around the hole can begin growing during the spring, potentially narrowing so much that the birds struggle to enter, or even become trapped inside. This risk is very real, and there is an observation from Białowieża, told to me by Marta Maziarz, of a dead Marsh Tit trapped by a narrowing cavity entrance. At Monks Wood one hole narrowed so much as the spring progressed that the male could no longer enter to feed the chicks, while the female could only enter with a struggle.

Re-use of nest-sites

Marsh Tits often nest in a cavity that has been used before, sometimes by the same female in multiple years, or by one of her predecessors. Some cavities are almost 'traditional', with tree holes remaining viable for decades and being used on and off by successive females (Wesołowski 2006, 2012). In Monks Wood 13 per cent of Marsh Tits were breeding in cavities that had been used the previous year (Broughton *et al.* 2011). In the Białowieża National Park 25 per cent of Marsh Tit nest-cavities were used in two consecutive years, although 22 per cent of cavities were lost between breeding seasons due to trees falling, entrances narrowing or holes filling with debris (Wesołowski 2006). Cavities are more likely to be re-used if the female survives and the previous nest was successful. At Monks Wood we had four cavities that were used in three consecutive breeding seasons, involving the same female at one cavity but another used by a different female each year. Another cavity was used four times in five years by two females.

These observations show that it is always worth checking previous nest-cavities when searching for Marsh Tit nests. Often the cavity is empty or occupied by Blue Tits, but they are reoccupied by Marsh Tits often enough to make checks worthwhile. The pattern of re-use suggests that certain cavities are particularly attractive to Marsh Tits for some reason, perhaps because of their dimensions or location among good habitat.

Willow Tits very rarely re-use a nest-cavity from a previous year, with many being destroyed over the winter by decay or foraging woodpeckers. Ludescher (1973) recorded only 3 per cent of nests in cavities that were initially excavated the previous year. However, Willow Tits do often return to the same stump or snag to excavate a new cavity if enough sound deadwood remains intact. As with Marsh Tits, repeated breeding around the same location hints at something particularly attractive about certain parts of the territory.

Marsh Tit nest-cavities

A typical Marsh Tit nest is a small cavity in a hole in the trunk, stem or branch of a living tree or mature shrub. Nests are occasionally in a dead stump or branch, or in the ground at the base of a tree or shrub. There are records of nests in holes in walls, and even an open nest between branches of a small spruce tree (Cramp & Perrins 1993).

Trees and shrubs used by nesting Marsh Tits

Marsh Tits use a wide variety of trees and shrubs for nesting, depending on their availability and tendency to form suitable cavities. In Monks Wood, two-thirds of nests are in Common Ash, which is used more often than its availability would predict (Broughton *et al.* 2011). Other frequent nest-trees at Monks Wood include Field Maple, European Aspen and young elms (Table 6.1). In the Oxford studies at Bagley and Wytham Woods, mature Common Elder was the favoured species despite being quite patchy in occurrence, with other nest-trees including Common Ash, Sycamore and Common Hazel (Morley 1953, Carpenter 2008). Silver Birch and Black Alder are also commonly used in Europe (Mildenberger 1984, Markovets & Visotsky 1993), and in the Białowieża National Park the most favoured nest-trees are Common Hornbeam and Small-leaved Lime (Wesołowski 1996).

Oak trees are rarely used by nesting Marsh Tits in Europe, despite being common in most study areas. Pedunculate Oaks and Sessile Oaks do not readily form the sort of cavities that Marsh Tits like, and they accounted for only up to 2 per cent of nests in the Białowieża and Monks Wood studies (Wesołowski 1996, Broughton *et al.* 2011). Nests in conifers, such as Scots Pine or Norway Spruce, are also unusual, as they form fewer cavities than most deciduous trees (Wesołowski 1996).

The tree or shrub stems used by nesting Marsh Tits can be as slender as 10–11cm in diameter. Most Marsh Tit nests in Monks Wood are in immature trees with a trunk diameter at breast/chest height (DBH) of 10–30cm (Broughton *et al.* 2011). A third of nests are in larger trees with trunks over 30cm DBH, and these tend to be higher above the ground in thinner limbs of the tree. Marsh Tits in Białowieża also commonly nest in the upper and thinner parts of larger trees, with the average DBH of nest-trees being 34cm, but at the cavity entrance the trunk or limb diameter averaged 27cm (Wesołowski 1996). For shrubs, stems of a suitable size for nesting can only be found in the thicker parts of quite mature specimens, such as old Common Elder, Common Hazel or Crab Apple.

Table 6.1. Trees and shrubs used for 222 Marsh Tit nests in England.

Tree/shrub species	Nests (%)
Common Ash	63.1
Common Elder	8.1
Field Maple	6.3
Elm	5.0
Pedunculate Oak	3.6
European Aspen	3.2
Common Hazel	2.7
Birch	2.3
Crab Apple	1.8
Hawthorn	1.4
Sycamore	1.4
Willow/Sallow	0.5
Common Dogwood	0.5
Common Beech	0.5

Values are pooled from Monks Wood (Broughton *et al.* 2011, and additional data) and Wytham Woods (Carpenter 2008).

Marsh Tits show an overwhelming preference for nesting in living wood, and largely avoid nesting in deadwood. Living wood is much firmer than dead or decaying wood, and so it is more secure from predators that may try to break in. In Monks Wood only 12 per cent of 192 nests were in deadwood, and only 8 per cent of 359 nests in Białowieża (Wesołowski 1996).

Old woodpecker holes are strongly avoided by nesting Marsh Tits, especially those of Great Spotted Woodpeckers, as the holes are obviously unsafe if these predators can already enter. Just 1–2 per cent of 554 natural nests in Monks Wood and Białowieża were in old woodpecker holes (Wesołowski 1996, my data). Marsh Tits do sometimes take over the smaller holes of Lesser Spotted Woodpeckers, even though they are in deadwood. At Monks Wood we found a

Marsh Tit nest in a new Lesser Spotted Woodpecker cavity in a large Common Dogwood. We do not know if it was already abandoned when the Marsh Tits took it over, or whether they actively stole it from the woodpeckers. In Norfolk, a pair of Marsh Tits were observed evicting Lesser Spotted Woodpeckers from their cavity soon after it had been excavated (Cleverley *et al.* 2019). The female Marsh Tit brought in nest material and defended the cavity from inside as the male woodpecker tried to enter, with lots of bickering at the entrance hole before the woodpeckers gave up. Marsh Tits will also evict Willow Tits from their nest excavations, and will use Willow Tit holes from previous years (Ludescher 1973).

Although Marsh Tits are secondary cavity-nesters, females are capable of enlarging a small cavity to create a suitable nest-chamber. Ludescher (1973) saw a Marsh Tit excavating a hollow that was only 3cm deep to begin with, and there were another eight cases of Marsh Tits taking over and extending Willow Tit excavations that had reached 5–10cm in depth. This behaviour can lead to misidentification for the unwary observer, who might think they've found a nesting Willow Tit.

Female Marsh Tits can spend a lot of time deepening a cavity or removing detritus from the bottom, typically in bouts of 20 minutes at a time. She carries away the debris in her bill and throat to discard nearby by shaking her head and spitting it out. Males do not take part in excavating or cleaning holes, but will often accompany the female and sing nearby. Sometimes the cavity is not used for nesting, even after lots of excavation, maybe because it could not be enlarged to the required size.

MARSH TIT CAVITY TYPE

Marsh Tits have a reputation for nesting low down or close to the ground, but this is biased by lower nests being easier to find. The nest heights in various studies range from ground level to 27m, but the average for natural nest-sites ranges over 1–6m (Table 6.2). Average heights of Marsh Tit nests are always lower than for Blue Tits and Great Tits in the same area (Van Balen *et al.* 1982, Wesołowski & Rowiński 2012, Maziarz *et al.* 2015). Occasionally a Marsh Tit's nest is underground, as with several nests at Monks Wood where the entrance hole was at the base of a young tree and the nest itself was in a cavity that extended for 10–20cm below the ground surface.

Table 6.2. Average heights (in metres) for natural nests of Marsh Tits.

Site	Mean (s.d.)	Minimum	Maximum	No. of nests	Source
Monks Wood	3.1 (2.5)	0.0	10.0	214	Broughton *et al.* 2011 *
Wytham	2.5 (3.0)	0.3	9.0	21	Carpenter 2008 *
Netherlands	3.4 (1.1)	1.6	4.7	6	Van Balen *et al.* 1982
Germany	3.7	1.0	8.6	33	Ludescher 1973
Baltic coast	2.5 (1.8)	<1.0	7.0	57	Markovets & Visotsky 1993
Białowieża	5.6 (4.8)	0.5	27.0	411	Wesołowski 1996
Sweden	1.1	<1.0	11.0	71	Nilsson 1984
China	2.7 (1.1)	—	—	14	Wang *et al.* 2003

All values were not available for some studies. Sources marked * include author's additional data.

Figure 6.2. Examples of typical Marsh Tit nest-sites in knotholes at Monks Wood in England. The knotholes are in Common Ash trees and the diameter of the entrances is about 3cm. (© Richard K. Broughton)

A common type of cavity used by Marsh Tits is a *knothole*, which forms where a side-branch breaks from a trunk and fungal decay takes hold (Figure 6.2). Knotholes were by far the most common cavity type used in the Monks Wood study, accounting for three-quarters of natural nest-sites (Table 6.3).

Table 6.3. Types of cavity used by nesting Marsh Tits in the Monks Wood study.

Cavity type	All nests (%)	Natural nests (%)
Knothole	67.8	74.4
Rotten sapwood/heartwood	11.7	12.8
Nest-box	8.9	—
Chimney	6.1	6.7
Slit	3.7	4.1
Woodpecker cavity	0.9	1.0
Fallen/leaning deadwood	0.9	1.0

Percentages calculated from 214 nest-sites comprising 195 natural sites and 19 nest-boxes.

Hollows in vertical fissures, or *slits*, are also regularly used by Marsh Tits, and elongated slit cavities are the most common type used in the Białowieża Forest (Figure 6.3). Many of these slits within 2–3m of the ground may develop from damage caused by large herbivores, such as Red Deer or European Bison, which strip the bark in winter and gouge it with their antlers or horns, allowing decay to set in (Broughton *et al.* 2022a). Other nest-cavity types include *chimneys*, which are tube-shaped cavities with an upward-facing entrance in a snapped stem, and also rotten sapwood or heartwood behind loose bark or damaged wood (Wesołowski 1996). Occasionally a nest-cavity is in the end of fallen or leaning deadwood suspended above the ground.

169

Figure 6.3. Marsh Tit nest-site in an elongated slit-shaped cavity in the Białowieża Forest. The cavity entrance is at the lower edge of the slit, from where the female is emerging in the right-hand picture. (© Richard K. Broughton)

Willow Tit nest excavations

Willow Tits usually excavate their nest-cavity in an upright dead stump, snag or young tree, in a damaged or dead branch on a larger tree, and sometimes in a fallen log lying horizontally. Nests have also been recorded in rotten fenceposts. The species used reflects the local availability and also their tendency to form deadwood that Willow Tits can excavate. Very soft and crumbling deadwood is avoided, as this could collapse during excavation or nesting. Instead, the ideal deadwood is often surprisingly firm, being able to be scored by a fingernail but not dented by pressing with a finger.

Willow Tit excavations in deadwood are often begun on the site of an existing break, scrape or shallow hollow, which seems to encourage them to dig (Figure 6.4). Deeper

Figure 6.4. A Willow Tit nest-site that has been excavated in the snapped branch of a Pedunculate Oak tree in England. (© Richard K. Broughton)

natural cavities are sometimes used, including those in living wood, which could probably function as nest-sites without much modification, but they are always excavated further. In Siberia and Mongolia 9–14 per cent of nests were in existing natural cavities (Pravosudov & Pravosudova 1996, Bai *et al.* 2005).

TREES AND SHRUBS USED BY NESTING WILLOW TITS

Birches are perhaps the most important nesting tree for Willow Tits across their range in the temperate and boreal forests (Figure 6.5). In Fennoscandia and Asia, birches account for at least three-quarters of nests, which probably reflects their abundance (Table 6.4; Pravosudov & Pravosudova 1996, Bai *et al.* 2005). Willow Tits will also readily nest in conifers, including pines, spruces and larches, and in the Białowieża National Park they nest in very large snags of dead Norway Spruces (Figure 6.6).

Figure 6.5. Examples of Willow Tit nest sites in birch snags in breeding habitat in northwest England (left: © Allan Rustell) and northern Finland (right: © Anne Laine).

Figure 6.6. Willow Tits emerging from their nests in conifers: a Corsican Pine snag in Swaffham Forest, Norfolk, England (left: © Ashley Banwell), and a 4m Norway Spruce snag in the Białowieża Forest, Poland (right: © Richard K. Broughton).

Many other trees and shrubs are used, however, and in British studies willows and Common Elder are popular (Table 6.4). Most British Willow Tit nests are in mature shrubs with thick dead stems, or in young and slender trees at the pole stage. The stems average 11–12cm in diameter at the nest height, but can be as thin as 5–7cm. Suitable Silver Birch, willow and Black Alder of this size can be abundant in relatively young wooded or wetland habitats, where dense stands of competing trees promote lots of standing deadwood. In unmanaged ancient woodlands Willow Tits will also nest in old Common Hazels.

Table 6.4. Trees and shrubs used by Willow Tits for 328 nests in Great Britain and 99 nests in Finland.

Tree/shrub species	Nests (%)	
	Great Britain	Finland
Willow/Sallow	33.0	7.1
Birch	31.5	75.8
Common Elder	18.0	0.0
Black Alder	9.2	14.1
Pine	3.7	0.0
Common Ash	0.9	0.0
Hawthorn	0.9	0.0
Common Hazel	0.6	0.0
White Poplar	0.6	0.0
Wild Cherry	0.6	0.0
Elm	0.3	0.0
Sycamore	0.3	0.0
Rowan	0.3	0.0
Pedunculate Oak	0.3	0.0
European Aspen	0.0	2.0
Spruce	0.0	1.0

British Willow Tit values are pooled from Foster & Godfrey 1950, Lewis *et al.* 2009a, Rustell 2015, Parry & Broughton 2018 and author's own data. Finland data are recalculated from Orell & Ojanen 1983.

Table 6.5. Average nest heights (in metres) of Willow Tits.

Site	Mean (s.d.)	Minimum	Maximum	No. of nests	Source
Wytham	2.1 (0.4)	1.8	2.6	3	Foster & Godfrey 1950
NW England	1.7 (0.7)	0.4	4.0	92	Rustell 2015 *
Wigan	1.2 (0.5)	0.5	4.0	103	Parry & Broughton 2018
Scotland	1.5	—	10.5	9	Maxwell 2002
Britain	2.2 (2.0)	0.2	10.0	72	Stewart 2010
Germany	3.1	0.1	9.5	65	Ludescher 1973
Mongolia	7.3 (4.6)	—	—	64	Bai *et al.* 2005

All values were not available for some studies. Source marked * includes additional data.

The range of Willow Tit nest heights in western Europe is broadly similar to the bulk of Marsh Tit nests, from near ground level to around 10m high (Table 6.5). Average heights tend to be lower than for Marsh Tits, however, at 1–3m. In Great Britain most nests are within a few metres of the ground, and they average lower than elsewhere.

WILLOW TITS AS NEST-SITE EXCAVATORS

Once a Willow Tit pair has settled on a final nest-site in early April, both sexes excavate the cavity, with the male contributing up to half of the effort. In an early description from Chobham Common in southern England, Witherby (1934) reported how it took eight days for a pair of Willow Tits to fully excavate a cavity. In our Wigan study the cavities were excavated in 6–8 days (Parry & Broughton 2018), although one pair completed it in only two days. In Ludescher's (1973) study most Willow Tits took 12–13 days to excavate a cavity, but with a full range of 7–18 days depending on the toughness of the wood.

The birds excavate throughout the day, and both sexes alternate during bouts lasting around 15 minutes. Initially the bird clings to the entrance and pecks at the wood, discarding fragments over its shoulder. This leaves a telltale scattering of small wood chips below an excavation, which can be useful for finding nests. Once the cavity is deep enough for the birds to enter, they carry away the wood chips for up to 20m before discarding.

Willow Tits may abandon their excavation at any stage for a variety of reasons, such as the wood becoming too hard or too unstable as they progress, or if the side of the cavity breaks open. Cavities can also be abandoned because of interference from predators or competitors, such as Great Spotted Woodpeckers, Great Tits, Marsh Tits and especially Blue Tits. If this happens, the pair will usually switch to an earlier trial excavation within the territory.

Nest-cavity dimensions

MARSH TIT NEST-CHAMBERS

The natural nest-sites used by Marsh Tits are usually quite narrow cavities with small entrances (Figure 6.7). The entrance holes can vary from round or oval to slit-shaped. Data from the Białowieża National Park (189 nests; Wesołowski 1996) and Monks Wood (10 nests) give average dimensions of 32–33mm across the horizontal width of the entrance hole and 49–81mm diameter across the vertical axis.

Before the nest material is brought in, a typical Marsh Tit cavity is 14–22cm deep from the entrance hole to the cavity floor, with extremes of 2–53cm in Białowieża and Monks Wood. The floor area of empty cavities averaged 79cm² for Białowieża nests, which is approximately 9 × 9cm wide at the bottom of the cavity, with a range of 5–18cm diameter. In Germany the cavity floor area averaged 45cm², with an absolute minimum of 23cm² (Ludescher 1973). The cavity depth is partly limited by the birds' ability to see in the dark. Only 1 per cent of the daylight at the nest-entrance reaches the nest-cup at a depth of 8–14cm (Wesołowski & Maziarz 2012). Marsh Tits are unlikely to be able to see colours at these light levels, and so the conditions within the nest-cavity are effectively nocturnal, similar to outside on a moonlit night.

Figure 6.7. The view inside a Marsh Tit nest in a natural cavity in the Białowieża Forest. The nest contains four chicks and is briefly illuminated by a torch during an inspection. (© Richard K. Broughton)

Willow Tit nest-chambers

The cavity that Willow Tits excavate is flask- or bottle-shaped, being narrow at the top and wider at the bottom, and broadly cylindrical. Initially the birds excavate horizontally from the entrance for a few centimetres before turning downwards to create the nest-chamber. The internal walls are clean and dimpled where wood chips have been peeled away (Figure 6.8).

The Willow Tit's cavity entrance is usually rounded or oval in shape, with its longest axis in the vertical plane. Nests in pre-existing cavities can sometimes have elongated slit-like entrances. The average horizontal diameter of the entrance is around 29mm in Great Britain, with a range of 17–38mm for 51 nests (Parry & Broughton 2018), and up to 35mm for larger birds in Asia (Bai *et al.* 2005). The average depth of an empty nest-cavity is 15–22cm, but is quite variable. These depths are similar to Marsh Tit cavities, but Ludescher (1973) found that Marsh Tits taking over old Willow Tit cavities always excavated up to 3.5cm further. Ludescher also measured the internal floor diameter of Willow Tit nest-chambers as 7–8cm, giving an average floor area of 40cm^2 (23–68cm^2) for 52 nests. In the Wigan study, the internal volume of 50 Willow Tit nest-chambers averaged 567cm^3 (425–765cm^3) (Parry & Broughton 2018). Compared to Marsh Tits, the Willow Tit's nest-cavity dimensions are a bit smaller, even though Willow Tits often have larger broods.

CAVITY ORIENTATION

The direction faced by the nest-entrance may influence nest-site choice if the birds avoid prevailing weather or sun exposure. How much control they have over the entrance direction differs between the species. Marsh Tits are limited to pre-existing cavities, whereas excavating Willow Tits can engineer which direction their cavity faces. However, if wood decay or cavity formation is biased in certain directions, such as the north-facing side of a tree, then this will also limit the choice of nest orientation.

For Marsh Tits, however, studies at Białowieża and the Baltic coast found no clear pattern of nest orientation, with entrances facing in all directions (Markovets & Visotsky 1993, Wesołowski 1996). For 212 nests at Monks Wood there was also no clear pattern, but Ludescher (1973) found that German Marsh Tits strongly avoided cavities facing south and west, where the prevailing weather came from.

Willow Tits in Ludescher's (1973) study also avoided excavating on southerly or westerly aspects, and in our Wigan study there was avoidance of excavating nests

Figure 6.8. Internal view of a Willow Tit nest-chamber, excavated from a block of deadwood inside a nest-box. The front panel of the box has been removed to show the cavity excavation. The entrance hole is near the top of the image and the pad of nest material is visible at the bottom. (© Richard K. Broughton)

facing southwest through to north (Parry & Broughton 2018). The Wigan birds may be trying to avoid the prevailing westerly wind and rain in the wet climate of northern England. Observations of nest orientation can have a practical purpose if certain directions are avoided, as this can help guide the placement of artificial nest-sites to maximise uptake. In western Europe it would seem prudent to avoid southerly or westerly directions.

Nest-boxes and artificial cavities

Studies of cavity-nesting birds often rely on nest-boxes. Checking convenient nest-boxes is far easier than trying to find natural nest-sites that the birds do not want to reveal. Blue Tits and Great Tits readily take to nest-boxes, but Marsh Tits and Willow Tits are notorious for their low uptake, at least in western Europe (Perrins 1979).

In Great Britain, even where nest-boxes are tailored to mimic the natural cavities used by Marsh Tits or Willow Tits, the uptake is usually so low that it is not worth providing them. Indeed, nest-boxes provided for these species tend to result in a high uptake by Blue Tits instead (Broughton & Hinsley 2014, Parry & Broughton 2018). This can be counterproductive by inflating the local numbers of dominant Blue Tits, which may then compete with Marsh Tits and Willow Tits for foraging areas and natural nest-sites (Parry & Broughton 2018, Maziarz *et al.* 2023).

The uptake of nest-boxes by Marsh Tits is generally under 15 per cent of those available, and this is often achieved only by providing large numbers of boxes. In Scandinavia, over 500 nest-boxes had an average annual uptake of just 8 per cent (Andreasson *et al.* 2023). In Wytham Woods almost a thousand nest-boxes are available but the uptake by Marsh Tits is typically under 1 per cent. Studies in the Balkans and China have had occupation rates of 10–15 per cent at best (Dolenec 2006, Zhang *et al.* 2021).

Willow Tits use nest-boxes even less frequently than Marsh Tits. Where nest-boxes have been targeted at Willow Tit territories and specifically designed for them, with a substrate to excavate, they have attracted just 6–10 per cent of local Willow Tit pairs in Germany, Finland and England (Ludescher 1973, Orell & Ojanen 1983, Parry & Broughton 2018). Empty nest-boxes provided for other tits are very occasionally used by Willow Tits, without any excavation. This seems to be more common in east Asia, and a study in northeast China had Willow Tits occupying an average 8 per cent of standard nest-boxes that were available each year, despite these having a spacious internal diameter of 14–17cm and depth of 34cm (Zhang *et al.* 2020). One major reason for the low uptake of nest-boxes by Marsh Tits and Willow Tits is their low population densities, where few pairs are present in relatively large areas of habitat. Another possible reason for low uptake is if the birds are selective about where they nest within their large territories, possibly due to local habitat quality. With plenty of natural sites available, any success in enticing Marsh Tits or Willow Tits into nest-boxes would be a happy coincidence.

An important conclusion from studies involving artificial nest-sites for Marsh Tits or Willow Tits is that they are clearly of limited use as a conservation tool. In natural or semi-natural habitats these species are constrained by the amount of available woodland for their large breeding territories, not by the number of nest-sites. The only exceptions are perhaps forestry plantations and heavily managed woodlands where deadwood and cavities can be scarce (Kumpula *et al.* 2023). In British woods, providing nest-boxes for Marsh Tits or Willow Tits, or creating dead stumps for Willow Tits to excavate, is unlikely to make a tangible difference. Instead, nest-boxes might actually make life more difficult for Marsh Tits and Willow Tits by increasing the local abundance of competing Blue Tits and Great Tits (Broughton *et al.* 2022c, Maziarz *et al.* 2023).

Marsh Tit nest-boxes

Nest-box designs for Marsh Tits have varied over time. Standard wooden nest-boxes with small (25–26mm) entrance holes have been placed low down on trees in the hope of dissuading dominant Blue Tits and Great Tits (Cromack 2018). Placing nest-boxes in pairs or trios within a few metres of each other is another tactic to ensure that a box remains available if Blue Tits take one (Nilsson & Smith 1988a). Where Blue Tits are abundant, however, then this tactic often fails as they occupy many of the paired boxes anyway (Broughton & Hinsley 2014).

At Monks Wood we trialled pairs of standard nest-boxes, but these had a Marsh Tit occupation rate of just 1 per cent while Blue Tits occupied 60–87 per cent of box pairs (Broughton & Hinsley 2014). In the same study we did a second trial with a new Marsh Tit nest-box design that mimicked natural nest-cavities, with a narrow diameter of 78 × 78mm and depth of 15cm below a 26mm hole (Figure 6.9). We placed pairs of the new boxes in known territories and near to previous nest-sites to maximise the likelihood of Marsh Tit uptake. Despite all of this effort, the average uptake of boxes was only 9 per cent during the trial. In the first year almost a third of the Marsh Tit pairs did use a nest-box, but this quickly fell to 5 per cent within a few years. This might suggest an initial attraction to new boxes that faded over time.

The Monks Wood trial underlined that providing nest-boxes for Marsh Tits is not very successful in Great Britain. Even with careful targeting, the birds prefer to use natural cavities. The general avoidance of nest-boxes might also be because they are essentially deadwood, and Marsh Tits prefer cavities in living trees. A cavity in living wood has a different microclimate to inside a nest-box, with the latter showing greater temperature fluctuations, poorer thermal insulation and far lower humidity (Maziarz *et al.* 2017). Whether Marsh Tits select for this kind of moist, insulated microclimate in natural cavities is unclear, but more stable temperatures and higher humidity could potentially be important for them.

WILLOW TIT NEST-BOXES

Willow Tit nest-boxes made of wood, woodcrete or hollowed logs have been filled with sawdust, blocks of deadwood or a mix of wood chips and flour to mimic natural nest-sites, providing something for the birds to excavate (Ludescher 1973, Orell 1983, Thessing 1999, Last & Burgess 2015). Polystyrene blocks have also been used as an infill, but this isn't a great idea as the excavated chips will become plastic pollution in woodland soils.

Hollowed sections of birch logs with a floor, lid and entrance hole drilled into the side were an early design, but Jimmy Maxwell (2002) used narrow (11cm diameter) plastic piping that was filled with wood shavings and camouflaged with tree bark to look like a dead tree stem with rotten heartwood. This followed a similar pipe design for chickadees used in Ohio (Grubb & Bronson 1995). Other researchers have tried wooden nest-boxes, with one design having a

Figure 6.9. Nest-boxes designed for Marsh Tits (left) and Willow Tits (right) to mimic the natural nests of both species, while providing easier access for monitoring. The Marsh Tit nest-box was trialled at Monks Wood. The Willow Tit nest-box, used in the Wigan study, contains a block of deadwood that has been excavated by the birds. (© Richard K. Broughton)

Figure 6.10. A brood of 10 Blue Tit nestlings in a nest-box designed for Marsh Tits. (© Richard K. Broughton)

narrow 7–8cm diameter internal cavity and a longer frontage to mimic standing deadwood (Revill *et al.* 2010, Last & Burgess 2015). A similar nest-box, minus the longer frontage, was used in the Wigan study (Figure 6.9; Parry 2017), but none have had high occupation rates. The best results have been where many boxes were provided in known territories with few natural alternatives. For example, Last & Burgess (2015) provided an average of four boxes per territory in a managed conifer plantation in England, and in some years every pair of local Willow Tits (up to nine territories) used a nest-box, although in other years only one box was occupied.

As with Marsh Tits, providing nest-boxes for Willow Tits may be counterproductive if competitors are common. In the Wigan study the few Willow Tits using nest-boxes had a high loss rate during excavation and egg-laying due to the boxes being spotted and taken over by Blue Tits (Parry & Broughton 2018). Blue Tits were even observed completing half-finished Willow Tit cavities and excavating the nest-boxes themselves. For Willow Tits and Marsh Tits alike, providing any kind of nest-box tends to benefit Blue Tits far more than the target species (Figure 6.10).

Stumps and logs for Willow Tits

Rather than go to the effort of making nest-boxes, a simpler approach is to provide more standing deadwood for Willow Tits to excavate. One method involves strapping suitable upright logs of dead birch, willow or alder onto living young trees (Orell & Ojanen 1983, Parry & Broughton 2018). These logs are about 11–15cm thick and up to 1.5m long, and are simply picked up off the woodland floor as fallen deadwood and attached to living trees with twine or (less sustainably) plastic cable-ties (Figure 6.11). Another approach has been to kill a small number of living trees by ring-barking at the base, or cutting high stumps or snags around 2m tall (Last & Burgess 2015, Pinder & Carr 2021).

In Finland, birch logs strapped to trees held 29 per cent of local Willow Tit nests in one seven-year study, compared to just 7 per cent in nest-boxes (Orell & Ojanen 1983). In the Wigan study around three strapped logs were available in each Willow Tit territory, and these held 9 per cent of nests, compared to 10 per cent in nest-boxes. There are no detailed studies of how readily Willow Tits use stumps and standing deadwood created by ring-barking or felling, but where existing deadwood is scarce then it can only improve their options.

As with nest-boxes, providing lots of deadwood cannot increase the density of Willow Tits, which are limited by the extent of breeding habitat for their large territories, and not by the availability of deadwood. The exception is areas of intensive forestry management where deadwood has been removed. In particular, the retention and promotion of standing deadwood is important for Willow Tits in managed forests in Finland (Kumpula *et al.* 2023). This can be achieved by simply not removing dead or dying trees in the first place.

Figure 6.11. A dead Silver Birch log tied to a living Common Ash tree to provide a potential nesting substrate for Willow Tits, which may last for several years. (© Richard K. Broughton)

Nest-site competition

NEST COMPETITION FOR MARSH TITS

Marsh Tits are part of the *secondary hole-nesting guild* of small woodland passerines, which includes other tits and Pied Flycatchers, Collared Flycatchers, Common Starlings and Eurasian Nuthatches, which can all breed in tree cavities with overlapping features. Despite this overlap, competition for nest-sites seems to be quite low for Marsh Tits. Tree cavities are fairly common in natural and semi-natural woodlands, and Marsh Tits favour the smaller cavities with narrow entrances at relatively low heights, which limits competition with larger species like the Eurasian Nuthatch (Wesołowski 1989).

The main nest competitor for European Marsh Tits is the Blue Tit. Being of a similar size and able to enter the same small holes, Blue Tits are socially dominant to Marsh Tits and also far more abundant in many regions. In the natural forest of the Białowieża National Park, where holes are abundant and Blue Tits are at natural densities (only about twice as common as Marsh Tits), there is no obvious competition for nest-cavities (Wesołowski 1989, 2023). In Great Britain, however, Blue Tits are much more abundant and can outnumber Marsh Tits in woodland by 5–10 to one, which has been suggested as a potential problem via nest-site competition (Smith *et al.* 1992, Fuller *et al.* 2005, Fuller 2022). We tested this in the Monks

Wood study but found that competition for nest-sites was actually very low, with only one of 117 Marsh Tit nests being taken over by Blue Tits, and another by Great Tits (Broughton *et al.* 2011). We did observe Blue Tits trying to take over four more nest-cavities, but the Marsh Tits successfully defended them and defeated the Blue Tits.

The main tactic of Blue Tits attempting to take over a cavity is to pull out the Marsh Tits' nest material, despite this moss and fur being the sort of thing that Blue Tits themselves would use. Instead, the aim is to destroy the nest, and to let the Marsh Tits see this, to make them abandon it. Marsh Tits are ferocious in fighting back, however, and there can be several days of intense conflict around the cavity. The Marsh Tits and Blue Tits will fly at each other and engage in rapid chases, sometimes grappling with claws and pecking, even falling to the ground together in a ball of fury. The low rate of Blue Tit success shows that Marsh Tits are obviously capable of defending their nest in most cases.

Nest competition for Willow Tits

Studies in Switzerland and Germany found the major competitors for Willow Tit nest-cavities were Marsh Tits (Thönen 1962, Ludescher 1973). In the German study Marsh Tits took up to 14 per cent of newly excavated Willow Tit cavities, and where takeover battles were observed they won 29 per cent of them. Coal Tits and Blue Tits were also common competitors, and Thönen (1962) even suggested that Willow Tits could only persist where competitors were scarce. Other species recorded usurping Willow Tits from their nests include Crested Tits, Pied Flycatchers and Eurasian Wrynecks (Orell & Ojanen 1983, Eggers & Low 2014).

In Great Britain, cavities being taken over by Blue Tits or Great Tits is a major cause of Willow Tit nest failure. As early as the 1930s a letter to *British Birds* magazine reported the 'menace' of Blue Tits and Great Tits ousting Willow Tits from newly excavated cavities in Cumbria (Johnston 1936). In Scotland's River Clyde catchment, Maxwell (2002) reported severe nest-site competition for a struggling Willow Tit population in the 1990s and 2000s, with 67 per cent of 30 cavities being taken over by Blue Tits and occasionally Great Tits. That Willow Tit population was extinct by 2010.

In the Wigan study, Blue Tits took 23 per cent of Willow Tit nests from the latter's first breeding attempts each spring, and usurped 17 per cent of first and repeated attempts overall (Parry & Broughton 2018). Blue Tits were responsible for 40 per cent of all Willow Tit nest failures in this study. Elsewhere in northern England another study found Blue Tits and Great Tits had successfully evicted Willow Tits from at least 7 per cent of nest-cavities, but 25 per cent of nests were abandoned due to competition, interference or other unknown causes (Rustell 2015). In southern England Willow Tits were recorded being evicted from nest-boxes by Blue Tits, Great Tits, Marsh Tits and Coal Tits (Last & Burgess 2015).

Blue Tits watch Willow Tits as they excavate a cavity, and show increasing interest as it nears completion. At this point Blue Tits start becoming aggressive and bringing in green moss, which is rarely used by Willow Tits and so is a good marker that the nest is being taken over. The conflict can stretch over a couple of days with intense chasing, grappling and pecking. Blue Tits will also pull out the Willow Tits' nest material, as they do at Marsh Tit nests, and will even take over after the Willow Tits have begun egg-laying or incubating, building their own nest on top of the Willow Tit eggs.

Willow Tits are less effective than Marsh Tits in defending their nest, and in the Wigan study they lost six out of eight observed conflicts with Blue Tits. Willow Tit pairs have been recorded losing up to five successive nest excavations in a single spring, with different Blue Tits evicting them from one cavity after another (Maxwell 2002, Parry & Broughton 2018). This waste of time and energy delays breeding by several weeks, and may ultimately leave the Willow Tits trying to nest in a suboptimal site, becoming out of step with food availability, or completely failing to breed. Late-hatching young from these nests probably also have a lower chance of settling after dispersal (Nilsson & Smith 1988a).

Avoiding nest-site competition

The similarity in body size and overlap in habitats means there is little that Marsh Tits or Willow Tits can do to avoid nest-site competition from Blue Tits across much of their European range. Nesting in cavities at lower heights might be a tactic to limit direct competition, and Marsh Tits seem able to physically repel Blue Tit attacks, but Willow Tits are clearly much more vulnerable.

Compared to mature deciduous woodland, the relatively lower densities of Blue Tits in mixed conifer–birch forest, boggy carr, large hedgerows, shrubland and early-successional woodland means that Willow Tits can sometimes avoid the largest numbers of competitors. In parts of Switzerland Willow Tits seem largely restricted to pure conifer forest and high mountainous woodland precisely because of the lower numbers of nest competitors in these habitats (Thönen 1962).

Problems arise when Blue Tits and Great Tits become very abundant in their main deciduous woodland habitats and then spill over into the more marginal conifer woodlands, shrublands and wetlands. Such increases have been happening in Great Britain over many decades, and Blue Tits and Great Tits are now very common almost everywhere (Hewson *et al.* 2007, Massimino *et al.* 2023). These marginal habitats for Blue Tits and Great Tits are where Willow Tits could previously escape from intense competition, but are now increasingly under pressure from dominant tits looking for nest-sites.

It is tempting to suppose that nest-site competition could be alleviated by providing lots of nest-boxes for Blue Tits and Great Tits so that they leave Willow Tits alone. Unfortunately this just intensifies the problem, because the nest-boxes produce lots of young Blue Tits and Great Tits that will settle locally and further increase the population (Perrins 1979, Dhondt & Adriaensen 1999), leading to even more competition. There is no easy answer, as the only mechanism to reduce competition is to discourage population growth of the competitors in the wider landscape.

Nest-building

Only female Marsh Tits and Willow Tits collect nest material and build the nest. Both species use different sets of materials, and so their nests are quite distinct from each other. This difference in nest construction can be useful, as the eggs and nestlings of both species can be difficult to tell apart.

Marsh Tit nest construction

Nest-building is probably prompted by increasing day length and temperature, and Marsh Tits need a minimum of 4–5 days to complete a nest, but more typically 8–14 days (Morley 1953, Wesołowski 2013). Sometimes there is a break in nest-building, or between nest completion

Figure 6.12. Marsh Tit nests in nest-boxes in Monks Wood, containing full clutches of nine and eight eggs. The nests have a base of moss to support the lining of mammal hair and fur, which is compressed into a felt-like cup to hold the eggs. (© Richard K. Broughton)

and egg-laying, especially if the weather turns cold and wet. These breaks can last for a week or more, and birds may also switch nest-sites during building if they are seriously disturbed.

Marsh Tit nests have a base of fresh green moss with a nest-cup embedded on top, large enough to hold the eggs, which is lined with animal fur, wool and hair (Figure 6.12). Unlike Blue Tit nests, the moss base does not contain bast (fine strips of bark fibres), although the occasional piece of dry grass may be present, probably picked up with the moss. Marsh Tits are apparently selective in the moss they use, and in the Białowieża National Park they utilised an average of 5–6 species for the nest-base, particularly Plait-moss and Flat Neckera moss plucked from tree trunks (Wesołowski & Wierzcholska 2018). These mosses are common across Europe and are probably widely used by Marsh Tits. The moss base provides structural support and also absorbs any water that gets into the nest-cavity or seeps from the surrounding walls of the living tree. Moss might also provide antimicrobial properties to limit bacteria, but this has not been tested (Wesołowski & Wierzcholska 2018).

Marsh Tit nests at Monks Wood are commonly lined with the hair and fur of Roe Deer, Muntjac Deer, Rabbit and Brown Hare, compressed into a felt-like pad. In the Białowieża Forest Marsh Tits also use the woolly fur of European Bison. Dog hair or sheep's wool may also be used, and all of these fibres are collected from the ground, trees or fences where the hair has snagged. Marsh Tits will also collect fur from dead animals, even if they are decomposing and smelly.

Artificial woolly materials are sometimes included in the nest-lining, such as the bright yellow polyester fibres from tennis balls that are brought into woods by dog-walkers. Monks Wood Marsh Tits have also been recorded using white polyester stuffing from a soft toy, and one female travelled 320m to collect roof insulation material (glass or mineral wool) from a nearby house under construction. At Oxford, Morley (1953) recorded female Marsh Tits travelling at least 250m to collect sheep's wool. Feathers are very rarely present in nests, with usually just a couple of small contour feathers from the female herself. This lack of feathers in the lining and the lack of bast or dry grass in the base easily distinguish Marsh Tit nests from those of Blue Tits. The female continues to bring nest-lining into the cavity after egg-laying has begun, and for up to five days into incubation.

Marsh Tits adjust the distance of their nest-cup from the cavity entrance by varying the amount of moss in the base. In very shallow cavities or old Willow Tit holes there may be only a rudimentary moss base of as little as 1cm, whereas in deep and damp cavities the moss could be 15–20cm thick (Ludescher 1973, Wesołowski 1996). There is a compromise between safety and visibility in placing the nest-cup out of reach of predators but not so deep that the birds cannot see (Wesołowski & Maziarz 2012). The *danger distance* for cavity-nesting birds is the safe distance beyond which larger predators like Pine Martens and Great Spotted Woodpeckers cannot reach eggs or chicks through the entrance (Wesołowski 1996). For Marsh Tits the minimum danger distance is about 18cm, and most build their nests at a depth at or beyond this limit.

WILLOW TIT NEST CONSTRUCTION

There is very little information on how long Willow Tits take to build their nest after cavity excavation. Foster & Godfrey (1950) described nest-building over 8–10 days, with further material added throughout the egg-laying period. Another old record from northeast England described a female Willow Tit incubating eggs within nine days of the cavity excavation, so the nest-building itself must have taken only a few days (Temperley 1934).

Willow Tit nests are distinctive and unlike those of Marsh Tits or other European tits (Figure 6.13). Moss is rarely used in nest-building, and instead there is a base of plant fibres from seed heads, such as Great Reedmace or thistles, and other plant materials including bast, dried grass

Figure 6.13. A Willow Tit nest in a nest-box in Wiltshire, southern England, containing a full clutch of nine eggs. The nest is a mixture of plant fibres (bast and grass), bud casings, wood chips, animal hair and small feathers. Note the lack of moss or a clearly defined structure of a base supporting a lined nest-cup. (© Matt Prior)

or scales from bud casings or fern rachises. The indistinct lining includes fur such as rabbit, deer hair and small feathers. The loose nest structure means the cup is barely distinguishable from the base material. Artificial fibres are uncommon, but may include pieces of wool, cotton or string.

On rare occasions Willow Tit nests can contain more than a few tufts of moss. Ludescher (1973) described a nest-box intended for Common Starlings that a Willow Tit had filled with a homogeneous mix of moss, bast and deer hair. In northern England green moss was only found in Willow Tit nests when Blue Tits or Great Tits were attempting to take them over and bringing their own nest material (Rustell 2015, Parry & Broughton 2018). Why Willow Tits don't build a mossy base for their nest, unlike other tits, is probably because their excavated nest-chambers in deadwood are relatively dry environments, and not surrounded by the humid walls of a living tree. So Willow Tits have little need of a barrier between the nest-cup and a soggy cavity floor, and the orientation of their entrances means they are unlikely to collect rainwater.

Mate guarding and 'demand' behaviour

During cavity excavation and nest-building the male Marsh Tit or Willow Tit *mate guards* the female as she enters her fertile period, often staying within a few metres for much of the day (Koivula *et al.* 1991). Mate guarding reduces opportunities for neighbouring males to get to the female for mating, but also limits the female's ability to seek out these extra-pair copulations. Mate guarding might also protect both members of the pair through extra vigilance. Losing a mate to a predator immediately before breeding would be disastrous for the survivor.

During nest-building, and from about a week before egg-laying, males begin providing extra food to the female. This gives the female extra energy as she develops the eggs, and the feeding becomes more intense and noisy as egg-laying begins. This behaviour used to be called 'courtship feeding' or 'begging' by the female (Morley 1949), but is way beyond courtship as the pair already have a firm bond. Susan Smith (1991) referred to this activity as *demand behaviour*, and it is also known as *food soliciting* (Foster & Godfrey 1950, Koivula *et al.* 1991).

The feeding exchanges between the pair are initially so rapid and infrequent that they are easily missed. The male makes a sudden and silent dash to the female as they both forage, places something in her bill (usually a caterpillar) and then flies off again. The earliest that I saw this silent feeding between Marsh Tits at Monks Wood during the 2000s was 4 April, about a week before the first eggs appeared. The full demand behaviour of the female begins around the time of egg-laying, which in England is about mid to late April. The female quivers her wings as the male approaches, and gives the persistent and characteristic call that Morley (1949) named the 'food cry' but is now known as the *broken dee* call in tits and chickadees. The Marsh Tit's broken dee call is a squeaking series of intermittent notes, while that of the Willow Tit is a more musical series of two or three descending notes (see Chapter 2).

The pair can be very conspicuous at this time as the female calls for extended periods, fluttering between twigs and branches and often allowing very close approach. Meanwhile, the male searches for small caterpillars to bring to the female, one at a time, and her calls rise to an excited high-pitched trill as she receives the food while crouching and shivering her wings. The feeding rate increases towards the time of egg-laying, with the male passing a food item to the female once or twice per minute. This was always one of my favourite times of the year in the Monks Wood study, finding and following Marsh Tits as they moved through the fresh leaves of the understorey on sunny April mornings.

The demand behaviour of females lasts throughout the incubation period and after the nestlings have hatched, and is a clear sign that a breeding attempt is under way. The pair may forage anywhere in the territory at this time, or even outside of it, so the demand behaviour does not necessarily mean that a nest is very close by, but it does mean that there is a nest somewhere within the territory.

Copulation

The behaviour around copulation is quite similar for both species and was described by Morley (1949), Koivula *et al.* (1991) and Haftorn (1994, 1995). Copulation often occurs when the female is showing demand behaviour by calling persistently and fluttering through the vegetation. She suddenly adopts a crouching posture with rapid quivering of the wings, and gives a very high-pitched trilling call. The male approaches with a distinctive slow flight, lands beside the female and may also quiver his wings, sometimes giving a short, deep, rapid trill. The female then bows her head and gently vibrates her tail, the male jumps on her back, and both birds twist their tails to touch their cloacas and pass the sperm. The male concludes with a bit of a flourish, giving a strong quiver of the wings before jumping off the female. The copulation takes only a few seconds.

The pair copulate multiple times in the days just before and during egg-laying, but it is not often seen. Despite closely observing 124 Willow Tit pairs over four years in Finland, Welling *et al.* (1997) saw only 16 copulations and 13 further attempts that were probably unsuccessful. Most copulations occurred from 11 days before the first egg was laid until the day of the penultimate egg, which was the female's fertile period. Pairs mostly copulated in the early morning, especially when they were reunited after roosting in separate holes overnight (Welling *et al.* 1995). Occasionally a male may try to solicit copulation by quivering his wings and chasing the female, but he is not always successful (Welling *et al.* 1997).

EXTRA-PAIR COPULATION

Although Marsh Tits and Willow Tits are both socially monogamous and have a very strong pair bond, both sexes sometimes copulate with other birds outside the pair, especially close neighbours. It is difficult to judge how common these *extra-pair copulations* might be, because they are secretive and involve a bird sneaking into a neighbouring territory.

There are no documented observations of extra-pair copulations in Marsh Tits, but in Monks Wood I have seen paired females visiting neighbouring males during their fertile period. The birds are suspiciously quiet during these meetings, and it's hard not to conclude that they're trying to avoid detection by their respective mates, who would chase them back into their own territory. A genetic study in China found that 15 per cent of Marsh Tit nestlings in 46 per cent of nests were the result of extra-pair copulations, fathered by a male who was not their mother's social partner (Wang *et al.* 2021). In other words, nearly half of the nests contained chicks with more than one father. Around half of these nestlings were fathered by a near neighbour but, astonishingly, some of the genetic fathers lived almost 3km away.

Haftorn (1995) documented an extra-pair copulation in Norway involving a Willow Tit pair confronting an unpaired male at their territory border. The males sang against each other from 25m apart, when the female suddenly appeared in the neighbour's territory and solicited a copulation. He happily obliged, but then the female's mate came over and began a ferocious fight with him. Ironically, the pair already had nestlings so there was no paternity risk by this point, although maybe it wasn't the first time that the female had copulated with the neighbour.

Although direct observation is rare, genetic evidence shows that extra-pair copulations must sometimes be relatively common in Willow Tits, as with Marsh Tits. A study of 117 Willow Tit broods in Finland found 21 per cent of nests contained some chicks that were fathered by a different male, comprising 7 per cent of all nestlings in the population (Lampila *et al.* 2011). This is maybe not surprising, as a study of the related Siberian Tit in Norway estimated extra-pair paternity to average 17 per cent of chicks per brood in 38 per cent of nests (Oddmund *et al.* 2020). Extra-pair young have also been found in at least a third of Black-capped Chickadee and Mountain Chickadee broods in Canada, comprising around one in six chicks (Mennill *et al.* 2004, Bonderud *et al.* 2018). The outwardly strong social bonds between pairs of Marsh Tits, Willow Tits and other *Poecile* species can therefore belie a high rate of infidelity. The reasons for this may involve females trying to secure better genes for their young, or maybe an insurance against infertility of their mate. Males can also gain the same benefits and have a better chance of reproducing if they don't have all of their progeny in one nest.

Eggs and egg-laying

After copulating, female Marsh Tits and Willow Tits may both begin egg-laying before the nest is finished, and will often continue to bring nest material. This is especially true if the birds are in a hurry after losing their first nest, or if the weather has quickly warmed up. There is often a delay of a few days before laying begins, however, and if the weather suddenly becomes cool or rainy then females can pause for up to 12 days between nest completion and laying the first egg (Wesołowski 2013).

Both species lay one egg per day until the clutch is complete. Svein Haftorn (1996) observed egg-laying using glass-sided nest-boxes and video cameras. Females began laying around dawn, from as early as 23 minutes before sunrise to as late as 49 minutes after, and the laying process lasted a minute or less. The female stood motionless in the nest-cup, breathing deeply and rhythmically, sometimes opening and closing her bill. The rump and tail began downward movement that increased in strength until the rump was finally raised and the egg was laid. During the final pushes, the female closed her eyes and sometimes turned her head under apparent strain. After the egg had been laid the female stood motionless during a recovery phase that lasted 30–90 seconds, but sometimes up to 7 minutes. She then preened or rearranged the nest material, and it could take 12 minutes or more before she left the nest, and occasionally more than an hour (Haftorn 1996).

Marsh Tit and Willow Tit eggs are similar to those of other tits and chickadees, being white with irregular brownish, reddish or purple-grey speckles (Figure 6.14; see also Figures 6.12 and 6.13). The speckles may be denser at the rounded end of the egg, but there is a lot of variation, with some eggs having very few speckles and others being heavily marked. The eggs of both species are of a similar size, averaging approximately 16–17mm long by 12mm wide for British birds and weighing around 1.1–1.3g, which is at least 10 per cent of the female's body weight. The egg size varies a millimetre or two between individuals and species, and across their respective ranges (Cramp & Perrins 1993).

After laying, the female usually covers the eggs with nest-lining material, so it is impossible for a predator (or researcher) to see them by looking inside the cavity. After leaving the nest for the day the female generally forages elsewhere in the territory and rarely comes near the nest, unless she brings some additional nest material. At dusk she returns to roost and may remove the material covering the eggs before settling on them for several hours, raising their temperature above the 25–27°C threshold at which embryos begin to develop (Haftorn 1979). This partial incubation happens even before the clutch is complete. During the night the female moves off the eggs to roost standing above them, so as not to incubate them further. In the morning she will lay another egg and leave the nest again.

Figure 6.14. Two Marsh Tit eggs from the same clutch in an abandoned nest at Monks Wood. The eggs measure 17.7mm and 17.4mm on the longest axis. Note the variation in the density of speckling. (© Richard K. Broughton)

FIRST-EGG DATES

The onset of egg-laying varies between years and regions. This is further complicated by the advancement of laying dates over time due to climate change. As such, it is difficult to compare laying dates between studies conducted at different times and in different regions. What is typical for Willow Tits in Finland is not typical in Greece, and what was normal for Marsh Tits in the 1970s will not be normal in the 2020s.

Nevertheless, there are broad trends in laying dates, and birds at southern latitudes begin egg-laying earlier than birds further north. In recent decades southern Marsh Tits in the Balkans can begin egg-laying in late March, whereas in mid-latitude England and Poland they begin around mid April, and further north in Sweden the average laying date is 3–5 days later still (Dolenec 2006, Massimino *et al.* 2023, Andreasson *et al.* 2023, Wesołowski 2023). There is also a northward pattern of later egg-laying in east Asian populations. In an urban forest in South Korea, Marsh Tits begin egg-laying in early to mid April, whereas further north on the Japanese island of Hokkaido the recent average onset of laying is in early May (Rhim *et al.* 2011, Nomi *et al.* 2017).

Willow Tits begin egg-laying at a similar time to Marsh Tits at similar latitudes. In England this is around mid April in recent decades, and in the Wigan study the annual first-egg dates varied by only a week during 2014–18, ranging between 7 and 13 April (Parry & Broughton 2018, Massimino *et al.* 2023). In colder climates Willow Tits begin egg-laying later in the spring. In northern Finland and Siberia during the late 1980s the first eggs did not appear until well into May or the first week of June (Orell *et al.* 1996, Pravosudov & Pravosudova 1996), and this has advanced by only a few days in the Oulu population in Finland (Vatka *et al.* 2011).

A remarkable feature of Marsh Tits and Willow Tits is their synchronised egg-laying within a population. For Marsh Tits in the Białowieża National Park, 90 per cent of females begin laying within 10 days of the first bird producing an egg (Wesołowski 1998). At Monks Wood an average 90 per cent of local Marsh Tits begin egg-laying within 10 days of the first bird (8–18 nests per year in 2005–2011). In Sweden 81 per cent of Marsh Tits in the Lund population hatched within 10 days of each other (Nilsson & Smith 1988a). Similar synchronisation is seen in first clutches of Willow Tits, although the overall laying dates become more scattered as a result of higher nest losses and repeat attempts (Pravosudov & Pravosudova 1996, Parry & Broughton 2018).

Individual females can show great flexibility between years, shifting their first-egg date by several weeks in different breeding seasons, depending on annual conditions. In two consecutive springs in Białowieża the same group of 16 female Marsh Tits all shifted their egg-laying by 13–23 days between years (Wesołowski *et al.* 2016). In Monks Wood we recorded 15 females shifting their individual egg-laying dates by 3–16 days in different breeding seasons. These adjustments by females are a response to the spring weather and phenology of vegetation and caterpillars. In colder springs the females delay their breeding for a few weeks until conditions improve, and in warmer springs they can take advantage of good foraging to advance their egg-laying. This gives individuals some flexibility in coping with unpredictable spring weather during their lifetime, but also means that the population as a whole has some ability to cope with earlier springs as a result of climate change (Wesołowski *et al.* 2016). This flexibility is unlikely to be limitless, however, and it is unknown at what point the birds might become unable to match their timing of breeding to the changing climate.

Clutch size

The typical clutch size for European Marsh Tits is 7–9 eggs for clutches in natural cavities (Table 6.6). Larger clutches often occur in bigger nest-cavities, suggesting that females are making subtle calculations when producing their eggs (Ludescher 1973, Wesołowski 2003). Clutches in nest-boxes average slightly larger, and with the largest numbering up to 11–12 eggs, which may be an effect of cavity size (Ludescher 1973). One-year-old females in their first breeding season usually lay slightly smaller clutches (1–2 fewer eggs) than older females, but experienced females also lay larger clutches than inexperienced birds of the same age (Smith 1993a). Late Marsh Tit clutches are smaller than those in early nests, with a reduction of around one egg for every week of delay (Nilsson 1991, Wesołowski 2000). Replacement and second breeding attempts also contain fewer eggs than first attempts (Delmée *et al.* 1972). Late birds might lay fewer eggs to try and catch up with their neighbours, or perhaps because foraging conditions deteriorate as spring progresses.

Willow Tit clutch sizes are similar to those of Marsh Tits, typically around 7–9 eggs for first attempts in natural cavities (Table 6.7). There is not much evidence that bigger clutches consistently occur in nest-boxes, as the very largest clutches of 12–13 eggs have been in natural cavities. Clutches laid early in the spring are larger than later or replacement clutches (Orell & Ojanen 1983). The repeat clutches recorded in the Wigan study after the loss of an earlier nest were almost one egg smaller on average, at 7.8 eggs for 28 nests, with a range of 3–9 eggs (Parry & Broughton 2018). In Finland, Willow Tit clutches laid towards the end of May were also around one egg smaller than clutches laid early in the month (Orell & Ojanen 1983).

Table 6.6. Breeding metrics for Marsh Tits in nest-boxes and natural cavities.

Location	Nest type	Clutch size Mean (range)	Clutch size No. of nests	Hatching success % eggs	Hatching success No. of nests	No. of fledglings Mean	No. of fledglings No. of broods	Reproductive success %	Reproductive success No. of nests	Nest failure %	Nest failure No. of nests
England: Monks Wood	Nest-box	8.4 (4–10)	19	97.8	16	8.4	15	79.5	18	21.1	19
England: Wytham	Nest-box	7.7	80	—	—	—	—	—	—	17.8	90
Belgium: Viroinval	Nest-box	8.8 (4–12)	61	—	—	7.7	<61	—	—	—	—
Germany: Radolfzell	Nest-box	8.3 (6–10)	199	—	—	—	—	78.9	199	—	—
Germany: Braunschweig	Nest-box	8.9 (6–12)	230	—	—	7.9	446	87.8	446	—	—
Denmark: Funen	Nest-box	8.8 (5–12)	49	—	—	—	—	—	—	—	—
Sweden: Lund	Nest-box	8.9 (5–11)	101	—	—	7.8	43	—	—	—	—
Sweden: central	Nest-box	7.9 (6–11)	45	98.9	45	7.8	45	98.3	45	0	45
Korea: Seoul	Nest-box	8.9 (6–11)	8	—	—	5.7	7	45.7	8	16.7	8
Japan: Hokkaido	Nest-box	9.6 (8–11)	30	97.1	26	—	—	—	—	25.8	31
England: Monks Wood	Cavity	7.5 (4–10)	101	97.8	82	7.0	61	73.6	73	16.0	144
England: Wytham	Cavity	7.1 (6–8)	16	99.2	16	6.5	12	68.7	16	25.0	16
Germany: Pfrunger Ried	Cavity	6.8 (4–9)	26	—	—	—	—	—	—	55.2	29
Baltic: Curonian Spit	Cavity	8.1 (7–10)	60	92.4	60	7.5	53	81.1	60	17.2	64
Poland: Białowieża	Cavity	7.4 (4–10)	202	—	—	—	—	—	—	25.9	968
China: Jilin	Cavity	6.9	12	92.4	11	7.1	9	77.1	14	35.7	14

Hatching success is calculated as the percentage of eggs that hatched (excluding depredated nests) across the population. The mean number of fledglings is for successful broods only (excluding nest failures). Reproductive success is the percentage of eggs laid that resulted in fledglings across the study population.

Sources: Delmée et al. 1972, Frederiksen et al. 1972, Johansson 1972, Ludescher 1973, Dunn 1976, 1977, Berndt & Winkel 1987, Nilsson & Smith 1988a, Nilsson 1991, Markovets & Visotsky 1993, Wesołowski 2000, 2023, Wang et al. 2003, Carpenter 2008, Rhim et al. 2011, Nomi et al. 2017, author's unpublished data.

Table 6.7. Breeding metrics for Willow Tits in nest-boxes and natural cavities.

Location	Nest type	Clutch size		Hatching success		No. of fledglings		Reproductive success		Nest failure	
		Mean (range)	No. of nests	% eggs	No. of nests	Mean	No. of broods	%	No. of nests	%	No. of nests
England: Wigan	Nest-box	8.6 (8–10)	8	100.0	8	8.5	8	98.6	8	0.0	8
England: Wiltshire	Nest-box	8.0	39	—	—	6.6	37	80.0	39	5.1	39
Scotland: central/south	Nest-box	—	—	—	—	6.4	19	—	—	15.6	32
Belgium: Viroinval	Nest-box	7.8 (5–11)	16	—	—	7.0	14	79.0	16	12.5	16
China: Hebei	Nest-box	—	—	—	—	—	—	—	—	3.2	31
England central/south	Cavity/box	—	—	—	—	—	—	—	—	34.4	81
Sweden: Gothenburg	Cavity/box	8.3	97	—	—	6.2	96	78.0	31	21.0	120
England: general	Cavity	8.3 (5–13)	37	94.7	3	5.7	3	89.5	3	0.0	3
England: northwest	Cavity	8.3 (6–12)	44	—	—	—	—	—	—	45.7	71
England: Wigan	Cavity	8.7 (6–10)	47	100.0	38	8.8	28	60.2	47	40.4	47
Germany: Pfrunger Ried	Cavity	7.2 (5–10)	29	—	—	—	—	—	—	61.1	36
Finland: Oulu	Cavity	7.6 (4–10)	138	96.0	125	6.2	125	84.6	125	21.1	152
Siberia: Magadan	Cavity	7.5 (6–10)	22	—	—	6.5	—	85.0	22	—	—

Hatching success is calculated as the percentage of eggs that hatched (excluding depredated nests) across the population. The mean number of fledglings is for successful broods only (excluding nest failures). Reproductive success is the percentage of eggs laid that resulted in fledglings across the study population.

Sources: Foster & Godfrey 1950, Delmée *et al.* 1972, Ludescher 1973, von Brömssen & Jansson 1980, Orell & Ojanen 1983, Ekman 1984, Pravosudov & Pravosudova 1996, Maxwell 2001, 2002, Stewart 2010, Last & Burgess 2015, Rustell 2015, personal communication, Parry & Broughton 2018, Zhang *et al.* 2020.

Incubation

Only females incubate the eggs, and the behaviour is very similar in both species. Prior to incubation the female develops a *brood patch* of thickened skin on the belly, which is rich in blood vessels to transfer heat to the eggs. The brood patch develops a couple of weeks before egg-laying begins, around late March or early April in Great Britain, and may still be visible until July before it feathers over during the annual moult. The temperature of the brood patch has been measured at 40–42°C in Marsh Tits, which compares with egg temperatures of 34–41°C for both species during steady incubation (Haftorn 1988, Deeming & du Feu 2008).

Incubation generally begins with the laying of the penultimate or last egg in the clutch, but there can sometimes be a couple of days' delay after clutch completion before incubation really starts (Nilsson 1993). Incubation may also begin gradually, with some nocturnal warming of the eggs before the clutch is complete, and then a build-up in daytime incubation over the first days after clutch completion (Haftorn 1979).

The start of full incubation is usually obvious, however, as the female spends much of her time inside the nest-cavity during the day. The female sits on the eggs and places her warm brood patch against them, with the feathers on her flanks and back splayed to provide an insulating cover over the nest-cup (Figure 6.15). At Monks Wood and Bagley Wood the female Marsh Tits usually incubated for periods of 15–45 minutes at a time, with frequent bouts of 20–25 minutes (Morley 1953). Nilsson & Smith (1988b) calculated that female Marsh Tits in Sweden spent 60–90 per cent of their day incubating the eggs. Haftorn (1979) recorded most incubation bouts of Willow Tits at 8–22 minutes, but they could be as long as 88 minutes.

Figure 6.15. A female Marsh Tit (left) and Willow Tit (right) incubating their eggs in nest-boxes in Monks Wood and Wiltshire, respectively. The birds themselves look very similar, but the nests are quite different in composition. (Marsh Tit © Richard K. Broughton; Willow Tit © Matt Prior)

The female's periods off the nest commonly last for around 5–15 minutes, but in good weather in the middle of the day it can be as long as 25–30 minutes. During these breaks the female will feed, defecate and sometimes bathe. In contrast to the laying period, during incubation the female no longer covers the eggs with nest-lining when she leaves the nest. This makes it easy for a researcher to count them by looking inside the cavity with an endoscope, or with a small mirror and torch. Like many others, during fieldwork I have spent countless hours quietly sitting near nests, watching the entrance intently, waiting for an incubating female to leave so that I can count the eggs. If the female is disturbed on the nest she will just sit tight for longer, so the best option is to sit somewhere comfortable and wait patiently, enjoying the forest.

During incubation the male frequently visits the nest to feed the female. When he arrives carrying a food item (usually a caterpillar) he sits in a bush or tree some distance from the nest and calls softly to tell the female that he's there (Figure 6.16). For Marsh Tits this is a *pitchou* call, and for Willow Tits a soft *tsip*. If the female does not respond the male may fly to the nest-cavity and lean inside to feed her. Alternatively, on hearing the male calling, the female often leaves the nest for one of her incubation breaks, flying to the male and showing demand behaviour by fluttering her wings and giving food-soliciting calls. Sometimes the female leaves the nest when the male is not there and she will begin calling, which quickly brings him to her. The pair then move away to feed, with constant demand behaviour from the female. During the incubation period the male can feed the female an average of 14 times per hour, either on or off the nest (Nilsson & Smith 1988b). If the female is alone,

Figure 6.16. A male Marsh Tit in the Białowieża Forest bringing a food item to a female, who is incubating eggs inside a nest. The male is perched a few metres from the nest and calling to the female to encourage her to come and receive the food. (© Richard K. Broughton)

however, perhaps having been widowed, then she will not get extra feeds and will silently forage alone.

The pair or lone female may forage over 100m from the nest. Pairs are easily located by their frequent calling, which helps a researcher to track down the nest. When the female decides to return to incubation the calling and feeding suddenly stops and she makes a direct flight straight back to the nest from wherever she is in the territory. During nest-finding fieldwork this is a cue to run after the female as she moves through the trees, trying to keep up until she stops near the nest before quickly going inside.

The incubation period lasts 13–15 days for Marsh Tits, and exceptionally 12–19 days before the eggs hatch. For Willow Tits the typical incubation period is a very similar 13–15 days with extremes of 11–19 days. Females with late clutches can shorten the incubation period by sitting on the eggs for longer periods and getting extra help from the male, who increases his feeding rate to offset her loss of foraging time (Nilsson & Smith 1988b, Wesołowski 2000).

While the female is inside the nest-cavity the male is conspicuously alone elsewhere within the territory. He can be quite easy to find during this time by listening for occasional calls, especially when he is near the nest. When the male is foraging, before long he will find a caterpillar but not eat it, instead removing the head and then carrying it in his bill, sometimes collecting one or two more. He then flies directly to the nest to feed the female, which can reveal its location to the patient researcher.

Hatching and nestlings

At the end of incubation the eggs hatch over one or two days, occasionally over three days, with each nestling taking up to several hours to emerge. Across a local population over half of the nests will hatch within a week of the first one. The eggshells are eaten by the female or carried out of the nest and dropped some distance away. During hatching the female's routine is superficially similar to during incubation, with extended periods on the nest brooding the chicks while the male regularly brings food.

If nests are not predated, then the hatching success of Marsh Tits and Willow Tits is very high, with at least 92 per cent of eggs hatching (Tables 6.6 and 6.7). The hatched chicks are initially bright pink with large bluish-grey eyelids that are sealed closed, and yellow gape flanges that extend from the bill and below the eyes (Figure 6.17). The wings and feet are short and poorly developed, but the chicks can actively move around the nest-cup by wriggling. Both species have a few greyish wispy tufts of natal down on their head and back, which is initially wet when they emerge from the eggs, but soon dries and becomes fluffy.

During breaks in brooding when the female leaves the nest she returns with very small caterpillars to feed to the chicks. The male also brings small food items for the chicks, usually passing them to the female in the first few days after hatching. This change in behaviour is clear evidence that eggs have hatched. The chicks begin demanding food within an hour of hatching by raising their heads and opening their bills to reveal a pale yellow-orange gape and mouth (Figure 6.17). Nestling development is very similar in Marsh Tits and Willow Tits. On the day of hatching the chicks weigh around 1g, but this increases by about 40–50 per cent daily in the first few days of life. The rate of weight increase steadily declines as nestlings approach their maximum of 10–12g by about 12–14 days old, similar to the adult weight (Foster & Godfrey 1950, Rheinwald 1975, Orell 1983, Thessing 1999). Nestlings maintain this weight until they leave the nest.

Figure 6.17. Marsh Tit nestlings during hatching in a nest-box at Monks Wood. The tufts of down are still wet on the backs and heads of the four nestlings that have hatched, and five unhatched eggs remain, along with a hatched eggshell that has not yet been removed or eaten by the female. (© Richard K. Broughton)

Young chicks have no ability to regulate their own temperature until they grow feathers, and so they rely on brooding by the female until they are more developed. The female also roosts in the nest-cavity with the chicks for at least the first 10 days, but later on she roosts elsewhere (Nilsson & Nord 2017). The increase in body mass as the nestlings grow is accompanied by rapid development of the feathers. Two days after hatching the feather tracts show below the skin as distinct greyish zones on the head, flanks, wings, back, rump and belly. Over the following days small whitish quills appear on the trailing edge of the wings, and the gape and mouth become bright yellow, deepening to orange in the throat (Figure 6.18). The eyes remain tightly closed until around day six after hatching, when they begin to open. By this stage, all of the feather tracts have emerged as short, spiky bluish-grey quills over the body, pushing out the natal down. The legs develop quickly, and after six days they begin to take on a bluish hue. Up to this age the nestlings are giving simple, soft *peep* calls when demanding food, but these become more complex after six days of age (see Chapter 2).

By eight days after hatching the nestlings' eyes are partly open and some of the growing feather quills are starting to split to reveal the brownish colour of the back and creamy buff of the underparts (Figure 6.19). The bill gets longer and darker but the wing and tail feathers remain mostly within the quills, with the emerging tips growing by about 3mm per day in a broadly linear fashion for the rest of the nestling period (Orell 1983, Pravosudov & Pravosudova 1996, Thessing 1999).

After 10 days most of the quills on the body and head have opened to reveal the feathery brown backs and black caps. The bluish-grey legs are fully grown by 11 days old but are a little thicker and more fleshy than the adults' (Orell 1983, Thessing 1999). Feathered nestlings can now thermoregulate on their own, so the female spends less time in the nest brooding them. By this stage the nestlings' food-demanding calls consist of *begging trills* (see Chapter 2), which make a high-pitched chorus from the whole brood when a parent arrives (Broughton 2019). Day 11–13 is the ideal time for a researcher to safely ring the nestlings, as they are developed enough yet still quite docile (Figure 6.20).

Figure 6.18. Willow Tit (left) and Marsh Tit (right) nestlings at four days after hatching, in Wiltshire and Monks Wood respectively. Both broods have grey fluffy down on their heads and backs, but dark quills of the feather tracks are developing under the skin. (Willow Tit © Matt Prior; Marsh Tit © Richard K. Broughton)

Figure 6.19. Willow Tit (left) and Marsh Tit (right) nestlings around eight days old, in Wiltshire and Monks Wood respectively. Dark feather quills have now broken through the skin and some are just beginning to open at the tips, revealing the plumage colouration. (Willow Tit © Matt Prior; Marsh Tit © Richard K. Broughton)

By 15–16 days after hatching the chicks are fully-feathered, wide-eyed and alert. The wing and tail feathers are still short, with the primaries barely extending beyond the secondary feathers on the closed wing, and the tail is only about 25mm long (half the adult length). At this stage the nestlings are capable of prematurely scrambling from the nest if they are attacked or frightened, even though they cannot yet fly, usually ending up on the ground and extremely vulnerable. Opening nest-boxes or intrusive nest checks can be quite risky after this point. The nestlings have become quite loud by day 15 after hatching and can be heard calling inside the nest from at least 5–10m away when an adult arrives to feed them. In the last few days before fledging the nestlings frequently exercise their wings by rapid flapping in short spurts.

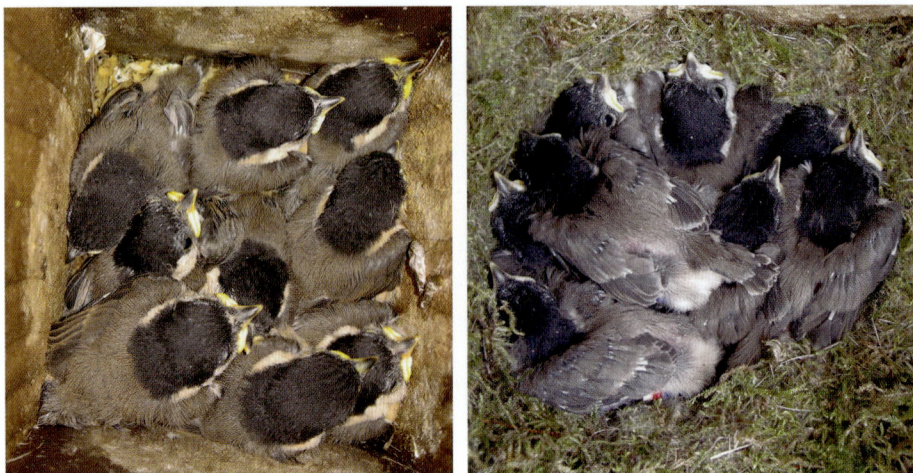

Figure 6.20. A brood of nine Willow Tit nestlings in Wiltshire (left) and eight Marsh Tits in Monks Wood (right), both around 13 days old. The feathers have emerged on the head and body, but the wing and tail feathers remain short yet growing. The nestlings of each species look virtually indistinguishable. (Willow Tit © Matt Prior; Marsh Tit © Richard K. Broughton)

A narrow nest-cavity will now have limited room for the large nestlings, which can be sat on top of each other and frequently changing position to be at the top.

SEX RATIO OF NESTLINGS

The ideal sex ratio in an adult population would be 50:50 for males and females. As such, it would seem straightforward for the primary sex ratio of the eggs in each clutch or across a population to contain an equal number of males and females. However, if the sexes have different chances of survival then the birds might want to skew their nestlings' sex ratio to favour the one that has most chance of being successful in entering the breeding population (Hasselquist & Kempenaers 2002).

Any advantage to producing chicks of either sex might vary over time and space. For example, larger males may be more costly to produce in tough years, but may also have higher settling success in fragmented habitat due to their shorter dispersal distances. Genetic evidence shows that some bird species do indeed have the ability to manipulate the sex ratio of their broods (Hasselquist & Kempenaers 2002), but there have been very few studies using genetic sexing of Marsh Tit broods, and none for Willow Tits.

A study in the Białowieża National Park found no evidence that Marsh Tits were manipulating the sex ratio of their nestlings (Czyż *et al.* 2012). In 66 broods the nestlings varied from all male in some nests to all female in others, but the most frequent, in 35 per cent of nests, was an equal sex ratio. Individual females were inconsistent with the offspring they produced, with one bird producing six daughters and two sons in one year, but seven sons and two daughters the following year. Overall, the sex ratio of nestlings in the Białowieża population was about even, with 53 per cent males and 47 per cent females across all nests.

We also looked at Marsh Tit brood sex ratios in Monks Wood and Wytham Woods, and got similar results to Białowieża (Broughton *et al.* 2018b). There were extreme ratios of some nests containing only females or almost all males, but overall the proportion of males across all broods was 51 per cent for Monks Wood and 46 per cent for Wytham Woods. The Monks Wood population is more fragmented than the larger woodland block of Wytham, which reduces the settlement chances of Monks Wood females trying to disperse over longer distances (Broughton *et al.* 2011). Nevertheless, we found no evidence that the Monks Wood Marsh Tits were manipulating their broods to favour males, which might have a better chance of survival in the fragmented landscape.

Despite the limited number of studies, the evidence so far is that Marsh Tit populations under varying conditions produce nestlings with a broadly equal sex ratio. Studies of Willow Tits would be useful but would probably show the same thing, which was also the case for Black-capped Chickadees (Ramsay *et al.* 2003).

NESTLING FEEDING RATES

The adults feed a variety of prey to the nestlings, which they glean from trees, shrubs and sometimes the herb layer or even the ground. By far the most important prey are caterpillars of butterflies and moths; these account for the vast bulk of the prey items (Figure 6.21), with spiders forming a small but significant component (Cholewa & Wesołowski 2011). More details of the nestling diet are given in Chapter 4.

Both parents feed the chicks throughout the daylight hours, and the initial feeding rate is approximately 10–15 feeds per hour in the first few days after hatching. This steadily increases to 18–30 visits per hour by seven days of age and is maintained at this rate until fledging (Markovets & Visotsky 1993, Pravosudov & Pravosudova 1996, Carpenter 2008). Initially the male makes

Figure 6.21. A Marsh Tit bringing a selection of caterpillars to its nestlings in the Białowieża Forest. (© Richard K. Broughton)

Figure 6.22. An adult Willow Tit removing a nestling's faecal sac from a nest in Norfolk, England. (© Ashley Banwell)

more feeding visits than the female, as she spends lots of time brooding rather than foraging, but as the chicks grow the parents' feeding rates become more similar.

What goes in must come out, and the nestlings produce *faecal sacs* that the parents remove during feeding visits. The white and black faecal sacs are semi-solid packages of waste products enclosed in a mucous membrane, which the adults can carry in their bill and drop some distance from the nest (Figure 6.22). In this way the adults keep the nest completely free of faeces until the day of fledging, when they reduce their feeding visits and stop removing faecal sacs. This shift in adult behaviour is probably their way of luring the youngsters to leave the nest, and a few telltale droppings left in the empty nest is a sign that at least some have fledged successfully.

Nest parasites

Cavity-nesting birds attract parasites, and Marsh Tits are no exception, but information is lacking for Willow Tits. The presence of an active nest elevates the temperature within the cavity (Maziarz 2019), promoting the development of invertebrates, and the nest and its occupants provide a food source. In the Białowieża National Park approximately half of Marsh Tit nests contain larvae of blowflies (*Protocalliphora falcozi*), which are maggots that bite nestlings to feed on their blood. The adult blowflies enter the tree cavities to lay their eggs onto the nest material after the chicks hatch, and the larvae pupate and fly off after the chicks have fledged (Wesołowski 2001).

Blowfly-infested Marsh Tit nests in Białowieża contained an average of three larvae per nestling. Larger broods had more larvae than smaller ones, and one nest had a remarkable 75 larvae. Despite this, there was no detectable effect of blowflies on the survival or development of nestlings. However, the parents did appear to work harder at infested nests, by increasing their feeding rate, which was associated with lower annual survival for these adults.

Other species of blowfly (*Protocalliphora azurea* and *P. maruyamensis*) have been found in Marsh Tit nests in Japan (Iwasa *et al.* 1995), and *Protocalliphora* larvae have been found in the nests of Chestnut-backed and Mountain Chickadees in California, with one nest harbouring a phenomenal 45 larvae per nestling (Gold & Dahlsten 1983). Again, these infestations did not seem to affect the nestlings. Blowflies probably parasitise Willow Tits in the same way, but there appear to be no records as yet.

The other major parasite of cavity-nesting birds is the Hen Flea, which can be abundant in European nest-boxes used by tits and flycatchers, alongside blowfly larvae (Blunsden & Goodenough 2023). Fleas feed on the blood of adults and nestlings, and nest-boxes seem to be particularly attractive due to the drier microclimate compared to natural cavities in living trees (Maziarz *et al.* 2017). Marsh Tits breeding in nest-boxes in Sweden had an average of just six fleas in their nests, which probably did little harm (Nilsson 2003). Where the number of fleas was experimentally tripled this resulted in a slower growth rate of nestlings, which had lower body weights and shorter wings than usual.

Hen Fleas are uncommon in Marsh Tit nests in natural cavities, with the Białowieża study finding only a few fleas in several nests (Wesołowski 2001). Interestingly, there was a higher prevalence of fleas in natural cavities used by Marsh Tits in areas where flea-infested nest-boxes were also present, even if Marsh Tits were not using the boxes (Wesołowski & Stańska 2001). Other species may have transferred fleas from boxes into cavities, perhaps when roosting or prospecting for nest-sites.

Fledging

Nestlings of both species usually leave the nest at 18–20 days after hatching, when they become fledglings. This can be as little as 16 days if the nest has been disturbed, or as long as 21–23 days if the weather is poor and they wait for better conditions. In the days before fledging the nestlings will peer outside the nest entrance, calling to be fed if an adult is nearby, and they can become quite noisy and noticeable.

Fledging usually takes place during the early morning, giving them the rest of the day to regroup outside the nest, but they can sometimes leave during the late morning or afternoon. All of the nestlings typically leave on the same day within an hour or two of each other, and often in quick succession. Fledging is initiated by the largest nestlings, meaning that if any of

Figure 6.23. Fledglings leaving the nest can sometimes become grounded, where they are very vulnerable to predators, like this young Marsh Tit in Poland's Białowieża Forest. Grounded fledglings must scramble up a tree to reach relative safety in the canopy. (© Marta Maziarz)

the siblings are behind in their development then they have to fledge before they might be fully ready, or get left behind (Nilsson & Svensson 1996).

I have watched Marsh Tits fledging on several occasions, including from quite low nests, and it follows lots of calling by the nestlings as they peer outside the entrance. The adults call from nearby and are possibly directing the nestlings to fly towards them to reach suitable cover. One by one the nestlings fly strongly out of the cavity and directly into the shrubs or trees, soon making their way higher into the sub-canopy of the woodland.

Willow Tits in northern England have been observed fledging in a similar manner to Marsh Tits, with lots of calling and peering out of the nest before a fairly competent and direct flight straight into surrounding bushes or trees. Fledging can sometimes be more gradual, especially if a low nest is surrounded by dense vegetation. Rather than a direct flight into the trees, the nestlings may clamber out of the nest and into adjacent tangles of foliage. Like Marsh Tits, they are safer if they can get up into the shrub layer or tree canopy, and the instinct of both species is to make their way upwards into the vegetation. Sometimes nestlings fall onto the woodland floor (Figure 6.23), and they will try to climb up shrubs and tree trunks to get off the ground.

Post-fledging

The post-fledging period lasts for one or two weeks until the fledglings disperse and leave the parental territory. Until then, the family spends the first few days within about 50m of the nest, scattered in a loose group among the sub-canopy of trees or tall shrubs (Figure 6.24). At this stage they can fly well but are not very mobile, making only short flights. Both of the adults forage for caterpillars in the trees around the fledglings and bring the food to them. As the adults approach, the fledglings demand to be fed by repeatedly calling, opening their bills wide and thrusting their heads towards the parent, and rapidly fluttering their wings. When I have watched Marsh Tit families the parents seem to feed the fledglings quite equally.

In the first days the fledglings and their parents must be quite vulnerable to attacks by Eurasian Sparrowhawks, Eurasian Jays or Pygmy Owls when calling and sitting in the same few trees all day long. These fledged families are quite easily located by a researcher, so they must be very obvious to a predator. In Wytham Woods a newly fledged Willow Tit was caught by a Eurasian Sparrowhawk within a day of leaving the nest, its ring being found beneath a plucking post (Foster & Godfrey 1950). Fledgling survival seems quite good, however, at least in Scandinavia, with Nilsson & Smith (1985) finding that 90 per cent of Marsh Tit fledglings from 20 broods survived for at least 11 days after leaving the nest. Hogstad (1990b) monitored six Willow Tit broods containing 34 fledglings, and found that only one had died before they began dispersing 14–15 days after leaving the nest.

During the fledglings' time within the family group they become increasingly mobile as the adults lead them to different foraging trees, which can be quite far from the nest. At Monks Wood we found fledged families up to 400m away within a week of leaving the nest. The adults' territory boundaries are not being defended at this time, so the families are free to wander.

The primary wing and tail feathers continue to grow for around a week after fledging, gaining a further 20mm in length (Thessing 1999). After 7–10 days the fledglings are largely feeding themselves but still demanding some food from the adults, although one or both parents may now start spending time away from the family. The fledglings' vocal repertoire also develops, with adult-type calls beginning to appear within a few days of leaving the nest

Figure 6.24. A Willow Tit fledgling in Norfolk, in eastern England, shortly after leaving the nest. Note the short tail, clean-looking plumage and yellow gape flange. (© Ashley Banwell)

(Haftorn 1993a, Broughton 2019; see Chapter 2). At Monks Wood I saw a fledged Marsh Tit family interacting with a newly dispersed independent juvenile from another family, which was already singing in the area where the fledglings were foraging. The independent juvenile showed aggression to the fledglings by chasing and posturing, but the fledglings were asking it for food, shivering their wings and giving demanding calls when it came near. This very different behaviour from birds only a few days apart in age underlined the sharp transition between fledgling and independent juvenile.

Finding fledged families during the short window between leaving the nest and family break-up can be used to confirm breeding success for nests that were not found or closely monitored. Sage *et al.* (2011) showed a 94 per cent probability of detecting a family group of fledged Marsh Tits if the territory area was searched at least twice per week during the post-fledging period, with four visits virtually guaranteeing to find the family. Similar methods could probably be applied to Willow Tits to monitor breeding success without necessarily finding the nest.

Breeding success

Despite fledglings still relying on their parents for a few weeks until they become independent, *reproductive success* is usually measured by how many fledglings have left the nest in relation to how many eggs were first laid. This includes all of the losses to predation or desertion, eggs that did not hatch, and chicks that starved or otherwise died. As nest-boxes can differ from natural cavities in clutch sizes, parasite loads and failure rates, this needs to be borne in mind when considering breeding success (Møller 1989, Wesołowski & Stańska 2001, Maziarz *et al.* 2017).

For Marsh Tits the overall reproductive success is usually around 70–80 per cent, meaning that for every five eggs laid, approximately three or four result in fledglings (Table 6.6). Notable extremes are seen in nest-boxes, with success as low as 46 per cent in a Korean study and as high as 98 per cent in Sweden. Across all nests, the number of fledglings produced from successful broods is typically 7–8 per nest, which reflects the very high hatching rate and low chick mortality for broods that escape predation. Some partial predation and starvation is usually responsible for the observed brood reduction, where the average number of fledglings is lower than the average clutch size (Wesołowski 2017, 2023).

The relatively low levels of total nest failure mean that the proportion of successful Marsh Tit nests producing at least one fledgling is around 70–84 per cent for most studies (Table 6.6). Interestingly, nest failures in the declining British population are lower than in the more stable population at Białowieża, which is a good indicator that nest failure is not a driver of the decline in Great Britain. The productivity of British Marsh Tits has been quite stable throughout the sharp decline in national abundance since the 1970s, with an average 5–6 fledglings produced from all breeding attempts when pooling successful and failed nests (Siriwardena 2006).

Willow Tits also have a high reproductive success in most studies, with 78–90 per cent of eggs producing fledglings (Table 6.7). In northern England, however, the reproductive success in natural cavities is much lower, due to a high rate of nest failure. These failure rates of British Willow Tit nests in recent studies are double those of Marsh Tits, although there is still some uncertainty over their significance as a factor in the Willow Tit's population decline (Siriwardena 2004). Where Willow Tits are successful, the number of fledglings produced is quite high, at around seven per nest. There is no clear trend over time in the average number of fledglings produced from breeding attempts in Great Britain, although small sample sizes hamper the analyses (Massimino *et al.* 2023).

Causes of breeding losses

There are multiple reasons for the loss of eggs or nestlings, or the total failure of a breeding attempt, but these can differ between the two species and their locations. Compared to most Marsh Tit nests in living wood, the vast majority of Willow Tit nests in deadwood are more vulnerable to predation or simply collapsing. Total failure can also result from events beyond the nest itself, such as poor weather reducing foraging success, or the death of one or both parents. Some causes of nest failure leave obvious field signs that can identify the reason, including the nest-tree falling down, a predator's mode of attack or the ingress of water, but in some cases the cause is a mystery.

It is also important to bear in mind that reported rates of nest failure in Tables 6.6 and 6.7 are usually minimum values. This is because losses in the early nesting stages are less likely to be detected, with failure occurring before the nest could be found by a researcher. Nest studies are therefore generally biased towards more successful nests that last long enough to be monitored. There are increasingly sophisticated statistical methods to deal with this bias (Weiser 2021), but for many studies only the number of nests observed and the proportion of failures are known, meaning they need to be treated with caution. Nevertheless, for some intensive studies of colour-marked populations the breeding status of every pair is sometimes known, and this can provide very accurate information for nest losses. Reasonable or very good data for causes of nest failure in some selected studies are shown in Table 6.8.

Table 6.8. Causes of nest failure and the percentage of all nests affected in major studies of Marsh Tits and Willow Tits from Monks Wood (England), Białowieża National Park (Poland), various sites in northern England, and Oulu (northern Finland).

Cause of nest failure	Marsh Tit (% of nests)		Willow Tit (% of nests)	
	Monks Wood	Białowieża	Northern England	Oulu
Competition/usurping	1.2	0	8.8	2.6
Desertion/adult died	3.7	4.6	0	5.3
Brood starvation	0	0	0	5.9
Flooding/soaking	1.8	2.6	0	0
Wind/collapse	0	0	0	0.7
Human forestry/destruction	0	0	0.4	1.3
Unknown	0	0	3.2	0
Great Spotted Woodpecker	2.5	4.3	17.3	4.6
Eurasian Jay	0	—	0.4	0
Forest Dormouse	—	1.9	—	—
Mice *Apodemus* spp.	1.2	?	?	0
Grey Squirrel	1.2	—	5.6 *	—
Red Squirrel	—	?	—	0.7
Eurasian Badger	0.6	0	0.4	0
Common Weasel	4.3	?	0.4	0
Total avian predation	2.5	4.3	17.7	4.6
Total mammalian predation	7.4	18.2	6.8 *	0.7
Total predation	9.8	23.2	27.7	6.6
Total nests monitored	163	1,051	249	152

A small minority of nest-boxes are included in calculations, excluding Białowieża. Only nests that reached the egg-laying stage are included. '?' denotes unknown proportions, '—' denotes that the predator does not occur in the study area. The Grey Squirrel predation marked * in northern England was unconfirmed, and some or all may have been Great Spotted Woodpeckers instead. *Apodemus* mice are Wood Mouse and Yellow-necked Mouse.

Sources: Monks Wood: Broughton *et al.* 2011 plus additional data; Białowieża: Wesołowski 2023; northern England: pooled from Parry & Broughton 2018, Lewis *et al.* 2009b, Rustell 2015 plus additional data; Oulu: Orell & Ojanen 1983.

NEST DESERTION

Marsh Tits and Willow Tits occasionally abandon their eggs or chicks if they feel that their own survival is seriously under threat. These threats might involve a very near miss with a predator or insufficient food to support a brood of chicks as well as themselves. Only once in Monks Wood did I ever see a pair of Marsh Tits abandon their nest, when a female gave up incubation in the very wet and cold spring of 2012 during terrible foraging conditions. In the Białowieża study, however, desertion accounted for a quarter of clutch losses and a tenth of brood losses (Wesołowski 2023). The likelihood of desertion was higher in colder weather, suggesting that the adults were struggling to find food and balance their energy needs.

In Scotland, there is a record of nest abandonment by Willow Tits due to unexpected heavy snow in May that covered the nest (Maxwell 2001). Abandonment was common in a Finnish study, accounting for 35 per cent of Willow Tit egg losses and 29 per cent of nestlings, totalling around 5 per cent of all nest failures (Orell & Ojanen 1983). These losses seemed to be weather-related in the boreal climate, which contrasted with no desertions recorded in the milder climate of northern England (Parry & Broughton 2018).

Nest desertion is usually obvious, because the parents stop coming and the eggs or chicks die, with no sign of disturbance by a predator. If the male parent disappears then the female can rear the young alone. However, if the female disappears before the eggs hatch or the chicks are fully feathered then the nest will fail, as the male cannot incubate eggs or brood nestlings until they can thermoregulate on their own.

A problem with identifying desertion is that parents often disappear because they have been killed elsewhere in the territory. Unless the parents are colour-ringed and seen again after the nest has failed, then genuine abandonment is hard to confirm. At Monks Wood, for example, 19 per cent of Marsh Tit nest losses were due to females disappearing and presumed killed because they were never seen again, as opposed to a nest confirmed as deserted when the parents were observed afterwards. Most of the birds killed away from the nest are probably caught by Eurasian Sparrowhawks, at least in temperate woodlands, where this predator eats a significant number of breeding tits (Geer 1978). At one nest in Monks Wood a female Marsh Tit was killed when the chicks were only a week old, but the male kept bringing food even as the chicks died from hypothermia over the course of a warm, sunny morning. It was tragic to watch the confused male continuing to bring caterpillars to the nestlings even after they had all died and lay still in the nest.

Females are sometimes abandoned by males to rear the chicks on their own. This can happen when the male is polygynous and breeding with two females. He will abandon the secondary female as soon as the chicks have hatched in the nest of his primary female, usually his long-term mate. Lone females can manage after being deserted by their partner, however, and one such female Marsh Tit at Monks Wood reared six fledglings after being abandoned by her polygynous male during incubation (Broughton 2006).

NEST SOAKING AND FLOODING

After heavy or persistent rainfall a nest-cavity can become soaked or fully flooded as a result of water running down tree stems (stemflow) and into the cavity, or seepage from the cavity walls. This is rarely a problem for Willow Tits nesting in dry deadwood and stumps, but for Marsh Tits it can cause nest failure.

The absorbent moss base of Marsh Tit nests gives them some buffering from water ingress into the cavity, but this can be overwhelmed by the amount of rainfall. In Białowieża, soaking or flooding was the second most important cause of nest failure, after predation (Wesołowski 2023). In some years a sudden downpour or a day of persistent rain resulted in up to 11 per cent of Marsh Tit nests being soaked and lost (Wesołowski et al. 2002). Overall, soaking and waterlogging accounted for 8 per cent of total clutch losses and a further 8 per cent of full brood losses in the Białowieża population, but partial losses of nestlings in other broods also occurred.

Nest soaking or flooding affected a few Marsh Tit nests at Monks Wood, accounting for 11 per cent of total nest losses (Table 6.8). In one case a nest in an upward-facing 'chimney' cavity was slowly filling with rainwater, and I saw the small chicks climbing onto the back of the brooding female to try and escape the rising water. There is a remarkable record from Białowieża of a single Eurasian Nuthatch nestling that survived a similar ordeal by climbing onto the bodies of its dead siblings and using them as a raft (Wesołowski *et al.* 2002). The likelihood of extreme weather and intense rainfall is increasing due to climate change, which will elevate the risk of these kinds of nest losses.

NEST PREDATION

The single biggest cause of nest failure for both Marsh Tits and Willow Tits is predation (Table 6.8). Under the near-natural conditions of the Białowieża National Park, where a full range of native predators occur at natural densities, predation accounts for three-quarters of Marsh Tit nest losses and almost a quarter of all nests. At Monks Wood, nest predators were responsible for 59 per cent of Marsh Tit nest losses but less than 10 per cent of all nests. In the German study at Pfrunger Ried, all of the nest losses of both species (except for those caused by competition) were due to predation (Ludescher 1973). This study was unusual, however, in finding 55 per cent nest losses for Marsh Tits, many of which were nesting in old Willow Tit holes in vulnerable deadwood. The nest losses for Willow Tits in this study were even higher, at 61 per cent. Willow Tit nests in Great Britain also have high losses during the incubation and chick stages, which are almost all due to predation (Tables 6.7 and 6.8). In studies across northern England, predators were responsible for most losses of eggs and chicks, and the nest predation rates in Great Britain are much higher than for Willow Tits breeding in Fennoscandia (Table 6.8).

Across all European studies, one of the main predators of Marsh Tit and Willow Tit nests is the Great Spotted Woodpecker (Figure 6.25). This woodpecker is a particularly serious nest predator of Willow Tits in western and central Europe, being responsible for the majority of nest predation. Losses to woodpeckers in some studies can represent around a quarter of all Willow Tit nesting attempts and over half of all clutches and broods being monitored. Great Spotted Woodpeckers are usually less of a problem for Marsh Tits, because they nest in the harder wood of living trees, which are safer than Willow Tit nests in deadwood. Wooden nest-boxes also seem to be more vulnerable to woodpeckers, possibly because they are more obvious than natural cavities (Nilsson

Figure 6.25. A Great Spotted Woodpecker in Norfolk: a major predator of cavity-nesting birds such as Marsh Tits, and especially of Willow Tits. (© www.garthpeacock.co.uk)

1984). However, Great Spotted Woodpeckers are well able to break into natural nests in living trees, and they often do so. In the Białowieża National Park the rate of nest failure increased with the density of Great Spotted Woodpeckers (Wesołowski 2023), and at Monks Wood Great Spotted Woodpeckers sometimes spent a lot of effort breaking into nests in thick living trees, but they only destroyed a small fraction of nests (Table 6.8).

The other main nest predators of Marsh Tits and Willow Tits are mustelids, especially Common Weasels, Pine Martens and Stoats, and very occasionally Eurasian Badgers. At Monks Wood, Common Weasels were the most significant predators of Marsh Tit nests, but this was still only a small proportion of breeding attempts. Particularly high weasel predation has been recorded for Marsh Tits in nest-boxes at Wytham Woods, where 18 per cent were depredated in one study (Dunn 1977). Abundant nest-boxes occupied by various tits may have given the weasels a 'search image' for a boxy structure that would be rewarding to investigate. Nests in natural cavities are far more cryptic and so a hunting weasel has to work much harder to find a nest in the thousands of available trees and shrubs.

Around three-quarters of Marsh Tit nest predation in the Białowieża National Park is attributed to mammals, with Pine Martens and Common Weasels thought to be the main species (Wesołowski 2002). Another major predator in Białowieża is the Forest Dormouse (Figure 6.26), which accounted for at least 16 per cent of clutch predation and another 6 per cent of brood mortality (Wesołowski 2023). Other rodents recorded or suspected as occasional nest predators in Europe include Red Squirrels, Grey Squirrels, Bank Voles, Wood Mice and Yellow-necked Mice. These species seem to be opportunistic rather than active hunters, destroying only a few nests.

Predation by non-native Grey Squirrels has been suggested as a potential factor in the decline of British woodland birds (Fuller *et al.* 2005). However, studies of Marsh Tits, Willow Tits and other species have since shown that Grey Squirrels are not very important nest

Figure 6.26. A Forest Dormouse, one of the most important predators of small cavity-nesting birds in the Białowieża Forest. (© Grzegorz Hebda)

predators (Broughton 2020). On the two occasions when squirrels destroyed Marsh Tit nests in Monks Wood they seemed to be interested in the cavity rather than predation, as they pulled out the nest contents but didn't seem to eat anything. In the Wigan study, Grey Squirrels were initially suspected of being significant predators of Willow Tit nests (Parry & Broughton 2018), but nest-cameras failed to show any attacks at all (Table 6.8). As far as Marsh Tits and Willow Tits are concerned, Grey Squirrels do not seem to be any more dangerous than the native Red Squirrels that they have replaced across much of Great Britain.

In east Asian populations, the predators of Marsh Tit and Willow Tit nests include the Large-billed Crow, Eurasian Jay, Black-billed Magpie, Azure-winged Magpie and Red Squirrel (Nomi *et al.* 2017, Yoon *et al.* 2017). In a Chinese study some nests were depredated by mice, but most attacks were attributed to Korean Rat-snakes or Steppe Rat-snakes (Zhang *et al.* 2020).

Insects can sometimes cause nest failure when ants, bumblebees or wasps take over the nest-cavity. This competition is far more common in nest-boxes than in natural cavities, probably because of the drier microclimate. Tree Bumblebees, Saxon Wasps and Common Wasps are the species most frequently involved in causing nest desertions by tits in Europe (Broughton *et al.* 2015b), and this has been noted in the Wigan study when Tree Bumblebees evicted nest-building Willow Tits from a natural cavity. Nestlings may also be killed by ants invading the nest, and this has been recorded for both Marsh Tits and Willow Tits (Ludescher 1973, Markovets & Visotsky 1993, Stewart 2010).

PARTIAL NEST LOSSES

Nest predation is not binary, and if a predator finds a nest it does not always take all of the contents, or not in a single episode. Quite often only some of the eggs or chicks are taken, and the predator may return another time, or the remaining nestlings can survive to fledge. These *partial predation* events are difficult to detect in standard nest-monitoring, as it is not always obvious what has happened between visits.

In many cases of partial predation the only sign will be missing eggs or chicks, and virtually every study of Marsh Tits and Willow Tits shows a reduction in the total number of nestlings across all successful nests compared to the number of eggs that were laid or hatched (Tables 6.6 and 6.7). In other words, many nests lose some eggs or nestlings. It is difficult to know whether these are losses from partial predation or because the eggs and chicks died from other causes and were removed by the parents. At Białowieża, however, years with lots of partial brood losses are correlated with years of increased nest predation, implying that predators are involved in many partial losses (Wesołowski 2023).

For Marsh Tits at least, Wesołowski (2017) calculated that 30 per cent of nests that were attacked by a predator actually survived, either partially or completely. Older nestlings are more likely to survive an attack, as they are able to hunker down in the nest material and hide. Nestlings can also give a hissing or growling call to deter a predator (Broughton 2005), and older nestlings from 16 days of age may jump out of a nest that's under attack, effectively fledging early.

Some predators are more lethal than others, with Marsh Tits in Białowieża surviving 40 per cent of Great Spotted Woodpecker attacks but none by Forest Dormice. Nestlings can survive woodpecker attacks because they are removed one at a time before the woodpecker kills them, so the attack may end before the whole brood is taken. For Forest Dormice, however, and also for Common Weasels and Stoats, the mode of attack is to enter the cavity and kill everything before removing them, or eating them *in situ*, so it's very unlikely that any eggs or nestlings will survive.

Besides predation, the other major causes of partial nest losses are chilling and starvation. Some eggs may fail to hatch in poorly insulated cavities, and in poor weather the weaker or younger nestlings may get insufficient food and die. In studies of Marsh Tits in Białowieża and Willow Tits in Finland, increased nestling mortality was associated with lower spring temperatures (Orell & Ojanen 1983, Wesołowski 2023).

IDENTIFYING NEST PREDATORS

The attacks by individual predators can often be identified by distinctive traces that they leave behind. Great Spotted Woodpeckers characteristically make a new hole at the level of the nest to pull out the material and take the eggs or chicks, also leaving distinctive peck marks in the wood (Figure 6.27). Sometimes, however, the woodpecker just hacks open the entrance and whole front of the cavity. Great Spotted Woodpeckers also have a grisly habit of wedging nestlings into a nearby nook and pummelling them with their bill to break them apart, and these remains can sometimes be found around the nest (Figure 6.28).

Common Weasels can climb to any height to enter very small holes without damaging them. Once inside they will eat eggs, nestlings and, if they can, catch the brooding female, leaving a mess of blood, feathers, body parts and often a telltale scat inside. Stoats do the same when the hole is large enough to enter, as they can also climb well, as can Pine Martens, who will bite and pull apart the nest-cavity to access the nest-chamber (Figure 6.29). Meanwhile, Eurasian Badgers can access low nests by simply ripping open decaying wood to leave a gaping hollow and distinctive claw marks. Mice can enter small entrances to eat eggs or nestlings inside, and

Figure 6.27. Great Spotted Woodpecker attacks on nests of a Marsh Tit (left) and Willow Tit (right). The woodpeckers have made new holes below the original entrances at the level of the nests to get to the contents. The Willow Tit nest has had the nest and eggs pulled out. (Marsh Tit © Richard K. Broughton; Willow Tit © Allan Rustell)

Figure 6.28. A Marsh Tit nestling taken from its nest and killed by a Great Spotted Woodpecker, which has wedged it into a crevice to peck and break apart. (© Richard K. Broughton)

Figure 6.29. A Willow Tit nest in a birch snag in the Oulu study that has been depredated, probably by a Pine Marten. The front of the nest has been ripped out from the lower rim of the entrance hole, and the contents removed. (© Markku Orell)

will often bring in new nest material of their own, such as dead leaves. Squirrels access nests by neatly enlarging the entrance and leaving distinctive chisel-like tooth marks, pulling out the nest material but often not eating anything (Broughton 2020). In eastern Europe and western Asia, Forest Dormice are able to enter small nest-holes to eat eggs, nestlings and brooding females, similar to Common Weasels, although they are distinctive in favouring the brains and often leaving more remains than weasels, plus depositing their own characteristic scats inside (Wesołowski 2017).

In many cases of nest predation the culprit is not clear. This is especially true for nests in deadwood, which can get so severely damaged that field signs are obliterated. For a low Willow Tit nest in a dead birch stump, for example, a cavity pulled open into small pieces could be the work of a woodpecker, marten, badger, squirrel or even cat. Field signs are better preserved in solid live wood, where the marks made by teeth, claws or a bill can still be seen around a semi-intact nest-cavity.

FEMALES DEPREDATED IN THE NEST

Total nest losses due to predators often result in the death of the female parent too. Females can get trapped inside the cavity when a predator enters or breaks in during incubation, or in the first 10 days after hatching, when she spends lots of time on the nest. Small predators entering through the nest-hole are the most dangerous, especially Common Weasels and Forest Dormice, as the female has little chance of escape.

In the Białowieża National Park, female Marsh Tits were killed in approximately half of all nest predation events during the incubation period, which was very likely due to Forest Dormice and small mustelids (Wesołowski 2023). At Monks Wood around a third of weasel predation on Marsh Tit nests also resulted in the female being killed. The main defence that females have against being trapped inside the nest-cavity by a predator is to give the *hissing display* (Zhang *et al.* 2020; see Chapter 2). Females of both species give this display, which involves a loud hiss-like call with a broad frequency, like white noise, accompanied by a flap of the wings against the cavity walls and a lunge with the bill that closes with a snap. The result is a blast of noise and air coming from the cavity, which is meant to startle the predator and make it hesitate or back off, perhaps giving the female a chance to escape. All tits and chickadees give the hissing display, and it is familiar to researchers, who get the full treatment when looking inside nests. Even when you know it is probably coming, the force of the hissing display can still take you aback. The hissing display can really work against predators, too. At Monks Wood an exploring Stoat was watched as it poked its head into a nest with an incubating Marsh Tit inside, and the hissing display from the female was enough to make the Stoat beat a hasty retreat. The nest went on to be successful. Zub *et al.* (2017) showed that Yellow-necked Mice also react negatively to the hissing call.

Females have another tactic when trapped inside the cavity with their nestlings, which is in stark contrast to the apparent bravery of the hissing display. Instead of threatening to attack, the female hides beneath the nestlings and uses them as a shield (Figure 6.30). The aim is probably to make it likely that the predator will take a nestling or two instead of her, and she may survive the attack. It is certainly a more cynical ploy than the hissing display, but ultimately the female's own life is more valuable than those of her nestlings, as she may live to breed again.

Figure 6.30. A female Marsh Tit hiding beneath her nestlings during a nest-box check in the Lund study, in Sweden. The female's long tail can be seen poking out at the top left of the image, but otherwise she is completely hidden as an anti-predator strategy. (© Johan Nilsson)

Nest defence

Aside from hiding or giving the hissing display if trapped inside the nest, adult Marsh Tits and Willow Tits have other anti-predator strategies if they return to the nest from a foraging trip to find it under attack. Both birds will intensively mob the predator with constant alarm calls, distraction displays and even direct attacks. These behaviours become more aggressive as the nestlings age, reflecting the parents' increasing investment in the chicks and the rapidly diminishing time to try and breed again if the nest fails. Around half of the adult population in both species can only expect to survive for one breeding season (see Chapter 8), so this will be their only chance to breed, and they do not give up easily.

The *nest-site distraction display* has been well described for Marsh Tits, Black-capped Chickadees and to a lesser extent Willow Tits (see Chapter 2), and it appears to be very similar for all *Poecile* species (Rytkönen *et al.* 1990, Smith 1991, Broughton 2012b). On finding a predator at or very near the nest, the adults will approach within a few metres and alarm-call intensively, switching perches in a very agitated manner. This can build to a spectacular distraction display where the bird leans forward, fans its tail and lifts its wings, sometimes slowly waving them, while swaying from side to side and periodically producing a loud hissing call. If that fails the bird may dive right at the predator's head, making direct contact with a peck of its bill, before returning to its perch, from where it may attack again in quick succession. At Monks Wood I recorded a Marsh Tit attacking a taxidermy weasel like this, hitting it with so much force that it knocked it over (Broughton 2012b). In the Wigan study

Willow Tits physically attacked a predatory Great Spotted Woodpecker and also a Blue Tit trying to take over the cavity, saving their nest on both occasions. Ludescher (1973) saw a female Willow Tit attacking a Great Spotted Woodpecker by diving at it while giving a loud 'squawk', and also attacking Great Tits near the nest.

Failed territories and non-breeders

As well as nest failure there is another, more cryptic, group of birds in the population that also fail to breed. These are territorial birds that do not reach the nesting stage because they are unpaired, or because their mate has been killed just before breeding, or because they have both survived but have decided not to breed. These birds are easily missed if monitoring relies only on nest-finding, and many studies don't quantify this portion of the population.

At Monks Wood the non-breeders accounted for 12 per cent of the Marsh Tit spring territories (Broughton *et al.* 2011). Most non-breeders were unpaired males, a few were unpaired females, some had been divorced at the start of spring, but around a quarter of non-breeding territories involved pairs where one bird was killed just before nesting or egg-laying. Similar results were found in a Finnish population of Willow Tits, where an average 11 per cent of males and 6 per cent of females were non-breeding each spring (Orell *et al.* 1994b). As at Monks Wood, most of these non-breeders were unpaired young males, with some unpaired females, divorcees and recent widows.

Unlike the Monks Wood Marsh Tits, however, some non-breeding Willow Tits in Finland were pairs that had deliberately decided to skip breeding, particularly in years that were productive for the wider population. This seemed to involve mostly young, low-quality females deciding not to invest in rearing offspring that would be unlikely to be competitive among large numbers of other juveniles. Instead, the females prioritise their own survival to produce more successful young in the future, when competition among juveniles might be less intense. Non-breeding males in this population also had a higher annual survival than breeding birds, so it can sometimes pay to wait out the challenging years.

There are no records of Marsh Tit pairs deliberately skipping breeding, nor of Willow Tits doing so in temperate regions like Great Britain. In these environments any non-breeding is more likely to be due to the repeated loss of nest excavations to usurping Blue Tits, until the Willow Tits run out of time to breed (Maxwell 2002, Parry & Broughton 2018).

Summary

Marsh Tits and Willow Tits breed in the northern spring and are exclusively or overwhelmingly single-brooded. Willow Tits are primary cavity-nesters that excavate a nest-chamber in standing deadwood, but occasionally use or extend an existing cavity. Marsh Tits are secondary cavity-nesters that rely on existing holes in living trees and shrubs, which they may partially excavate. Nest-sites are plentiful in most habitats, but may be scarce in heavily managed forestry.

Nest-boxes cannot increase populations, as the birds are usually limited by the area of available woodland for their territories, and not by the number of nest-sites. Providing nest-boxes as a conservation tool is likely to be counterproductive in temperate regions, such as Great Britain, where they attract and benefit competing Blue Tits rather than Marsh Tits or Willow Tits. Maintaining or increasing standing deadwood and expanding diverse habitat is a better approach for Willow Tit conservation.

Populations are highly synchronised in their breeding, with most birds in a given locality beginning to lay eggs within 10 days of each other. The timing of breeding is affected by spring weather and the development of woodland foliage and invertebrate prey. Birds in northern latitudes breed later than southern birds, and the timing of breeding is advancing due to climate change. A typical clutch for both species is 7–9 eggs, which are incubated by the female for around two weeks.

The nestlings are fed by both parents for around three weeks until they fledge, after which the family remains together for another week or two until the fledglings become independent and disperse. Caterpillars are the major prey fed to nestlings of both species, but Willow Tit nestlings have a more diverse diet than Marsh Tits.

Nest failure rates are relatively low for Marsh Tits, who may attempt to breed again if the first nest is lost at an early stage. Nest failure rates for Willow Tits in temperate regions, such as Great Britain, are much higher than in boreal regions, and far higher than for Marsh Tits, largely due to increased nest-site competition by Blue Tits and predation by Great Spotted Woodpeckers. Individual pairs of Willow Tits may lose multiple nesting attempts each spring. Genuine second broods have been recorded for boreal Willow Tits after a successful first brood, but not in Great Britain.

Productivity in surviving nests is usually high, with most eggs resulting in fledglings. Marsh Tits and Willow Tits are very committed parents, and will try to defend their nests against attackers much larger than themselves. Parents are occasionally killed while breeding, either inside the nest or elsewhere in the territory, and this can be a cause of nest failure. Nest abandonment is uncommon, but some northern Willow Tits do not breed in years when there are low chances of their offspring joining the local population.

CHAPTER 7

Dispersal and movements

Marsh Tits and Willow Tits are resident and sedentary, meaning they do not migrate or undertake regular movements between different breeding and wintering sites. Nevertheless, both species do undertake significant movements during their lives, which fall into the category of *dispersal*. Dispersal is the permanent movement of a bird from a place where it has been resident for a while, to another place where it will settle for an extended period. Unlike migration, which is a seasonal movement back and forth, dispersal is usually a permanent, one-way movement. However, birds can sometimes disperse several times, and they can even disperse back to where they came from. Some movements of Willow Tits also fall into the category of *irruptions*, which are an extreme form of irregular mass dispersal. Irruptions are not annual but occur in some autumns in northern latitudes in response to high numbers of juveniles and/or a failure of the forest seed crop.

The major movements undertaken by Marsh Tits and Willow Tits are *natal dispersal* and *breeding dispersal*. Natal dispersal is the permanent movement of a juvenile from the place where it was hatched and raised to the place where it first breeds as an adult. Breeding dispersal is the movement of an adult between two breeding seasons, usually between nest-sites or breeding territories in consecutive years. Natal dispersal tends to involve much greater distances than breeding dispersal, as once a bird is settled in a breeding area it is likely to remain nearby for the rest of its life, having found a place that supports its needs.

Dispersal has been quite well studied in Marsh Tits, but there is surprisingly little information for Willow Tits. Despite this, there is a general pattern of dispersal among the *Poecile* species, and the additional studies of Black-capped Chickadees and Siberian Tits can help to fill in some of the gaps for Willow Tits in particular (Virkkala 1990, Smith 1991, Orell *et al.* 1999a). The available information gives a fairly good understanding of the drivers and mechanisms of dispersal in *Poecile* species, which is characterised by relatively short distances (only a few kilometres for most birds), narrow time windows during the year and a great sensitivity to habitat fragmentation. These factors can have major implications for Marsh Tits, Willow Tits and their conservation.

Natal dispersal

The biggest movement in the life of a Marsh Tit or Willow Tit is natal dispersal, which juveniles undertake within only a few weeks of leaving the nest. Natal dispersal is often the most challenging period of their lives, and many juveniles will not survive it, especially in very fragmented habitat. A few birds may appear in unusual places, tens or even hundreds of kilometres out of their usual range, sometimes reaching the coast.

Studying natal dispersal in Marsh Tits relies on ringing nestlings, preferably with colour-rings to allow later observation in the field without recapturing. This is because fledglings are virtually impossible to target for catching during their first weeks out of the nest, when they are mobile in the tree canopy, and they disperse soon afterwards. So the only opportunity to ring young birds before they disperse is when they are still in the nest (Figure 7.1). To know where

Figure 7.1. A Marsh Tit nestling in the Białowieża Forest that has been safely extracted from its nest for ringing, before being quickly returned. Only by ringing birds at this age can natal dispersal be studied. (© Richard K. Broughton)

Figure 7.2. A Willow Tit nestling briefly extracted from its nest for ringing in the Oulu study, Finland. Note the mosquitoes! (© Anne Laine)

a ringed nestling has later dispersed, it must also be re-caught or re-sighted within its first year, before it has finished breeding for the first time. For a reasonable study of natal dispersal it is necessary to find most nests within a large study area, colour-ring lots of nestlings and then search intensively across a wide area to find the juveniles after dispersal during their first year.

There are several major studies of Marsh Tit dispersal. In Sweden, the landmark research from Lund was critical in revealing the key drivers and consequences of natal dispersal (Nilsson & Smith 1985, 1988a, Nilsson 1989a, 1989b). Other important studies have come from Basel in Switzerland (Amann 1997), Monks Wood in Great Britain (Broughton *et al.* 2010) and the Białowieża National Park in Poland (Wesołowski 2015). Together, these studies have given us a good understanding of the Marsh Tit's behaviour immediately after fledging and around natal dispersal, and how this is affected by variables such as parental behaviour, individual quality and the woodland environment they live in.

Willow Tits are far less studied than Marsh Tits in the post-fledging period, and natal dispersal distances are still quite poorly understood, with only a few major studies. As with Marsh Tits, dispersal studies are only possible by ringing nestlings before fledging, which requires a lot of effort and skill to find the nests and safely extract the chicks (Figure 7.2). Almost all available information comes from Fennoscandia, especially from Oulu in Finland (Orell *et al.* 1999b), with more limited studies from Norway (Hogstad 1990b, Haftorn 1997b). Further inferences of local settling after dispersal come from Sweden (Ekman 1979a, 1984). In Great Britain there is very little published information from the post-fledging and dispersal periods, with a single study of two fledged families at Wytham Woods (Foster & Godfrey 1950) and dispersal distances for a few juveniles in Lanarkshire (Maxwell 2002).

DRIVERS OF NATAL DISPERSAL

Marsh Tit families generally remain together for 8–16 days after the fledglings leave the nest, and Willow Tits for around 12–21 days (Figure 7.3). During this time the fledglings develop physically, vocally and behaviourally, until they are ready for independent living (see Chapters 2 and 6). Eventually, the fledglings decide that it is time to leave the family group, and they disperse. This can occur quite suddenly and involve all of the fledglings in the family leaving on the same day, or it might take up to six days for all of them to go (Nilsson & Smith 1985, Hogstad 1990b). Once the fledglings have left their parents behind in the natal territory they have become dispersing juveniles.

Is leaving the parental territory the choice of the fledglings themselves, or do their parents drive them away when they think the time is right? Nilsson & Smith (1985) investigated what causes Marsh Tit fledglings to disperse from the family group, and Haftorn (1997b) looked at similar questions for Willow Tits. Both studies found very low levels of aggression between parents and fledglings, but Willow Tit siblings showed increasing amounts of aggression towards each other in the days leading up to dispersal, including sporadic chasing and supplanting attacks. At Monks Wood I also observed some aggression between Marsh Tit siblings, with agitated calling and posturing, but there was no obvious escalation towards the time of dispersal. In Black-capped Chickadees, Holleback (1974) did notice an escalation of posturing, calling and chasing between siblings, and from parents towards the fledglings, in the days before family break-up. However, Weise & Meyer (1979) did not see any increased aggression among Black-capped Chickadee families, and concluded that the fledglings' decision to disperse was innate. To look at this in more detail, Nilsson & Smith (1985) experimentally adjusted the size of Marsh Tit families to see if this increased conflict, but found no effect on the amount of aggression or the timing of dispersal. Given these findings, sibling aggression is probably not the major impetus for family break-up and dispersal, and parents do not seem to chase their youngsters out of the territory when they think it is time for them to leave. It would actually be a poor strategy for parents to force their offspring to disperse before they were ready. Instead, it is the fledglings that appear to decide when it is time to leave the natal territory and become independent juveniles. The trigger for this transition seems to be the ability to fend for

Figure 7.3. A young Marsh Tit in Monks Wood in mid June, approximately 10 days after fledging, but almost ready to disperse from its family group just out of sight in the bushes. (© Richard K. Broughton)

themselves by acquiring foraging skills and a sufficient vocabulary of calls and song. Becoming competent in these skills takes at least a week or two (Haftorn 1993a, Broughton 2019), but once they have been mastered the fledglings disperse as soon as they can.

The juveniles have a driving urge to rapidly find and acquire their own home-range and a partner of the opposite sex, before other competing juveniles get there first. The classic Marsh Tit studies at Lund showed that hatching and fledging date, and the density of local juveniles fledging around the same time, were critical factors driving the timing, distance and success of natal dispersal (Nilsson & Smith 1985, 1988a, Nilsson 1989a, 1989b). In years with a high density of other dispersing juveniles, the competition for vacant home-ranges is severe, and leaving early is a major advantage. For juvenile Marsh Tits the pressure for early dispersal and rapid settling is particularly strong for males, as being quick gives them a better chance to find and secure a vacant home-range of their own. Interestingly, dominant juveniles of both species that are heavier or larger as nestlings tend to disperse and settle first (Nilsson & Smith 1985, Haftorn 1997b). This might explain any size-related dominance in later social groups, which could be due to the seniority of larger or heavier birds dispersing first, settling early and then being dominant over later arrivals (Smith 1991). Late-hatched and late-fledging males of both species have the lowest chance of becoming established or achieving a high dominance rank in the population, and their life chances are quite poor (Nilsson 1989a, Hogstad 1990b).

THE JUVENILE SETTLING PROCESS

During settling, the juvenile males try to locate a vacant area of woodland which they can occupy, and they advertise their claim by singing and calling at other males that might arrive or settle nearby. Among all *Poecile* species the juveniles have a curious respect for the rule of *seniority*, by which they recognise and accept that prior occupancy determines ownership

(Hogstad 1987, Nilsson 1989b, Koivula *et al.* 1993). This rule works by the simple order of arrival: if a juvenile disperses into a patch of habitat and finds it already occupied, then the existing owner wins the confrontation. There will be lots of calling and posturing, and sometimes physical aggression, but the juveniles do not battle to determine if a stronger bird can displace a weaker one. Instead, prior occupancy and seniority establishes the dominance hierarchy among males arriving in any given home-range, and this rule is respected regardless of other factors, such as body size (see Chapter 5).

For juvenile females the competition to disperse early and become established in a home-range is less intense than for males. Early-hatched and early-fledging females are still more likely to be successful in finding a home-range after dispersal, and the seniority rule still applies, but they do have more leeway than males to assess the quality of a home-range, and also to assess any occupying males who got there first. This is because juvenile males will tolerate females that are passing through their home-range during dispersal. Indeed, the males are hoping that a female will join them to form a pair. Meanwhile, intruding males are chased away by settled males and females alike, so have less chance to assess habitat or mate quality (Nilsson 1989a, 1989b). Ultimately, this means that females have more opportunity to investigate the habitat quality, potential partners or winter flocks they might want to join (Nilsson 1989a).

Juvenile Willow Tits and Marsh Tits are probably under the same competitive pressures during dispersal and settling, but there are subtle differences in different parts of their ranges. In temperate woodland, Ludescher (1973) observed a rapid dispersal among both species at Pfrunger Ried in Germany, and settled birds generally remained in the same place for the rest of their lives. In Monks Wood and Switzerland, too, Marsh Tits quickly dispersed and settled into home-ranges and a basic flock structure (see Chapter 5). However, northern Willow Tits in Fennoscandia seem to have a slightly different strategy, first settling with adults or dominants in discrete flocks rather than in their own individual home-ranges. Before joining these discrete flocks, dispersing Willow Tits in northern latitudes may first spend a few weeks in loose temporary groups with other juveniles, and occasionally with adults or family groups that have not yet dispersed (Haftorn 1997b). These juveniles may be assessing each other and forming pairs, moving noisily around the forest with lots of antagonism. Crucially, during this period the juveniles come into contact with lots of other birds and multiple adult territories, before deciding whether to join them in a winter flock (Haftorn 1999). Aside from the German study (Ludescher 1973), information on Willow Tit dispersal and settling from temperate lower latitudes is severely lacking, but they probably behave more like the rapidly settling Marsh Tits from Monks Wood and Switzerland.

TIMING AND DURATION OF NATAL DISPERSAL

Most studies suggest that the actual movement period during dispersal is a rapid phase that many birds complete within a matter of days or weeks. At Monks Wood around 85 per cent of Marsh Tits had completed their full natal dispersal within the first couple of weeks after the family break-up (Broughton *et al.* 2010). For many birds, especially juvenile males, dispersal was completed within a few days, sometimes as little as a single day. During fieldwork I often found a bird within its family group one day, and the next day it would have dispersed a kilometre or more and be singing or calling in a new home-range, where it remained to breed the following spring.

In the Lund study in Sweden, Nilsson & Smith (1988a) showed that two-thirds of juvenile Marsh Tits were settled in home-ranges within 10 days of family break-up. The remainder settled soon afterwards. In Switzerland, Amann (1997) found that most juvenile Marsh Tits were paired and settled in home-ranges by the end of June, within a few weeks of dispersal. Bardin *et al.* (1992) recorded a regular surge of juvenile Marsh Tits dispersing along the Curonian Spit on the Baltic coast during late June and early July. Ringing recoveries showed that the median age of these dispersers was just 39 days, or about three weeks after fledging, with this surge of dispersing juveniles lasting around 10 days. A few birds appeared to disperse again in late September, corresponding to the end of the post-juvenile moult (Broughton *et al.* 2008b). We observed a similar phenomenon at Monks Wood, where some settled juveniles that were moulting during August or September later moved several kilometres away by October or November. It seems that some juveniles settle only long enough to complete their post-juvenile moult in late summer before later continuing to disperse in the autumn.

At the end of winter there is another important period of dispersal, with a significant minority of young Marsh Tits moving just before the breeding season. At Monks Wood, 7 per cent of first-year males and 5 per cent of females undertook significant new movements to a different territory between February and April. Immigrants into the study area were also appearing during this period. These juveniles seemed to be leaving their winter home-ranges after being evicted by resident adults or dominant juveniles ahead of the breeding season. Other youngsters may be dispersing because they have been widowed or unpaired over the winter, and they are now trying to find a breeding partner (Amann 1997, Broughton *et al.* 2010). This reshuffling of first-years at the end of winter and into early spring was very important at Monks Wood, as the new arrivals comprised up to 22 per cent of the breeding population in some years (Broughton *et al.* 2010). These birds filled late vacancies in breeding pairs and empty territories that had appeared due to winter mortality, and without their last-minute arrival the annual number of breeding pairs would have been significantly lower.

Juvenile Willow Tits in Fennoscandia settled in discrete flocks within one or two months of the family break-up, by August or September, although some birds took as little as 10 days to disperse and settle (Ekman 1979b, Hogstad 1990b, Haftorn 1997b). Ekman (1979b) found that around half of the juveniles during late September and October were still assessing potential vacancies in winter flocks, or were floaters (see Chapter 5). At the end of the northern winter, in April, the discrete flocks break down into pairs that establish breeding territories in the vicinity. For those birds that survive to breed, therefore, the dispersal distance to the winter flock is approximately the same as the overall natal dispersal distance to their first breeding site. Some birds will move to breed within the range of a neighbouring winter flock, which will complete their full natal dispersal, and this usually involves moving up to an additional 1km or so (Ekman 1979b).

NATAL DISPERSAL DISTANCES

Individual juveniles are extremely difficult to follow during the dispersal phase, and it is likely that some ringed birds will leave the study area and travel beyond the limits of detection. The further a bird travels, the larger the potential search area and the less likely it is to be found by the researcher or somebody else. Even within a defined study area it is difficult to know whether a dispersing bird has left or simply died, with predation or starvation being

Figure 7.4. Colour-ringing nestlings at Monks Wood showed that, after dispersal, most of the juvenile Marsh Tits settled within 2km of their birthplace. (© www.garthpeacock.co.uk)

obvious risks for inexperienced young birds. Consequently, there is a potential for bias and underestimation of dispersal distances, especially those that involve small sample sizes or limited study areas.

Nevertheless, where most of the re-sightings of dispersed birds are well within the maximum search radius, with very few near the outer limits of detection, then this will indicate that a negligible number of birds have moved beyond the search area. Luckily, most investigations into Marsh Tit dispersal have had relatively large study areas, and so the results should be very reliable, as few birds will have been missed. In the Monks Wood study we found that the great majority of colour-ringed juveniles settled within a short distance of their origin (Figures 7.4 and 7.5), far short of the maximum detection distance (Broughton *et al.* 2010). Only a very small number of dispersed juveniles were found near the search limit, or were reported further afield by other observers. The fact that the settling Marsh Tits were restricted to the woodland patches within a relatively hostile arable landscape helped to narrow down the search for colour-ringed youngsters in the weeks and months after dispersal (Broughton *et al.* 2010).

A summary of the natal dispersal distances detected in various Marsh Tit studies is given in Table 7.1. Several things are evident from these figures, such as the similarity of dispersal distances in different populations and the short distances involved, typically within just a few kilometres. Also notable is that juvenile females consistently disperse further than males. Among all studies, the dispersal of nearly every juvenile took it outside of the natal territory in which it was hatched. This is probably an innate tactic to ensure that a bird doesn't end up breeding with one of its parents or another close relative that has also settled very close

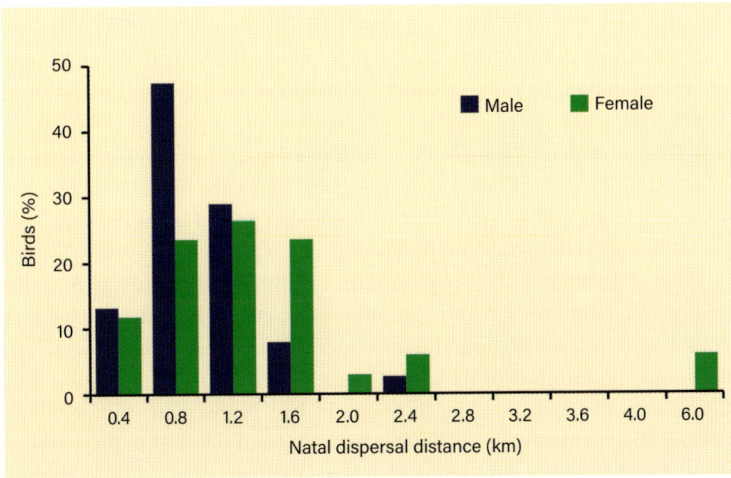

Figure 7.5. Natal dispersal distances of 34 female and 38 male Marsh Tits in the Monks Wood study. Note that the males are skewed towards shorter distances.

by. At Monks Wood, however, one bird did settle and overwinter in its natal territory after its parents had both disappeared, and it would probably have bred there if it had survived until the spring. Among the maximum dispersal distances, three of the Marsh Tit studies recorded birds moving 9–22km before settling, and these individuals were all detected by chance outside the main study areas.

Another way of looking at dispersal distances is by the number of territories that a juvenile has travelled across before it settles. This can be a better metric for comparison than standard distance measurements, as it controls for the gaps between woodland habitat in patchy landscapes, such as open fields. For the Marsh Tit studies in Table 7.1, juvenile

Table 7.1. Natal dispersal distances of juvenile Marsh Tits from European studies.

Study area	Dispersal distances (km)				Source
	Female	Male	Range	Extra-limital	
Monks Wood, England	1.3	0.8	0.2–5.8	10.5	Broughton et al. 2010, additional data
Lund, Sweden	1.4	1.0	0.2–7.3	—	Nilsson 1989a, Wesołowski 2015
Basel, Switzerland	1.1	0.6	0.3–1.6	2.2	Amann 1997
Lower Saxony, Germany	0.5	0.5	0.2–3.2	9.0	Berndt & Winkel 1987
Curonian Spit, Baltic coast	4.5	3.5	—	—	Markovets 2001
Białowieża, Poland	1.7	0.6	0.1–5.1	21.8	Wesołowski 2015

Values are means or medians, depending on the source. Extra-limital distances are those detected ad hoc beyond the study area. Additional data for Monks Wood gave total samples of 39 males and 34 females in 2004–2015.

females typically dispersed across 3–6 territories before settling, while males crossed 2–4 territories. These values are remarkably similar in different populations and underline that Marsh Tit dispersal distances are very short, whether the metric is territories or kilometres that were traversed.

For Willow Tits, almost all of the dispersal-distance data come from Fennoscandia. Despite relatively large search areas in these studies, they were generally insufficient to encompass all of the dispersing juveniles. Two studies from Finland reported natal dispersal between the nest-site and the first breeding location that averaged 1.6km for males and 1.7km for females (Orell *et al.* 1999b, Pakanen *et al.* 2016a). Another study found slightly greater distances, of 1.7km for males and 2.0km for females (Kumpula *et al.* 2023). Partial dispersal distances are available from other studies, detailing the movements between the nest in which the Willow Tit was raised and the location of its winter home-range, which is not quite the full natal dispersal distance to the first breeding location. In two Norwegian studies the averages of the partial dispersal distances were 0.8–0.9km for males and 0.9–1.2km for females (Hogstad 1990b, Haftorn 1997b), with a maximum of 1.5km. Pakanen *et al.* (2016a) showed that the distances moved at the end of winter from the home-range to the subsequent breeding site was only 0.4km, so a rough approximation of average natal dispersal would be around 1–2km in Fennoscandia. This was similar to the 1–2km dispersal distances for a handful of juvenile Willow Tits monitored in Scotland (Maxwell 2002), with the maximum distance for the Scottish birds being 6km.

Overall, most *Poecile* species seem to have broadly comparable natal dispersal distances, although Marsh Tits are probably at the lower end of the range. For Black-capped Chickadees the median natal dispersal distances have been reported as 1.1km (Weise & Meyer 1979), but some birds dispersed up to 11.2km. For Siberian Tits, median natal dispersal distances of 2.5–4.3km were reported by Orell *et al.* (1999a), with a maximum of 10.3km. Among other tit species, there are reported average natal dispersal distances of 0.6km for Coal Tits (Dietrich *et al.* 2003) and 0.5–1.1km for Great Tits, but these are known to be underestimates due to the limited study areas (Van Tienderen & Van Noordwijk 1988, Dingemanse *et al.* 2003, Szulkin & Sheldon 2008). Indeed, many juvenile Great Tits are known to disperse more than 2km from where they were hatched (Verhulst *et al.* 1997, van Overveld *et al.* 2014).

The available natal dispersal distances for Willow Tits in Fennoscandia are also acknowledged to be underestimates. This is not unusual, and underlines the difficulty of studying dispersal. Many studies have found that most of their ringed nestlings disappear from the study area after independence, only to be replaced by the arrival of many other unringed juveniles from outside. Ekman (1984) found that only 7 per cent of juvenile Willow Tits settling within a 6km² study site were born locally, meaning that nearly all of the local fledglings had dispersed beyond this area. Likewise, in Finland, Virkkala (1990) ringed 465 Siberian Tit nestlings but later found only a single bird had settled in the study area, which was 6km wide. In North America fewer than 14 per cent of dispersing Black-capped Chickadees born within study areas of 65–800ha actually settled within them, with most presumed to have dispersed beyond the study limits (Smith 1967, Weise & Meyer 1979).

Within this wider context, and despite the information being quite patchy and incomplete, it is clear that the natal dispersal distances of most juvenile Marsh Tits and Willow Tits are within a couple of kilometres of where they were hatched. Some birds do disperse over greater distances, but this is uncommon, and very few birds appear to disperse over 10km.

SEX BIAS IN NATAL DISPERSAL

There is a clear sex bias in natal dispersal, with juvenile females on average dispersing further than males (Table 7.1, Figure 7.5). This phenomenon is true of all tits and chickadees, and also many other birds (Smith 1991, Clarke *et al.* 1997). The drivers of this sex-biased natal dispersal could be competitive pressure on males to settle quickly in the first suitable place they find. Females can disperse further because they are under less pressure than males when assessing settling opportunities, so they can take their time (Nilsson 1989a). This fits with the suggestion by Greenwood *et al.* (1979) for juvenile Great Tits, where females disperse further than males because they move for a longer period before settling. Another explanation for the sex bias in dispersal distances is that the females are avoiding inbreeding with their male siblings, and so they instinctively try to move further away from their nest of origin (Pusey 1987). This makes sense, as if sisters disperse further than their brothers then they won't end up settling and breeding nearby.

One more way to avoid settling in the same places as your siblings could be to disperse in a different direction. There is no evidence for this, however, and the studies of Marsh Tits, Willow Tits and other tits and chickadees indicate that dispersal directions are essentially random for both sexes. This does mean that it can occasionally happen that siblings end up in the same place and form a breeding pair. In one case during the Białowieża study a pair of Marsh Tit siblings had each dispersed almost 1km before breeding together (Wesołowski 2015). In the Monks Wood study we recorded eight cases of siblings being found together after natal dispersal, including six male–female associations. All of these cases were first detected in autumn, involving dispersal distances of 0.6–1.7km. One pair of siblings stayed together from September until at least late January, when they appeared to be defending a territory together, although they were not monitored into the breeding season. In another case a male was closely associating with a female sibling in the autumn, but by the spring I was amazed to find that he was now paired and defending a territory with a different female, who was also his sibling. The original female had disappeared, but this means that three siblings from the same nest had all ended up in the same territory, which was 1.5km from where they had fledged. Breeding could not be confirmed for this pair, however, as they were not closely monitored.

Dispersing siblings are more likely to find themselves in the same place when the woodland habitat is fragmented, as in the Monks Wood study area, rather than in continuous forest, such as Białowieża. The apparent randomness of natal dispersal directions can be skewed or biased when patchy woodland limits the juveniles' options and channels them in certain directions (Gosler 1993, Alderman *et al.* 2011). The most unbiased studies of natal dispersal come from large expanses of intact forest where juveniles can disperse across the landscape in any direction without having to break cover, as in the vast Białowieża Forest (Wesołowski 2015). At Monks Wood, and in most other landscapes, the varying connectivity between different woods determines where the juveniles are able to disperse, and how easy it is to get there (Broughton *et al.* 2010, Alderman *et al.* 2011). It is perhaps not surprising, then, that siblings dispersing from the same territory might have to follow the same habitat corridors, and then find themselves settling in the same place. What is unknown is whether they recognise each other once they are away from the family group, and if there is any kind of kin avoidance to try and reduce the chance of inbreeding.

223

Breeding dispersal

In contrast to the natal dispersal of juveniles, breeding dispersal involves settled adults leaving their breeding territory and moving to another one in the following year. Breeding dispersal is infrequent and generally involves much shorter distances than natal dispersal, because adults usually have little reason to leave a territory or home-range that has already enabled them to survive for a year or more. Indeed, most Marsh Tits and Willow Tits will remain in the vicinity of their first breeding territory for the rest of their lives (Figure 7.6). Several studies have shown that 85–95 per cent of adult Marsh Tits stay in more or less the same breeding territory between years, though the borders and nest-site will shift by a few tens or hundreds of metres (Amann 2003, Broughton *et al.* 2010, Wesołowski 2015). There is surprisingly little information on breeding dispersal in Willow Tits, probably because defining the breeding-territory border is more difficult than for Marsh Tits. Nevertheless, studies from Fennoscandia indicate that territory switching by adult Willow Tits is uncommon (Orell *et al.* 1994a, 1999b). In the wider context, a study of Siberian Tits showed that breeding dispersal was rare (Virkkala 1990), but another study found that a third of adults switched breeding sites by at least 1km at some point during their life (Orell *et al.* 1999a).

CAUSES OF BREEDING DISPERSAL

There are a few circumstances when an adult Marsh Tit or Willow Tit of either sex will leave an established territory and switch to a new one. These circumstances can be divided into needs and preferences. For an adult to need to undertake breeding dispersal, a key resource must be missing or have been lost, such as a breeding partner. A preference to leave an established territory might be due to an improvement in circumstances, such as a better quality of mate or foraging habitat somewhere else.

In Monks Wood we identified the circumstances in which 10 female and 11 male Marsh Tits switched territories between breeding seasons (Broughton *et al.* 2010). For most of these birds the breeding dispersal was in response to the death of a neighbour in a nearby territory, allowing them to move in and replace it. Another factor for most of these birds was that their own mate had also disappeared, and so one bird could switch territories to enable the two widowed adults to form a new pair. In a few cases a bird divorced its mate and left it behind in the old territory so that it could join another bird to form a new pair (see Chapter 5). This kind of divorce was rare at Monks Wood, however, involving just 2–3 per cent of adult Marsh Tits each year. This compares with a 12 per cent divorce rate for Willow Tits in Finland (Orell *et al.* 1994a).

The Marsh Tit results suggest that breeding dispersal is mostly driven by the need to find a new breeding partner if a bird finds itself widowed or unpaired as the breeding season approaches. Most breeding dispersal occurs at the end of winter or in early spring, during the same dispersal period that involves some juveniles moving again, also in search of breeding partners. An unpaired adult has a dilemma of whether to wait and see if a breeding partner arrives into its own established territory, or to leave and go looking for a breeding vacancy elsewhere. Most dispersing adults move only into a nearby territory, so they will have been monitoring their neighbours and will already know when an opportunity exists. For adults with no local options, dispersal is clearly more of a gamble, with no guarantee of success. At Monks Wood I observed two unpaired adult males that both dispersed up to 1.8km to different woods, where they were unlikely to have known the situation regarding any opportunities.

Figure 7.6. An adult Willow Tit in northern England. Once they have settled into a breeding territory, adult Willow Tits and Marsh Tits usually remain in the vicinity for the rest of their lives. (© Philip Schofield)

One male did manage to find a breeding partner that he could join, but the other male was unsuccessful and was chased out of the wood by resident birds, eventually returning to his former territory, where he was unable to breed that year. I also recorded the arrival of a small number of unringed adults into Monks Wood in some years, showing that some Marsh Tits were dispersing into the study area and must have travelled at least several kilometres to do so. Occasionally a colour-ringed adult would disperse back into the study area, having being ringed as a nestling or juvenile in a previous year and then disappearing shortly afterwards. These birds must have initially moved just beyond the study area during natal dispersal, settled in a territory somewhere for a year or more, and then moved back as adults during breeding dispersal if their settled circumstances had taken a turn for the worse.

Breeding dispersal distances

For Marsh Tits there are only a few studies reporting breeding dispersal distances. At Monks Wood the adult dispersal distance was measured as the shift in the central point of the territory between years, whereas in the Białowieża study it was measured as the distance between consecutive nest-sites (Broughton *et al.* 2010, Wesołowski 2015). The median or mean values of breeding dispersal from both studies were very short, at around 50–100m, which would not take most birds outside the radius of a typical 4–6ha breeding territory. This basically revealed that most birds did not undertake any meaningful breeding dispersal at all, with fewer than 10 per cent of adults moving further than 300m between years. However, some larger movements were recorded, with one male at Monks Wood dispersing 1.8km between breeding seasons, which took him to a different wood. The previously ringed juveniles that returned as adults, mentioned above, also indicate some larger movements during breeding dispersal. Overwhelmingly, though, the dispersal distances show that most Marsh Tits remain in or around their previous breeding territories in consecutive years.

The only detailed information for the breeding dispersal distances of Willow Tits comes from the Oulu study in northern Finland, which also measured the distance between nest-sites (Orell *et al.* 1999b, Lampila *et al.* 2006, Kumpula *et al.* 2023). The median or mean breeding dispersal distances in this population are more than double those recorded for Marsh Tits, varying between 215m and 480m for females and males. This difference may be due to larger territories of Willow Tits, as their breeding density at Oulu was less than half that of the Marsh Tits at Monks Wood or Białowieża. For both species, any sex-biased differences in breeding dispersal between males and females are minor and inconsistent.

Irruptive movements

Natal and breeding dispersal are seasonal behaviours that take place every year. Another type of dispersal behaviour that is shown by some Willow Tits, but not Marsh Tits, consists of irregular *irruptions*. Irruptive movements occur in populations of many bird species across the boreal zone of northern Eurasia, and are sporadic mass emigrations in response to specific conditions. Birds relying on tree seeds and berries are particularly prone to irruptions, and in Europe these famously include Bohemian Waxwings and Bramblings. Willow Tits are also an irruptive species in northern latitudes, especially in Fennoscandia and probably further east into northern Eurasia, along with Coal Tits and Blue Tits, although curiously not Crested Tits. In North America, Black-capped Chickadees also undertake irruptions. Factors influencing the likelihood and magnitude of irruptions include population pressure after a particularly good breeding season, and a collapse of food availability when forest trees fail to produce a seed or berry crop. Harsh weather can also contribute to irruptions, such as heavy snowfall and prolonged freezing conditions. Irruptions of tits and chickadees are often synchronised with those of other species, such as finches, suggesting similar causes (Bock & Lepthien 1976, Koenig 2001).

A superb overview of the topic of bird irruptions is given by Ian Newton (2010), with good accounts of irruptive behaviour in tits and chickadees given by Christopher Perrins (1979) and Susan Smith (1991). Among the *Poecile* species, irruptive behaviour is best studied in the Black-capped Chickadee, whose irruptions from boreal forests seem to vary in the number of birds involved and also in the regions affected. Smith (1991) outlines how the mass movements of chickadees from northern forests to more southern regions coincide with years of scarce tree seeds, and that almost all of the birds involved are juveniles. Vast numbers of irrupting birds can be concentrated at bottlenecks on shorelines, with Bagg (1969) recording 36,000 Black-capped Chickadees passing along Lake Ontario during October 1961.

IRRUPTIONS OF WILLOW TITS

In Europe, irruptions of Willow Tits analogous to those of Black-capped Chickadees have been well documented at bird observatories and field stations in Fennoscandia and the Baltic states. These Willow Tit irruptions involve pale northern birds that are classified in the subspecies *borealis*. Some southerly movements of apparent *borealis* birds occur most years, but irruptions are distinguished by their scale, sometimes involving huge numbers of individuals. During irruption years birds can begin moving in August, with peak counts occurring at coastal watchpoints from mid September to mid October as birds heading south and west from the boreal forests reach the Baltic coast (Ehrenroth 1973, Aalto *et al*. 1995). On the western coast of Finland nearly 8,000 Willow Tits passed through the Tauvo field station in one week in mid September 1969, with up to 5,000 within a six-hour period on a single day (Hildén 1971). The number of Willow Tits ringed at coastal observatories in Finland and Estonia during irruption years can be in the thousands, whereas the totals in southern Sweden tend to be in the tens or low hundreds (Ehrenroth 1973, Aalto *et al*. 1995). These ringed birds are just a small sample of those actually passing through these locations during irruptions.

What do irruptions look like? Reports from the coastal field stations on the Baltic Sea give some idea of the drama (Hildén 1971, Ehrenroth 1973, Tiainen 1980). The birds move during daylight and mostly from a few hours after sunrise until the early afternoon, particularly in calm weather. Although irrupting Willow Tits can move singly, where larger numbers are involved they can travel in flocks ranging in size from 2–3 birds to as many as 250. Olavi Hildén (1971) described how flocks of 60–70 Willow Tits moved along the coast at Tauvo in Finland, flying just above the tops of the bushes and frequently dropping down into them, but some flocks flew much higher. In coastal Finland and Sweden, Tiainen (1980) and Ehrenroth (1973) watched flocks of up to 150 Willow Tits moving through the treetops before reaching the sea. Each flock showed lots of hesitation before flying high into the sky and out over the water, but some turned back after a few hundred metres, only continuing if they could see land on the other side. Very few Willow Tits appear to successfully cross significant expanses of open sea, even those as little as 10km wide, despite some observations of small flocks heading out to sea (Hildén 1971, Tiainen 1980, Saurola 1981). These observations show how open spaces and water are a huge barrier to movement, and also how coastlines can concentrate and funnel birds that are on the move.

Irruptions of Willow Tits overwhelmingly involve juvenile birds, presumably those that have failed to secure a home-range or establish in a winter flock after dispersal. Others may have abandoned their settled position due to food scarcity. Of 7,000 irrupting Willow Tits trapped in Finland during the autumns of 1979–1980, more than 99 per cent were juveniles (Saurola 1981). These birds tended to have low body weights, suggesting they were in poor condition (Tiainen 1980). A few adults are certainly involved in irruptions, and Smith (1991) wondered what would drive them to leave the security of the home-ranges that have obviously supported them over at least the past year. The risk of starvation was one potential explanation, if the seed crop had failed, but Smith (1991) suggested that habitat destruction could also force some adults to disperse. Indeed, extensive forestry and logging in Fennoscandia has been shown to reduce the habitat quality and carrying capacity of forests for Willow Tits (Siffczyk *et al*. 2003, Virkkala 2016).

There is no information from ringing or other observations to show that irrupting birds ever return to their original home-ranges, although Smith (1991) outlines the rather patchy evidence for some northward movement of returning Black-capped Chickadees in North America. These movements appear to be rather aimless and exploratory, rather than birds 'homing' back to their original locations, and very little is known about the eventual fate of irrupting tits and chickadees.

Ringing data show that some of the Willow Tits caught during autumn irruptions in Fennoscandia travel great distances to or from the coastal field stations. Birds can move over tens or hundreds of kilometres in a matter of days or weeks, in various directions. Aalto *et al*. (1995)

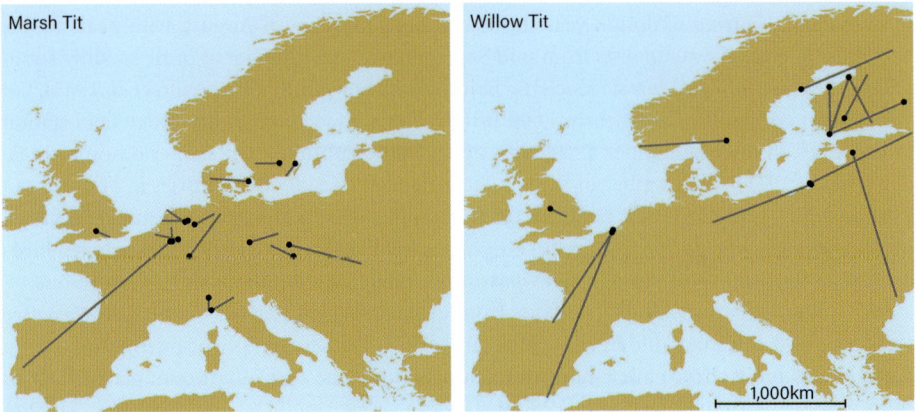

Figure 7.7. Ringing recoveries showing the largest recorded movements (100km or more in Great Britain, 150km or more elsewhere) of Marsh Tits and Willow Tits in the EURING Migration Atlas databank (Spina *et al.* 2022). Black dots are the ringing site and lines indicate the recovery location. EURING data provided by the BTO.

estimated that these Willow Tits travelled 20–30km per day, but one bird was documented as moving 70km in a day. There are multiple records of ringed birds being recovered after moving distances of over 150km around the Baltic region (Figure 7.7). These movements also clearly show how Willow Tits irrupting south and west through Fennoscandia are blocked by the Baltic Sea, however, with no birds recorded crossing the sea and heading further south or west into central Europe. Any Willow Tits moving west through Sweden and Norway are also blocked by the North Sea, which is an overwhelming barrier to further movement.

Some irrupting Willow Tits seem to take a southwesterly direction through the Baltic states, which avoids a sea crossing into the rest of Europe. Hundreds or thousands of Willow Tits are ringed on the southern Baltic coast during irruptions, particularly in Estonia and Lithuania (Aalto *et al.* 1995), but they may originate much further away to the northeast. However, the vast majority of these birds not appear to penetrate much further, with only low numbers recorded as far west as Poland (Aalto *et al.* 1995, Sokolov *et al.* 2002).

IRRUPTING WILLOW TITS IN WESTERN EUROPE

Willow Tit irruptions are clearly dramatic phenomena, but in Europe they barely register outside of Fennoscandia and the Baltic region. The distances travelled by these birds over hundreds of kilometres mean that some individuals potentially could reach locations well beyond their typical range. The distinctive pale appearance of northern Willow Tits compared to the darker birds in western Europe means that they should be readily identifiable if they get there. Figure 7.7 shows that some birds ringed around the Baltic coast, probably during irruptive movements, have travelled staggering distances. For example, a Willow Tit ringed at the beginning of September 2018 on the Lithuanian coast was found 73 days later between Berlin and Leipzig in Germany, 690km to the west. This ringing record shows that irrupting Willow Tits can very occasionally reach deep into central Europe at least.

In western Europe there have been several claims of Willow Tits of the *borealis* subspecies in Great Britain, but only one record is accepted by the British Birds Rarities Committee, which adjudicates on such records (Naylor 2021). The bird concerned was reportedly shot in March 1907 in Gloucestershire, western England. This inland county is not an obvious location for an irruptive Willow Tit, especially in March, but if genuine then it may have come via the southern Baltic route and crossed the English Channel to winter in Great Britain. What sounds a likely candidate is a record from November 1935 on the northern Scottish island of Fair Isle (Naylor 2021). The observer described a pale Willow Tit associating with an arrival of three immigrant Blue Tits, which sounds quite plausible for a group of birds irrupting from Norway. The specimen was not obtained, however, and so the record was not accepted by the committee.

Regardless of the veracity of the British records, the paucity of other observations highlights that irruptions of Willow Tits from Fennoscandia, northeast Europe and the Baltic states are very unlikely to reach western Europe on anything other than an extremely rare basis. If any *borealis* Willow Tits have genuinely ever reached Great Britain it must be a heroic feat, as the North Sea would be a formidable barrier, whichever crossing route was taken. It is possible that a Willow Tit could hitch a ride on a ship for part of the journey, but the substantial decline of Willow Tit populations in Fennoscandia in recent decades, driven by extensive logging of the boreal forests, means that the likelihood of one reaching Great Britain is probably becoming even more remote.

It is unclear why irruptive behaviour occurs in Willow Tits, Black-capped Chickadees or Coal Tits but not in Marsh Tits or Crested Tits. In the case of Marsh Tits, there are far fewer of them in the northern boreal forests than Willow Tits, which have a more northerly distribution. This could mean that some Marsh Tits will be overlooked among the Willow Tits, but this seems unlikely, as there are few records of Marsh Tits being ringed at coastal bird observatories around the Baltic during irruptions. There are also very few ringing recoveries of Marsh Tits moving long distances like irrupting Willow Tits. It seems that Marsh Tits genuinely do lack the irruptive habit.

Ringing recoveries

Of all of the birds of all species that are caught and ringed, only a small minority of those outside of intensive studies are ever recorded again. The finding or retrapping of a bird that has previously been ringed constitutes a *ringing recovery*. This may involve a bird being recaptured by the same or another ringer, or one that is found dead or injured by a member of the public and reported to the relevant ringing scheme. For Marsh Tits and Willow Tits there are tens of thousands of records of ringed birds that have been re-encountered by somebody in one way or another. These ringing recoveries contain a mixture of natal dispersal, breeding dispersal, irruptive movements and any other kind of wandering away from the ringing site.

Ringing recoveries are influenced by the behaviour of ringers and the public, as well as by the birds themselves. For example, the majority of birds are first ringed as full-grown individuals, as these are far more likely to be caught and ringed than nestlings. In Great Britain, of the 59,007 Marsh Tits ringed between 1931 and 2021, only 27 per cent were nestlings, most of which will have been ringed opportunistically in nest-boxes (Robinson *et al.* 2023). British Willow Tits use nest-boxes even less often than Marsh Tits, so there are only 3.9 per cent nestlings among the 54,209 Willow Tits ringed in Great Britain between 1940 and 2021. This bias in favour of full-grown birds means that most will have already finished dispersing by the time they are first caught. Consequently, general ringing recoveries largely fail to capture the dispersal movements of Marsh Tits and Willow Tits.

Robin Sellers analysed the finding circumstances of 75 ringed Marsh Tits and 86 Willow Tits that were found dead in Great Britain (Sellers 1984, Wernham *et al.* 2002). For both species, 40–43 per cent of recovered birds had been killed by domestic cats. A further 35–38 per cent were killed by some other human-related cause, such as struck by traffic or hitting a window. Just 7–10 per cent of the deaths were attributed to natural predators, with remains presumably found near Eurasian Sparrowhawk nests, in Tawny Owl pellets, or similar circumstances. This underlines how ringing recoveries largely reflect where and when people are likely to find dead birds rather than where Marsh Tits and Willow Tits occur or how they mostly die. In reality only a small minority of all Marsh Tits and Willow Tits will be killed by cats or traffic, as the vast majority of birds are unlikely to encounter them out in the woods. Similarly, a potential finder is unlikely to encounter a ringed Marsh Tit or Willow Tit when it has been killed by a Eurasian Sparrowhawk, Pygmy Owl or starvation.

These biases are very important to bear in mind when interpreting general ringing recoveries, but that is not to say that this information cannot tell us anything useful. The numbers of birds ringed during general ringing and at observatories are so much greater than those involved in detailed ecological studies that they can provide a different sample. The EURING databank (EURING 2023) includes ringing recoveries that have been gathered by national ringing schemes from across Europe, and these provide useful context for the more detailed dispersal studies of Marsh Tits and Willow Tits. The EURING databank contains ringing recoveries for 37,924 Willow Tits and 26,594 Marsh Tits from various countries up to 2022, and the recovery distances from the initial ringing site are mostly very short. Overall, 87–88 per cent of European Marsh Tits and Willow Tits had not moved from the ringing site before they were recovered, whether days or years later. Of the 12–13 per cent of other recoveries, most were encountered within just 5km of where they had been ringed. Only 1.4 per cent of Marsh Tits and 3.3 per cent of Willow Tits were recovered further than 10km from the ringing site. These figures reinforce the findings from the detailed population studies, that full-grown Willow Tits and especially Marsh Tits are extremely sedentary across their European range.

Although northern Willow Tits do undertake some long movements during irruptions, as shown in Figure 7.7, the extent of these is not very obvious in the EURING databank, as any long-distance recoveries are swamped by those of more sedentary birds, which have a higher chance of being re-encountered near the ringing site. It is also difficult to know if exceptional movements in the EURING databank are genuine or if they involve human influence, such as birds transported accidentally on vehicles, or errors during data input. An incredible recovery involved a Willow Tit ringed in Estonia in March 1971 that was found dead 1,431km south in Odesa, in Ukraine, more than five years later. Even more astonishing are two Willow Tits ringed in 1964 and 1968 in Belgium that were later recovered in Spain, at distances of 967km and 1,701km from the ringing site. These three records are so eccentric that they seem highly unlikely to be natural movements, and the EURING data suggest that any Willow Tit records beyond 600km probably require additional checks and validation.

There are also some long-distance recoveries among the Marsh Tit data in the EURING databank, despite this species not taking part in irruptions. Twenty records exist of Marsh Tits being recovered more than 150km from the ringing site, although one bird that supposedly moved 1,610km between Belgium and Portugal in 1971 seems as unreliable as the Willow Tit recoveries between Belgium and Spain (Figure 7.7). Another odd record is of a Marsh Tit ringed in December 1954 in what is now the Czech Republic, which was recovered 501km to the east in Ukraine just 79 days later. This puzzling movement was well outside of any dispersal period. A more interesting record involves a Danish Marsh Tit ringed on the island of Zealand during the post-fledging dispersal period in June 1954, which was recovered 217km away the following spring on the Danish mainland. This movement would have involved a sea crossing of at least 12km, which is a formidable barrier for a Marsh Tit, but the record seems genuine. There are several other seemingly reliable recoveries of Marsh Tits that had moved distances of over 125km in various parts of Europe, and so around 195–217km probably reflects the upper limit of natural movements for this species.

Even within the confines of an island like Great Britain, with a high density of people and also ringers, the data from the BTO ringing scheme include very few long-distance ringing recoveries (Figure 7.7). A Marsh Tit ringed in Wiltshire in December 1982 was recovered more than three years later after hitting a window in Surrey, 100km away (Robinson *et al.* 2023). Meanwhile, a Willow Tit ringed in Greater Manchester in October 1977 was found dead the following spring at the Carlsberg Brewery in Northampton, 167km away. This record sounds like an unfortunate Willow Tit might have been hit by a delivery truck near Manchester and then carried back to the brewery on the vehicle somehow, but we'll never know. A previous analysis by Robin Sellers (1984) found that just 4 per cent of Marsh Tit ringing recoveries in Great Britain were further than 10km from the ringing site, with a larger 10 per cent for Willow Tits. In my own study of 1,693 Marsh Tits colour-ringed at Monks Wood up to 2024, the furthest from the ringing location that any of them was ever detected was 10.5km. All of the others were under 6km, with the vast majority being under 2km.

Other observations of dispersal

For species that are as sedentary as Marsh Tits and Willow Tits, even general observations of birds outside of their normal ranges can provide some interesting insights into movements and dispersal capabilities. For well-watched bird observatories and nature reserves, outside of regular populations of Marsh Tits or Willow Tits, the timing and frequency of any observations can give an idea of how far and how often birds might disperse from the nearest breeding areas.

A good example of such observations comes from the East Riding of Yorkshire, in northeast England. Marsh Tits are resident in reasonable numbers in the western part of this county, on the hilly Yorkshire Wolds, but they are absent from a broad 40km-wide swathe of low-lying farmland called Holderness, which lies between these western populations and the North Sea coast to the east (Figure 7.8). The area's geography is a natural funnel that channels birds dispersing into Holderness down towards the Spurn Bird Observatory, where they have a very good chance of being detected. This gives an interesting case study of when and how often Marsh Tits manage to cross the 40km of arable farmland habitat via its sparse hedgerows, scattered thickets and very limited woodland. In total, there were just seven Marsh Tits reliably recorded at Spurn during a 33-year period, which is an average of one record per 4–5 years. Most of the records occurred in summer, between late June and August, coinciding with juvenile dispersal. The small number of Marsh Tit records at Spurn and its hinterland in Holderness, as shown in Figure 7.8, demonstrate the limited willingness or ability of Marsh Tits to disperse across expanses of arable farmland. It also hints at why colonisation and recruitment can be so difficult for them in highly fragmented habitat, as even 40km of farmland appears very difficult to cross with any regularity.

Some other observations from Great Britain point to the limits of a Marsh Tit's dispersal abilities. In addition to the above-mentioned 100km ringing recovery in Surrey, there is also a

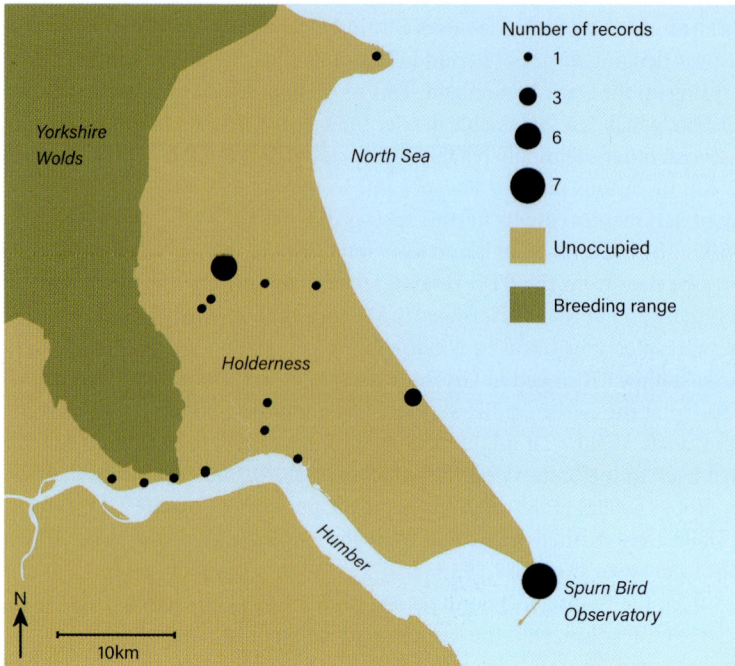

Figure 7.8. Marsh Tit records outside of the breeding range in the East Riding of Yorkshire, in northeast England, during 1978–2010. The geography funnels birds dispersing eastwards towards the Spurn Bird Observatory, but few complete that 40km journey from the breeding range. Data were collated from numerous published, online and personal records. Contains Ordnance Survey data, © Crown copyright and database right 2023.

reliable observation of an unringed Marsh Tit at Stevenston, in Ayrshire on the west coast of Scotland, 100km outside of their normal range. The bird was photographed on a garden bird-feeder by Iain Hamlin in February 2007. There is no doubt about the identification, but the nearest breeding Marsh Tits to Stevenston are in the Scottish Borders or northern Cumbria. An even more extralimital record was a bird at Arbroath in eastern Scotland in June 1996, which was at least 170km from the nearest breeding population, assuming a route that avoided crossing any extensive water (Forrester & Andrews 2007). In Wales there are only 12 records since 1954 of Marsh Tits dispersing from the mainland onto the offshore island of Anglesey, which involves a minimum sea crossing of 400m across the Menai Strait (Pritchard *et al.* 2021). The Irish Sea seems to present an almost insurmountable barrier to any dispersing Marsh Tits or Willow Tits departing from England, involving a minimum flight of at least 50km across open water, even if the Isle of Man is used as a stepping stone. There is a single record of a Marsh Tit on the official Irish list, with a bird reported by Oscar Merne from his garden in Bray, just south of Dublin, on 17 December 1990 (Merne 1993). This record is very close to ferry routes between Dublin and Great Britain, and may have involved a Marsh Tit hitching a ride, or some other human influence.

The fate of these wanderers is invariably going to be quite bleak. Having dispersed out of their populations and failed to find any suitable habitat or conspecifics to settle with, they keep

Figure 7.9. A dispersing Willow Tit at Paull on the Humber Estuary in northern England, in September 2023. This juvenile has dispersed far beyond the nearest breeding populations at least 35km away, and has wandered into an empty landscape where it has no real prospects of finding a partner and settling. (© Simon Brebner)

moving and searching until something stops them, such as the coast, starvation, a predator or an accident. Dispersing from the natal territory and heading outwards from dwindling populations can often mean heading into a landscape that no longer has other Marsh Tits or Willow Tits nearby. Such extralimital birds are nice to see for the observers who find them unexpectedly, perhaps at the coast or in the former breeding range (Figure 7.9), but the tragedy for these juveniles is that the further they disperse the less likely they are to meet another bird of the same species, let alone a bird of the opposite sex. In reality these wandering birds are lost from the breeding population and probably do not live very long.

The challenges of dispersal

All indications are that dispersal is an immensely challenging activity for Marsh Tits and Willow Tits. For natal dispersal, what is so remarkable is that the juveniles are facing this test when they are only just able to fend for themselves, at little more than five or six weeks old. Dispersing juveniles must have little idea of where they are going, as they have no prior knowledge of where vacant territories might be located and how to get there, particularly if this involves moving between different woodland patches. Presumably the juveniles find vacancies and conspecifics by random exploration.

A key problem in understanding the challenges of dispersal is that it is so rapid and rarely observed in action. Nobody has ever seen or documented the full process of an individual juvenile leaving its natal territory, moving through the landscape, and settling in a new place. We only get to observe brief snippets of an individual's dispersal journey. I have only occasionally seen birds in the act of dispersal, part way through their journey. On one occasion at Monks Wood, for example, a juvenile was seen in a family group one day but the next day was alone and moving purposely through the wood for several hundred metres in one direction, evidently in active dispersal. It was not seen again. Weise & Meyer (1979) had a similar encounter with a dispersing Black-capped Chickadee, which was followed for 30 minutes as it travelled through 500m of habitat, occasionally pausing to forage. When it came across two other juveniles they chased it away. It then carried on for another 200m in the same direction before being lost to view. Away from Monks Wood, I have sometimes seen juvenile Marsh Tits or Willow Tits in summer as they were apparently dispersing along hedgerows through farmland, frequently calling or foraging, but relentlessly moving through the landscape. It is impressive that any of them are successful, and the failure and mortality rate seems to be quite high (see Chapter 8).

Dispersal in fragmented habitat

Habitat fragmentation is a major issue for Marsh Tits and Willow Tits during dispersal. The woodland patches that the birds want to find, explore and settle in are typically located within a more hostile landscape matrix, such as intensive farmland (Figure 3.40). This means that woodland patches are separated by large gaps that the birds must cross, using woody corridors like hedgerows or tree lines, if they exist. Often, however, there is no obvious or direct corridor or connectivity between woodland patches, leaving some woods more inaccessible than others.

When coming to a large gap or the edge of a woodland patch, there is ample evidence to show that juvenile Marsh Tits and Willow Tits are very unwilling to break cover and fly across them. The larger the gap, the more of a barrier it represents. In the Monks Wood study, Paul Bellamy found a juvenile Marsh Tit struggling during active dispersal, 2km from its natal territory. The bird was watched as it tried to cross 416m of open fields between small woodland patches in the arable landscape. On three occasions it was seen to fly upwards and out over the field for about 100m before abandoning the attempt and turning back. It was not seen again, and must have later made the crossing and continued its journey, found an alternative route, or died. This is reminiscent of Tiainen's (1980) observations of irrupting Willow Tits reaching the Finnish coast, where they flew high and headed out to sea before turning back after a few hundred metres.

The Marsh Tits in the Monks Wood study can see some of the neighbouring woods across the arable fields, and they can even hear other Marsh Tits singing in them, which probably encourages dispersing birds to try and head in those directions. Without an obvious corridor, however, a dispersing bird might well be thwarted (Figure 7.10). Desrochers & Hannon (1997) showed that dispersing Black-capped Chickadees were attracted to experimental mobbing calls of other chickadees across a 50m forest gap. However, the chickadees preferred to avoid crossing the open space if possible, and would go around the gap via the surrounding forest even if that meant the journey was three times longer. This experiment neatly demonstrated that juveniles are attracted to calls of their own species during the dispersal period, but crossing even small gaps in the woodland habitat can be a problem.

Figure 7.10. A Marsh Tit arriving here at the edge of Monks Wood has to cross a significant gap of 780m to reach the next wood, visible in the distance. Just a single tree lies as a stepping stone across arable land. Most Marsh Tits will be unwilling to cross such large gaps. (© Richard K. Broughton)

How big does a gap have to be before Marsh Tits and Willow Tits are unwilling to cross it? That probably depends on push and pull factors, such as hunger or an aggressive bird chasing them away, or the sight and sound of good habitat and a potential mate on the other side. Ehrenroth (1973) saw irrupting Willow Tits happily crossing gaps of 100m in a single flight, and at Monks Wood we observed Marsh Tits flying across gaps of 91–133m, with circumstantial evidence of 250m gaps being crossed in a single flight (Broughton & Hinsley 2015). However, gaps of around 400m are enough to make some Marsh Tits turn back (Broughton *et al.* 2010), and the barrier effect for most birds probably starts becoming important when faced with a gap of over 100–200m.

Habitat configuration can also have an effect on dispersal. Juveniles fledging from territories along the edge of woodland patches are less likely to disperse successfully and join the local population (Alderman *et al.* 2011). This is probably because these juveniles are more likely to leave the woodland via 'exits', such as hedgerows and tree lines, which can channel them out of the woodland but do not necessarily lead to other suitable habitat, at least not directly. There is then a greater chance of such birds becoming lost in the landscape and exposed to greater risks.

The fragmentation effect is illustrated by the pattern of Marsh Tit dispersal in the Monks Wood study (Figure 7.11). This graphically shows how many Marsh Tits were colour-ringed in one wood within the study area and were later seen again in another wood after dispersing. For those birds that moved between woods, the great majority only dispersed relatively short distances to the next woodland nearby. Very few birds dispersed successfully between the larger populations in Monks Wood and the bigger outlying woodlands, despite being within only a 5km range. Some of the smaller and more isolated woods had no birds moving to them at all. Furthermore, the 102 birds that moved between different woods represented only 7 per cent of the colour-ringed Marsh Tits whose movements were monitored in the Monks Wood study. The other 93 per cent of birds ringed as nestlings, juveniles and adults were not recorded ever leaving the wood in which they were ringed. These data illustrate just how much of a barrier the gaps between woodlands really are, even gaps of just a few fields.

Tomasz Wesołowski examined Marsh Tit dispersal in the extensive Białowieża Forest in Poland and compared his findings to those of studies from fragmented landscapes in Sweden, Switzerland and Monks Wood in England (Wesołowski 2015). These comparisons, similar to Table 7.1, showed how the absolute dispersal distances for females are slightly greater in Białowieża than elsewhere, such as at Monks Wood. However, when looking at the number of territories that the juveniles dispersed across, in Białowieża the females crossed an average of 6.4 territories, which was more than double the 3.1 traversed at Monks Wood. The values for males were similar between the studies, at 2–3 territories crossed. The females want to disperse much further than the males, however, and so Wesołowski's (2015) comparison shows that this ability was severely curtailed in the fragmented woodland at Monks Wood. The females faced greater problems because of the need to cross large gaps between the woods just to try and cover the same sort of dispersal distance as in the continuous forest of Białowieża. Because of these gaps the Monks Wood females actually achieved far less real dispersal in terms of the number of territories they moved across before being able to settle.

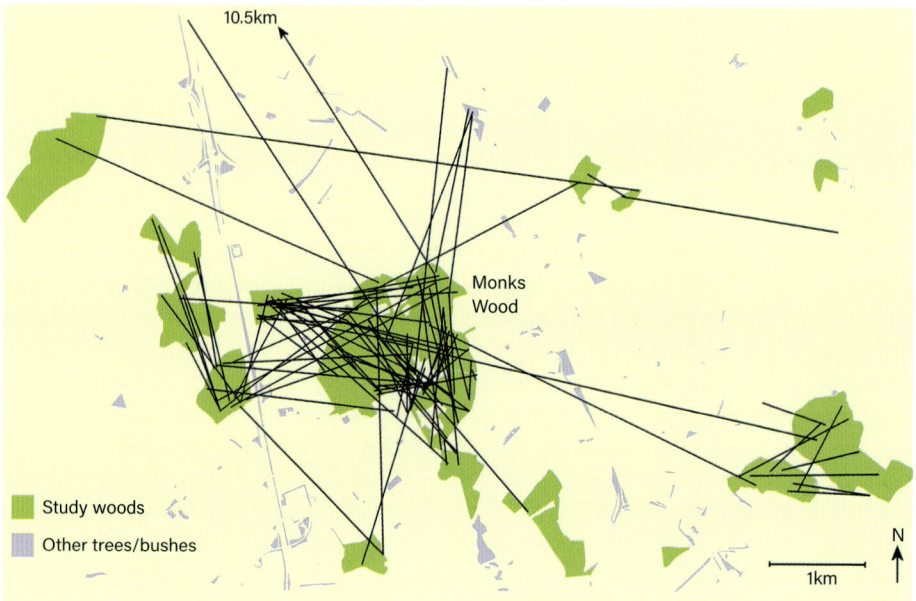

Figure 7.11. Marsh Tit dispersal in the Monks Wood study area. The lines connect the origin and destination of 102 colour-ringed birds that moved between different woodland patches. The actual routes taken by each bird were probably more indirect, using hedgerows and woody corridors across the open farmland. Very few birds moved between distant or isolated woods, showing the severe impact of habitat fragmentation on dispersal. One bird was found outside the study area at 10.5km distance.

The upshot of these findings is that young female Marsh Tits at Monks Wood have a limited dispersal ability, which has important and far-reaching impacts on the population. It means that many females are unable to successfully disperse between woods to find all of the territories that need recruits, and so some areas remain unoccupied, or tenanted by lone males (Broughton *et al.* 2010). Ultimately, we found that there were fewer breeding pairs across all of the woods than there otherwise could have been, which meant fewer juveniles fledging in the following summer and dispersing between these woods after the breeding season. It's easy to see how this could lead to a downward spiral in fragmented landscapes of poor recruitment, fewer pairs, leading to fewer fledglings and even poorer recruitment, and so on. The potential impacts of truncated dispersal on inbreeding are also obvious.

There is little comparative information for Willow Tits, as the available studies have tended to be in extensive habitat rather than in fragmented landscapes. For example, in a Finnish population the immigration rate was found to be remarkably high in an area of 22km², with 63 per cent of male and 76 per cent of female breeders originating from outside the local population (Orell *et al.* 1999b). The higher proportion of immigrant females will reflect their longer dispersal distances compared to males, with relatively more locally born females leaving the study area and being replaced by immigrant females dispersing inwards. The extensively forested landscape clearly facilitated these movements, enabling birds to freely disperse over large distances, and the number of unpaired individuals was low due to plenty of immigrants filling the vacancies (Orell *et al.* 1994b).

SETTLING PATTERNS IN FRAGMENTED WOODLANDS

As mentioned previously, there are difficulties in directly tracking the dispersal of small birds like Marsh Tits and Willow Tits. However, the consequences of dispersal in fragmented landscapes can be seen in the pattern of bird communities in the woodland patches. There is a good collection of studies showing that tits and chickadees have a higher turnover and lower survival in smaller woodland patches than in larger ones. This makes populations in small woods much more heavily reliant on regular recruits arriving during dispersal. Yet, as we have seen from the studies of dispersal distances, the more fragmented and isolated woodlands are more difficult for juveniles to find. This is compounded by the fact that just one bird finding an isolated wood is not enough, as forming a pair requires two birds of opposite sexes to have found the wood and each other at the same time.

Shelley Hinsley, Paul Bellamy and Ian Newton looked at rates of bird species turnover, random extinction and recolonisation in 146 woods in an agricultural landscape in eastern England (Hinsley *et al.* 1995b). Of the 59 woodland birds examined, Marsh Tits had one of the highest rates of turnover, meaning that their presence in the woods was one of the least stable. Marsh Tits could be present in an individual wood one year but not the next, and then recolonising a later year. This only applied to larger woods of around 4–30ha, as the smallest woods had no Marsh Tits at all. Even where Marsh Tits were present, the typical size of the woods meant that there were just a few birds present in each. If one or two of those birds died within a year then there might be no intact breeding pairs left. What the instability and high turnover revealed was that Marsh Tits often did die in these small woods, and, crucially, they were not always replaced by other birds arriving during dispersal periods. Because the woodland was so fragmented, in small and scattered patches, dispersing Marsh Tits just didn't arrive in enough numbers or with sufficient reliability to maintain a stable, constant presence over time.

A very similar effect of woodland size and isolation was found for Marsh Tits in agricultural landscapes in the Netherlands (Opdam *et al.* 1985, van Dorp & Opdam 1987). In these studies, Marsh Tits were far less likely to occur in smaller woodland patches, those that were most isolated, and those least connected to other woods. Again, this largely reflects the increased difficulty of dispersing birds in reaching and finding small, isolated woods and persisting in them to maintain a constant presence.

The negative relationship between woodland fragmentation and dispersal is common to all *Poecile* species, as shown by an experimental study of Carolina Chickadees in Ohio (Groom & Grubb 2006). Chickadees in this study were removed from a selection of small woodlands to simulate extinction, and then the process of recolonisation by dispersing birds was monitored. The more heavily fragmented and isolated woodlands were recolonised more slowly than better-connected woods, and they were also less likely to remain occupied. In another study of Carolina Chickadees, also in Ohio, Doherty & Grubb (2002) showed that individuals in small woods had a lower survival rate than in larger woods, echoing the high turnover of Marsh Tits in woodland patches in England (Hinsley *et al.* 1995b).

The clear lesson is that habitat fragmentation intensifies the challenges of dispersal by hindering the settling success of birds that are trying to find a vacant territory and a breeding partner. Dispersal is most straightforward and successful in large, continuous expanses of wooded habitat. Where woodland is fragmented into small and scattered patches in a hostile matrix, such as intensive farmland or urbanisation, then dispersal is more difficult and less successful, leading to lower recruitment that eventually manifests as local extinctions and empty patches of habitat.

THE PRESSURE OF TIME

Even in intact forest, where movements are unconstrained by habitat fragmentation and gaps, there are severe pressures on dispersing juveniles, and also on their parents. The need for juveniles to disperse early imposes a huge demand on the parents to breed as early as possible, so that their young have the best chance of dispersing ahead of the competition and becoming established in the population. Tomasz Wesołowski (2023) concluded that Marsh Tits adjust their breeding time each year to nest as soon as weather conditions and food availability allow, rather than to try and synchronise with the spring peak of caterpillar availability (Figure 7.12). The reason behind this is to enable adults to give their fledglings the best chance of early dispersal and successful settling, as outlined in the studies by Nilsson & Smith (1985, 1988a) and Nilsson (1989a, 1989b). Essentially, the birds are racing against time, and against their neighbours, which explains the high level of breeding synchrony observed in all Marsh Tit studies. A delay of even a day in undertaking dispersal can put a juvenile bird at a distinct disadvantage.

A similar pattern appears to exist for Willow Tits, where recruitment is dependent on early breeding and dispersal in relation to other local Willow Tits, rather than synchrony with caterpillar abundance (Pakanen *et al.* 2016b). However, in years where the timing of breeding happens to coincide more closely with caterpillar abundance during the nestling or fledgling stage, then recruitment is also improved. As such, for both Marsh Tits and Willow Tits the timing of juvenile dispersal ultimately depends on their parents' ability to begin breeding as soon as possible, ahead of the local competition. The strong drive for juveniles to fledge and disperse as early as possible, and the relatively low chance of late juveniles settling in the

Figure 7.12. A female Marsh Tit brooding nestlings in the Białowieża Forest. Tomasz Wesołowski (2023) showed that these birds usually breed well before the peak availability of caterpillars, as soon as conditions allow, to give their offspring an advantage during dispersal. (© Richard K. Broughton)

population, probably also explains why Marsh Tits and Willow Tits are single-brooded, and why brood losses are rarely replaced. Fundamentally, the successful dispersal and recruitment of juveniles owes a lot to the decisions of their parents and the configuration of the landscape.

Summary

Marsh Tits and Willow Tits undertake several types of dispersal, leaving one settled location for another place. Natal dispersal is the single most important and challenging movement of their life, in which they leave the parental territory and try to find a new home-range of their own. Juveniles begin this journey when only around five or six weeks old.

In temperate regions the juveniles immediately seek a home-range to occupy as a pair. In boreal regions the juveniles settle in discrete flocks in a communal home-range or territory, which they later partition into breeding territories. Successful establishment relies on dispersing earlier than other competing juveniles, which is largely determined by their parents' ability to breed as soon as conditions allow. Late-dispersing juveniles have little chance of finding a vacant home-range in which to settle.

Natal dispersal is a sudden and rapid process, sometimes taking as little as a day to accomplish, and juveniles are generally settled by the end of summer. Most juveniles disperse only within a few kilometres of their parents' territory. Some birds take part in a second dispersal phase at the end of winter and into early spring, in search of breeding partners, and these birds form an important component of breeding populations, by compensating for winter mortality.

Breeding dispersal involves settled adults leaving their established territories and moving to another one. This is much less common than natal dispersal and usually involves far shorter distances. Breeding dispersal is mostly undertaken by widowed or divorced adults moving to a neighbouring territory in order to form a new breeding pair, but longer distances can be involved. Dispersing adults help to fill vacancies created by mortality.

In some years large numbers of Willow Tits from northern boreal regions take part in irruptive movements in response to a high population pressure and/or a failure of seed crops. Irruptions occur between August and November and overwhelmingly involve juveniles, which undertake movements in any direction. Tens of thousands of birds pass through Fennoscandia, but the Baltic Sea represents a major barrier that they seem unable to cross. Others avoid a sea crossing by moving south through the Baltic states and into central Europe, with some travelling 600km or more. There is limited convincing evidence of northern Willow Tits reaching western Europe, which is probably extremely rare. It is unknown how many irrupting juveniles make it back to the boreal forests to breed.

Ringing recoveries offer useful insights into movements, but are biased by most full-grown birds being ringed after having completed natal dispersal. The vast majority of recoveries are within 5km of the place of ringing, showing how most birds are very sedentary after dispersal. Observations of unringed birds can provide further insights into dispersal abilities, especially where Marsh Tits and Willow Tits are observed outside of their known breeding range. Together, ringing and extralimital observations suggest that movements over 50km are very unusual, and distances of 100–200km are exceptional for birds in non-irrupting populations.

Dispersing Marsh Tits and Willow Tits are very sensitive to habitat fragmentation. Gaps of a few hundred metres between woodlands represent barriers that begin to inhibit effective movement through the landscape. Smaller and more isolated woods receive fewer dispersing immigrants than larger woods, resulting in fewer recruits, a higher turnover, and a greater chance of local extinction.

CHAPTER 8

Survival and longevity

Most of the Marsh Tits and Willow Tits that are born each spring can only expect to live for a few months. A significant minority die in the nest, a few are caught by predators after fledging, and many more do not survive the next few weeks of independence and dispersal. Those birds that do survive to settle in their own home-range and be recruited into the population must cope with changing food sources as summer turns into autumn, and then deteriorating weather into the winter. By the spring only a small percentage of the eggs that were laid in the previous year will have resulted in mature birds that have survived to enter the breeding population.

The breeding season is also a dangerous time for full-grown Marsh Tits and Willow Tits. The males are particularly vulnerable to predators when advertising their position by frequent singing, while females spending a lot of time inside the nest-cavity can also be trapped and killed by a predator. Having their activity tied to a fixed point at the nest for weeks on end also means that the movements of both parents become more predictable, giving predators a better chance to find and ambush them.

Nevertheless, once a Marsh Tit or Willow Tit has survived its first winter and entered the breeding population, then the odds of it surviving for a further year are usually slightly better than even. Compared to similar-sized mammals such as voles or mice, tits and chickadees can be very long-lived, with some birds surviving for over a decade. Overall, the two biggest causes of mortality throughout their lives are predation and starvation, but sometimes birds also die because of accidents, although deaths from disease seem rare.

There is a huge evolutionary pressure acting on Marsh Tits and Willow Tits to avoid or limit the causes of mortality, which has sharpened their behaviour and instincts, but luck also plays a big part. Sometimes a bird is just in the wrong place at the wrong time, or born into a cold and wet spring, or its environment becomes more hostile through reasons beyond the bird's control. In these situations there is not much a bird can do about it. As species, Marsh Tits and Willow Tits are becoming more unlucky in many regions, such as in Great Britain and Finland, where their populations are declining as the mortality rate exceeds the birds' ability to maintain their numbers.

Juvenile survival

As we saw in Chapter 6, in non-depredated nests almost all of the eggs that are laid by Marsh Tits and Willow Tits survive to hatch, meaning that the embryo mortality is very low (Tables 6.6 and 6.7). Some embryos may die within the egg after becoming chilled or because of congenital defects, or if the nests are taken over and built upon by competing Blue Tits or Great Tits, and this is probably under-recorded in many studies. The greatest cause of egg and nestling mortality is predation, and together with other causes it means that around a fifth of all Marsh Tits and Willow Tits die before they leave the nest, and some before they even hatch. For fledglings, despite their initial vulnerability in being unable to fly very well, and their frequent calling, the mortality in the weeks after leaving the nest appears to be very low. This is a difficult period of the life cycle to study, because tracking and observing fledged families is not always easy, but the available studies suggest that at least 90 per cent of fledglings regularly survive until they become independent and disperse (Nilsson & Smith 1985, Hogstad 1990b).

When the juveniles reach independence they enter the very risky period of dispersal, where they have to leave their parental territory and try to find their own home-range (see Chapter 7). The mortality rate in this period is much higher. The young birds have very little knowledge of the wider area, no knowledge of where suitable vacancies might be located, and they don't know where they might be going. They are also quite inexperienced at finding food, with only a few weeks of 'practice' within the family group since they left the nest. Their dispersal might take them through poor or hostile habitat where food may be scarce and exposure to predators is increased, especially when crossing open habitats between the cover of dense woodland. At this dangerous time, these inexperienced dispersing juveniles are at risk of starvation, predation, accidents (such as being hit by traffic) or simply getting lost in the landscape.

It comes as no surprise, then, that in the Monks Wood study we found that two-thirds of the juvenile Marsh Tits that successfully fledged did not survive to settle in their own home-range by the autumn (Broughton et al. 2010). A similar proportion is indicated in Sweden and Switzerland (Smith 1993b, Amann 1997). Considering that the survival of fledglings is initially quite high, this means that the bulk of juvenile mortality in the summer occurs during or around dispersal and attempted settling (Figure 8.1). Of those juveniles that do successfully navigate dispersal to find and settle in a home-range, there is a steady but quite low-level mortality rate over the autumn and winter. By the following spring an average of only 10–20 per cent of the original fledglings are left to enter the breeding population. At Monks Wood and in Sweden the annual proportions of Marsh Tit fledglings that survived to breed were as low as 5–7 per cent in some years.

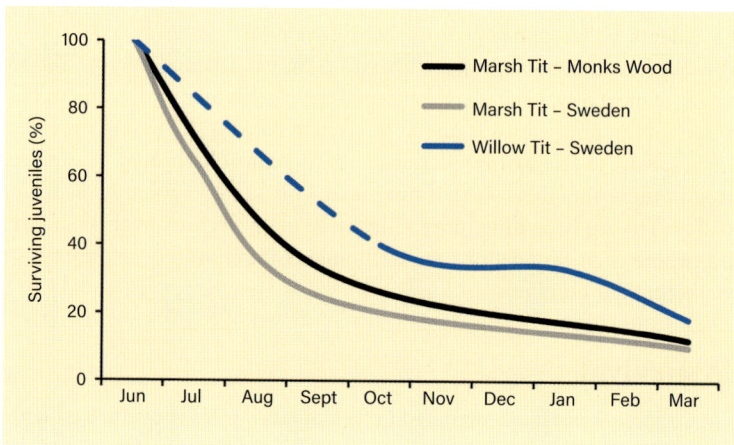

Figure 8.1. Estimates of juvenile survival over time for Marsh Tits at Monks Wood (averaged over 11 years, 614 colour-ringed birds, author's own data) and at Lund in Sweden (averaged over 4 years, 1,277 birds, estimated from Smith 1993b), and also for Willow Tits in southwest Sweden (averaged over 5 years, approximately 935 birds, estimated from Ekman 1984). The Willow Tit estimate includes counts of unringed birds during a period of dispersal and immigration (dashed line) and later counts of colour-ringed birds (solid line).

From these available studies, it seems that juvenile Marsh Tits and Willow Tits suffer their greatest levels of mortality shortly after becoming independent in the summer (Ekman 1984, Smith 1993b, Amann 1997, Schaub & Amann 2001, Broughton *et al.* 2010). If they are successful in finding and settling in a vacant home-range before their first autumn then their chances of survival improve significantly. However, the multiple risks and challenges in their first few months of life mean that only a small minority of fledglings will ever survive to breed.

Adult survival

Determining adult survival is more straightforward than for juveniles, as adults are more sedentary and less likely to disperse, so their local disappearance can only be due to death in the vast majority of cases. In all studies of Marsh Tits and Willow Tits that have good sample sizes, the annual survival rate usually shows that around half of the adult birds in a population survive from one year to the same point in the following year (Table 8.1). Usually these annual survival rates are measured between breeding seasons, as this is when territorial adults are most easily located.

Table 8.1. Annual survival of adult Marsh Tits and Willow Tits in colour-ringing studies. The survival percentages are averaged across the given study years and number of birds.

Location	Survival (%)	No. of birds	Years	Source
Marsh Tit				
Monks Wood, UK	55	486	2003–2023	Author's data
Bradfield Woods, UK	58	30	2015–2018	Author's data
Białowieża, Poland	59	253	1993–2000	Wesołowski 2001
Basel, Switzerland	47	117	1949–1953	Schaub & Amann 2001
Willow Tit				
Gothenburg, Sweden	53	c.200	1974–1981	Ekman & Askenmo 1986
Oulu, Finland	49	185	1978–1986	Orell & Koivula 1988
Oulu, Finland	59	1,230	1991–2000	Lampila et al. 2006
Budal, Norway	50	230	1999–2014	Hogstad & Slagsvold 2018

The annual survival rates for Marsh Tits and Willow Tits are similar to those of other tits and chickadees, although there is a critical lack of good survival information for Willow Tits outside of Fennoscandia, such as in Great Britain. Gavin Siriwardena (2004) analysed the BTO's general ringing data from Great Britain over about 45 years and calculated an annual survival of 42 per cent for adult Willow Tits, which is slightly lower than in the more intensive studies (Table 8.1), but this might simply reflect the less intensive method of data collection. There is also a particular lack of survival data for both species from their Asian populations. Only ringing studies can provide this information on individual fates and survival, using standard metal alloy rings, colour-rings or PIT tags, and whilst these activities are somewhat invasive they carry only a very small risk of injury or mortality for Marsh Tits and Willow Tits (Broughton 2015).

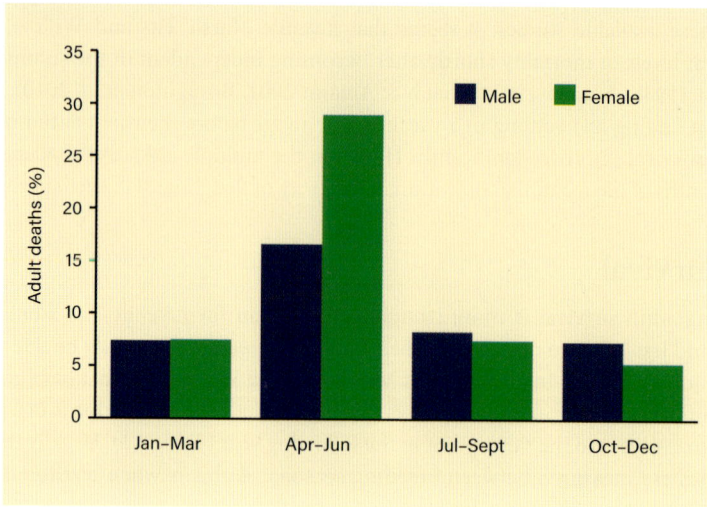

Figure 8.2. Timing of mortality for colour-ringed adult Marsh Tits at Monks Wood across the year, based on a bird's last sighting. The breeding season (April to June) has the highest mortality for both sexes. Data are from 2004–2008 for 108 males and 93 females, of which 89 birds died.

For adult Marsh Tits, the breeding season has the highest mortality rate of any time of year. Some adult females are killed inside the nest by predators, probably by mammals such as Stoats, Common Weasels, Pine Martens and Forest Dormice. This mortality appears to affect Marsh Tits much more than Willow Tits, although some female Willow Tits are also killed by predators inside the nest (Ekman 1984). In Monks Wood and the Białowieża National Park, however, nesting female Marsh Tits were killed on between a third and a half of the nest predation events during incubation (Broughton *et al.* 2011, Wesołowski 2023). Despite the apparently high proportion of female mortality during predator attacks on the nest, it actually accounts for relatively few breeding birds overall. In Monks Wood just 3 per cent of breeding female Marsh Tits were known to have been killed on the nest over an eight-year period, while another 5 per cent were killed away from the nest, presumably by Eurasian Sparrowhawks. Over the full breeding season of April to June, which includes the post-fledging period, more than a quarter of the female Marsh Tits at Monks Wood disappeared (Figure 8.2). Male mortality is also highest during the breeding season, but to a lesser extent than for females.

These seasonal mortality rates at Monks Wood differ from those calculated by Schaub & Amann (2001) for a Marsh Tit population in Switzerland, where adult survival rates were higher during the spring breeding season (March to June) than at any other time of the year. In Wytham Woods, meanwhile, Maziarz *et al.* (2023) found that fewer than 5 per cent of Marsh Tits died over the winter period, which means that the bulk of mortality must have taken place in other parts of the annual cycle. In the mild climate of Great Britain, then, winter does not appear to be the most challenging time for those Marsh Tits that are established in a home-range.

Over the full course of the year the annual survival rates for males and females are quite similar. Over 20 years in the Monks Wood study the survival rates of birds in all of the study

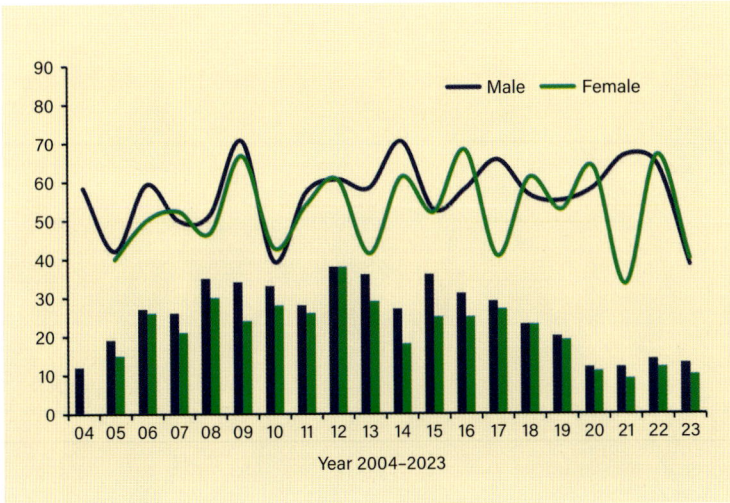

Figure 8.3. Fluctuation in the annual survival rates of adult male and female Marsh Tits in the Monks Wood study area, based on colour-ringing observations between spring breeding seasons. Lines are the percentage annual survival rates for each sex, and columns are the number of birds in the annual sample.

woods averaged 57 per cent for males and 53 per cent for females (Figure 8.3). There was huge annual variation in survival, however, particularly for females and at lower sample sizes. Despite this variation there was no significant long-term trend in adult survival rates for either sex.

Causes and consequences of mortality

Despite various predators attacking Marsh Tits when they are breeding, especially Great Spotted Woodpeckers, mustelids (Stoats, Common Weasels and Pine Martens) and Forest Dormice, the losses are not huge. In fact, the levels of total nest loss, which are mostly due to predation, are notably higher in the Białowieża National Park (28 per cent), where Marsh Tits are relatively stable, compared to the Monks Wood study (18 per cent) and elsewhere in Britain where Marsh Tits are in decline (Broughton *et al.* 2011, Wesołowski 2023). This is a strong sign that nest predators, and nest failure in general, are not the major driver of Marsh Tit population changes.

It is often difficult to disentangle other causes of mortality, as the actual death of a Marsh Tit or Willow Tit is rarely witnessed, and any remains of birds that die away from the nest are seldom found. The causes of mortality can also conflict with each other, as individuals that are killed by predators may have starved if they had lived longer. A bird on the point of starvation may take more desperate risks for food, which would make it more vulnerable to predators or accidents. General recoveries of ringed birds reported by the public are not very informative, as they are heavily biased to suburban gardens and other areas where people are present and are more likely to find a dead bird. For example, three-quarters of the recoveries of ringed Marsh Tits and Willow Tits from the British ringing scheme involve birds that were killed by cats or traffic (Wernham *et al.* 2002). Clearly, these recoveries are a very unrepresentative sample of what really kills most Marsh Tits and Willow Tits.

Figure 8.4. A dead juvenile Marsh Tit on the woodland floor at Monks Wood. It is unusual to find dead Marsh Tits or Willow Tits, and in many cases, such as this, the cause of death is not apparent. (© Richard K. Broughton)

In the more natural conditions of the Monks Wood study, aside from breeding females killed inside the nest, only six Marsh Tits from the study population have ever been found after dying, and it was not always obvious what was the cause. It is sometimes clear that an individual has been killed by a predator, such as remains that were found at the plucking posts of a Eurasian Sparrowhawk, or the rings found inside a Tawny Owl nest-box. Another bird was found dead on the side of a road, obviously killed by traffic. Other birds were simply found dead in the woodland, and the cause was unknown (Figure 8.4).

Across much of their European range, Eurasian Sparrowhawks and Pygmy Owls are the main predators that are likely to threaten Marsh Tits and Willow Tits on a regular basis throughout the year. Despite often being the most important predator of tits in western Europe, Eurasian Sparrowhawks are territorial and generally occur at low densities (Newton 1986), so their impact is probably limited. However, at Monks Wood we found that 12 per cent of Marsh Tit breeding territories failed to reach the nesting stage due to one or both of the occupants disappearing, which we suspected was largely due to sparrowhawk predation (Broughton *et al.* 2011). Sparrowhawks can also learn to target bird-feeders, and the very frequent visits by Marsh Tits and Willow Tits to collect food for caching can make them especially vulnerable, through their predictable behaviour. Of course, the issue here is the unnatural and semi-permanent concentration of food at feeding stations, which disrupts the normal behaviour of Marsh Tits, Willow Tits and the predatory sparrowhawks.

In Fennoscandia, Pygmy Owls also regularly ambush Willow Tits, and Ekman (1986) estimated that they foraged within an approximate 700m radius of their roost or food cache,

where they store the killed Willow Tits and other prey during winter (Figure 8.5). Although a lot of Willow Tits are eaten by Pygmy Owls, Ekman showed that the tits are by no means the owls' most favoured prey, and are usually not targeted disproportionately to their local abundance. Other ambush predators of Willow Tits in Fennoscandia include the Northern Hawk Owl, Great Grey Shrike and Siberian Jay (Hogstad & Slagsvold 2018).

Starvation is probably also a major cause of mortality, but this can be more than the simple lack of food. Food may be available, but interspecific competition from other tits might prevent Marsh Tits or Willow Tits from accessing enough of it. Several studies have demonstrated that Marsh Tits or Willow Tits are displaced from preferential foraging places by dominant Blue Tits, Great Tits or Crested Tits (Hogstad 1978, Krams 1996, Maziarz et al. 2023). This exclusion probably has survival consequences for some individuals, but these mortality effects have not yet been examined in relation to interspecific competition.

At the broader scale, two studies in Great Britain have looked for correlations between the abundance of Marsh Tits, Willow Tits and their competitors and predators, focusing on the breeding period (Siriwardena 2004, 2006). The analyses compared the presence or abundance of Marsh Tits or Willow Tits in the national survey data (Common Birds Census) with the co-abundance or annual fluctuations of the other competing tits and predatory woodpeckers or Eurasian Jays. Both studies failed to find any widespread negative relationships, and in fact the associations between Marsh Tits and potential competitors were positive. Comparisons were pooled across sites of varying quality, however, which could have smoothed any variation in the local trends. The analyses might also have been too limited in looking for annual increases of Blue Tits, Great Tits or Great Spotted Woodpeckers that coincided with annual declines or absences of Marsh Tits or Willow Tits, without considering the magnitude of any change. For example, all of the tits could increase in good breeding years and decline in bad ones, but for Marsh Tits or Willow Tits the increases could be smaller and the declines much greater than for the other species. Over time that would result in a divergence in their population trends, even if the annual ups and downs were synchronised between Marsh Tits, Willow Tits, and their competitors or their predators. It would be useful to look into these more detailed questions in the historical

Figure 8.5. Prey of Pygmy Owls cached in two nest-boxes in Finland. The caches include voles, Blue Tits, Great Tits, a Yellowhammer and several Willow Tits. (© Giulia Masoero)

data, although the national survey methods may not be very robust in representing the abundance of low-density Willow Tits and Marsh Tits (Siriwardena 2004).

In more intensive studies it is clear that other species can have strong local effects on Marsh Tits and Willow Tits, including foraging opportunities and nesting success (e.g. Parry & Broughton 2018, Maziarz *et al.* 2023), which were not detected in the broader analyses. This does not mean that wider impacts are not happening, as it would be surprising if big increases in major predators and competitors in woodland habitats did not make them more hostile for Marsh Tits or Willow Tits. In Great Britain the breeding abundance of Blue Tits has increased by more than a quarter, and Great Tits have almost doubled, over the 53 years that have seen Marsh Tits and Willow Tits declining by a respective 80 per cent and 94 per cent (Harris *et al.* 2022). Those trends do look suspicious for species that can all occur in the same habitats and have partially overlapping niches.

The impacts of competition on survival may also operate outside the breeding season, such as in the late summer or autumn. At this time there are peak numbers of juvenile tits all competing for the same foods, just as the invertebrate abundance is on the wane. In this scenario large numbers of dominant Great Tits and Blue Tits could outcompete the subordinate juveniles of Marsh Tits and Willow Tits, which could exacerbate the naturally high mortality during this period (Figure 8.1). Our data from the declining population at Monks Wood do suggest that juvenile Marsh Tits are under increasing pressure, as we have seen juvenile recruitment falling by 63 per cent over 20 years, despite breeding success and adult survival remaining buoyant. Clearly, something is piling greater pressure on the subordinate juveniles in late summer, which may well involve increased competition, declining food resources, or a combination of both.

Disease or parasites are other potential causes of mortality, but Marsh Tits and Willow Tits do not seem to be significantly affected. Blood parasites have been found in Willow Tits, including haemosporidian protozoa (Stanković *et al.* 2019, Rintamäki *et al.* 2000), and these infections may have fitness consequences, but they are not obvious. Avian pox virus (*Avipoxvirus* spp.), which has been reported in a few Marsh Tits or Willow Tits (Lawson *et al.* 2012), causes severe wart-like skin lesions. It has particularly affected Great Tits in Great Britain, and the lesions can become large growths on the head, legs, scapulars and wings that are so debilitating that the bird is unlikely to survive. I have seen only one Marsh Tit at Monks Wood with symptoms of avian pox virus, alongside very many infected Great Tits. The Marsh Tit had a small growth above the eye, which increased over a few weeks to about 5mm in size before regressing and disappearing. The bird survived for several more years with no obvious impact. Marsh Tits have also tested positive for other nasty bacteria, including *Chlamydia* spp. and antimicrobial-resistant *Escherichia coli*, although the birds appeared healthy at the time (Holzinger-Umlauf *et al.* 1997, Zurfluh *et al.* 2019).

SURVIVING THE NIGHT

Both species spend the night roosting, and so each bird needs to have enough energy reserves to see it through until the morning. Except for breeding females that spend the night with their chicks in the nest, each bird roosts singly throughout the year. Birds might sometimes roost in dense foliage in a bush or tree, but Marsh Tits mostly seem to use some kind of tree cavity throughout the year. Willow Tits may also roost in tree holes, and even hollows in deep snow, but can also roost in the open throughout the year (Reinertsen 1983). Sleeping inside a cavity or hollow gives some protection from the elements and reduces energy loss from wind chill, and also hides the bird from nocturnal predators.

While roosting, the birds appear to fall into a very deep sleep. For a few weeks I looked after an injured Marsh Tit from Monks Wood, which was kept indoors in a large wooden box-cage while it recovered, and I could see its roosting behaviour. Around sunset the bird would go to sleep on the highest perch in the cage, against the side of the box, with its body feathers puffed out and its head tilted upwards. A little later the head would be turned around and tucked into the feathers on the back, and the body feathers would be so puffed out that the bird looked almost spherical. It slept so deeply that it seemed oblivious when I checked on it, and it showed no alertness to any small sound that I made. Such roosting birds must be very vulnerable if discovered by a predator during the night, like a Common Weasel or Forest Dormouse.

When radio-tracking Marsh Tits at Monks Wood we followed one female to its roosting cavity behind loose bark on a dead birch, about 1.5m above the ground. This female went to bed quite early during the winter, about half an hour before dark. Marsh Tits and Willow Tits both emerge from their roosting sites shortly after dawn each morning. This means that Willow Tits at the northern edge of their range in the boreal forests may have less than six hours of full daylight during midwinter in which to feed and store enough fat reserves to avoid night-time starvation. They can achieve this by reducing their metabolism and lowering their body temperature by several degrees Celsius, effectively going into a nocturnal hypothermia. This can reduce their energy consumption by up to 10 per cent, extending their ability to withstand long nights before they can feed again in the morning (Reinertsen 1983).

Birds do sometimes die inside their roost holes. At Monks Wood one day I was searching for a particular male Marsh Tit in his territory, to check he was still present, but I couldn't find him. While searching I passed an interesting-looking cavity at head-height in a young elm tree, and I peeked inside with a small torch, as I often do with accessible tree cavities. I was amazed to find the Marsh Tit in there, in a sleeping posture but quite dead. There was no obvious injury or trauma, and I concluded that he had simply died in his sleep, perhaps through having insufficient fat reserves to get him through the February night. In Japan, Matsukoa & Kudo (2009) found a Marsh Tit that had apparently suffocated in its roost hole after snow and ice covered the entrance during the night while it was asleep, sealing it inside.

Tits roosting in tree cavities may also suffer feather damage that can reduce the integrity of their plumage, potentially increasing their energetic cost through reduced insulation. I see this feather damage occasionally on Marsh Tits in England, and strongly suspect the cause to be the larvae of moths and beetles that overwinter in tree cavities and feed on old nest material, such as feathers, fur and plant material. The belly and flanks of the roosting birds are presumably in contact with the cavity walls, allowing the larvae to graze on the feathers. Some birds are so heavily affected that they have large dark patches on the underparts where the tips of the feathers have been eaten away, exposing the dark bases (Figure 8.6). The numbers of affected Marsh Tits at Monks Wood have not been sufficient to test for a survival impact, and several birds have certainly survived the winter with severe feather loss, but in cold weather it must surely have an energetic cost.

Figure 8.6. A colour-ringed Marsh Tit caught for a brief examination at Monks Wood, showing feather damage on the underparts that is presumed to be caused by moth or beetle larvae in its roost hole, which graze on its feathers during the night. Marsh Tits with such dark, abraded patches on the underparts are seen occasionally during winter and spring. (© Richard K. Broughton)

SEX-BIASED MORTALITY

Habitat fragmentation can influence mortality and recruitment, and this can have serious consequences for declining populations. Juvenile female Marsh Tits and Willow Tits disperse further than males (see Chapter 7), and in landscapes that have small patches of heavily fragmented woodland, such as in Great Britain, this can lead to greater female mortality. Whereas juvenile males might settle within a neighbouring territory in the woodland patch in which they were born, females more often leave their natal woodland in order to disperse further than their male siblings. These females then have to cross fairly hostile habitat, such as open farmland, in order to try and find another wood in which to settle. Similarly, female immigrants have to cross the same hostile landscapes in order to replace the female emigrants and any other losses of breeding females. Unsurprisingly, juvenile females are less successful than males at dispersal and settling, due to the greater risks they face when moving between woodlands, such as exposure to predators, failure to find enough food, or getting lost. In the fragmented woodlands of the Monks Wood study area, despite an even sex ratio among fledglings, there was frequently a deficit of juvenile females recruited to the population. At Monks Wood itself, over the 20 years from 2004 to 2023 we found unpaired males present in more than half (60 per cent) of the breeding seasons, averaging 14 per cent of territorial males and ranging up to 23 per cent each spring. By contrast, unpaired females were only found on a few occasions, and this clearly showed that by the time of breeding not enough females had been recruited or immigrated to balance the population.

A male bias due to higher female mortality and lower recruitment during dispersal is a distinctive feature of declining and increasingly isolated bird populations that are becoming vulnerable to local extinction (Dale 2001). In Great Britain, including at Monks Wood, Marsh Tit populations in several woodlands have been shown to have a male bias of around 55–60 per cent males, with only 40–45 per cent females, whereas stable populations elsewhere in Europe typically have an even sex ratio (Broughton & Hinsley 2015). The unpaired males at Monks Wood are therefore a symptom of the decline that the population is experiencing. Similar imbalances in the sex ratio are highly likely to be present within Great Britain's rapidly declining Willow Tit populations, resulting from the same processes of female-biased dispersal and mortality, but analysing this has been hampered by a critical lack of demographic data.

Longevity

An annual survival rate of around 50 per cent for adult Marsh Tits and Willow Tits means that from a cohort of 100 breeding birds only three individuals (3 per cent) will be expected to be alive five years later. At Monks Wood, however, the Marsh Tits slightly outperformed this projection, with five-year-olds making up an average 4.6 per cent of adults within the main study population (Figure 8.7). This is because birds with a few years of experience seem to have a better survival rate than first-years. For example, while the overall proportion of two-year-olds in the population was almost half that of the one-year olds, suggesting a virtually 50 per cent mortality rate among first-years, the proportion of two-year-olds did not halve the following year, but instead the attrition was only 40 per cent. This means that two-year-olds had a higher survival rate into the next year than did one-year-olds, on average. By the time birds get to five years or older, however, the attrition rate increases back up to 50 per cent or more, suggesting that older birds are past their peak fitness. In Monks Wood, then, Marsh Tits aged 2–4 years old seem to have an optimum mix of experience and vigour, giving them a better than even chance of surviving into the next year. Only 8 per cent of birds lived longer than this and, of the 1,693 individuals ringed in the wider Monks Wood study over 22 years, the oldest individuals survived for nine years.

Hogstad & Slagsvold (2018) found very similar rates of survival and longevity in a Norwegian population of Willow Tits, where 51 per cent of adults survived over one winter, 25 per cent survived for two winters, 9 per cent for three winters and just 6 per cent of birds survived four winters. The oldest birds in this 15-year study survived for at least nine years. Annual survival rates were similar for males and females and were unrelated to body size, so larger birds did not have any better chances than small ones. Instead, the main factor that underpinned survival was getting established in a dominant position in a winter flock during their first autumn, which led to a permanent home-range and a breeding territory. As with Marsh Tits, therefore, the key to surviving and having the potential for a long life is to acquire a home-range immediately after dispersal, which sets the course and duration

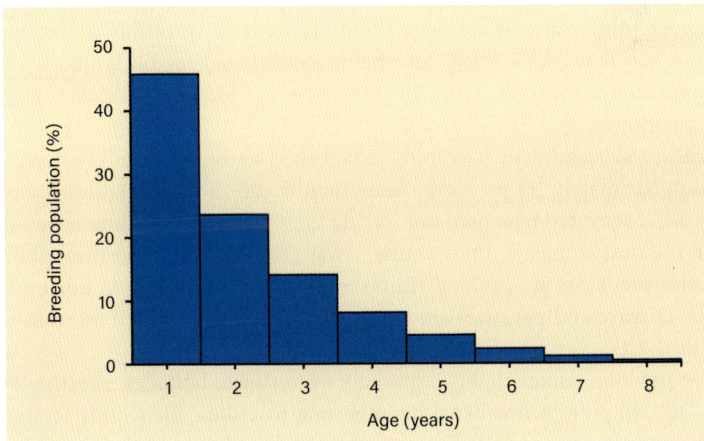

Figure 8.7. The age distribution of the Marsh Tit breeding population in Monks Wood, averaged over 2005–2020. Data are based on an average of 36 birds per year.

of the rest of their lives. Most birds that do not find a vacancy by the autumn or end the winter as a territory-holder do not live much longer. Those that succeed might live for over a decade more.

The oldest recorded individuals from national ringing schemes in Europe include a male Marsh Tit from Sweden that survived for at least 13 years and two months after ringing, when it was recaptured by a ringer, and a Marsh Tit from Switzerland that was found dead 13 years and 8 months after being ringed. A Willow Tit from Finland survived for at least 11 years and 4 months, when it was recaptured by a ringer (Fransson *et al.* 2023). These birds would have been older than they appeared, as they were ringed when already full-grown. In Great Britain the oldest recorded Marsh Tit was ringed as a juvenile at Catterick Garrison, in North Yorkshire, in July 2004. It was recaptured in the same place 11 years and three months later. This bird would have hatched in May 2004, so it could be quite accurately aged at 11 years and five months at the time of recapture, but it was not seen again. The oldest recorded Willow Tit in Great Britain was ringed as an adult in June 1974 at Attenborough Nature Reserve in Nottinghamshire, and was recaptured there 10 years and four months later, meaning that the bird was over 11 years old (Robinson *et al.* 2023). These ringing data are useful by indicating the maximum likely lifespan of Marsh Tits and Willow Tits, which seems to be around 12–14 years. Other similar-sized tits and chickadees have much the same longevities from ringing records (Smith 1991, Fransson *et al.* 2023), showing that this is probably the limit for parids of this size. Generally, however, almost all individuals survive for far shorter periods than this. At Monks Wood, in a typical year, 70 per cent of the Marsh Tit breeding population was just 1–2 years old, and that is how long the majority of individuals can expect to live before they succumb to a predator or starvation.

Summary

The period of greatest mortality for Marsh Tits and Willow Tits is in their first few months of life. A significant minority die as nestlings, mostly due to predators such as Great Spotted Woodpeckers or Common Weasels, but around two-thirds of juveniles die around the time of independence, dispersal and settling. These juveniles are probably killed by predators, starvation and accidents, with these risks being exacerbated for birds that try to disperse between fragmented woodlands. Only 10–20 per cent of local juveniles survive to join the breeding population.

Birds that are successful in becoming established as breeding adults have an annual survival of approximately 50 per cent. Many adults die in the breeding season, such as females that are discovered by a predator during incubation or brooding, when they can be killed inside the nest alongside their young. Away from the nest and throughout the year the other major predators are probably Eurasian Sparrowhawks and, in northern latitudes, Pygmy Owls. Disease and parasites are not currently major causes of mortality for Marsh Tits and Willow Tits.

To survive the long winter nights, especially in northern latitudes, roosting Willow Tits and Marsh Tits can enter a nocturnal hypothermia to reduce their body temperature and energy needs until the morning. Birds are sometimes found dead in their roost holes, due to starvation or accidents, but may also be vulnerable to nocturnal predators.

In stable populations the adult mortality is compensated by the juvenile recruits each year. As a result, around half of the annual breeding population is made up of birds in their first year, and typically more than two-thirds of birds are 1–2 years old. Marsh Tits and Willow Tits can potentially live for 12–14 years, but vanishingly few birds reach such an age. Birds that reach 6–8 years old have done very well.

Stable populations generally have an even sex ratio where mortality is balanced by recruitment for both males and females. However, female-biased dispersal and higher mortality mean that populations in fragmented habitats can suffer from an imbalance, with more males than females in the breeding population. This male-biased sex ratio is a feature of declining populations in patchy woodland, where females are less successful in dispersing between woods to be recruited into the population.

There is a critical lack of demographic data for many regions and populations, especially for Willow Tits in temperate woodlands, including Great Britain, and for both species in Asia. Marking and monitoring individuals by ringing and tagging can provide this high-quality information, which is urgently needed to better understand the population declines of both species.

CHAPTER 9

Conservation conclusions

A dominant theme through this book has been the widespread decline and dire conservation status of Marsh Tits and Willow Tits across parts of their ranges. Compounding these trends are the significant gaps in knowledge, which can limit our ability to formulate coherent and effective conservation polices. In Great Britain, Marsh Tit abundance is now just a fifth of what it was in the 1970s, while Willow Tit abundance has fallen to less than a tenth (Massimino *et al.* 2023). Both species have been lost from large parts of the country, including almost all of Scotland, while the Willow Tit has also been lost from most of southern and eastern England. With no sign of stabilisation in the declines, the British Willow Tit is at very real risk of extinction in the coming decades (Figure 9.1). Here, I summarise the relevant science from earlier chapters to outline what we know about the declines of both species, and what actions could be taken to conserve them, with a focus on the British context.

Habitat change

Chapter 3 outlined how both species rely on extensive wooded habitats. Marsh Tits are specialists of mature woodland with a complex, multi-layered structure and high species diversity, typified by old-growth forest and ancient woodland. Intensively managed woodland and plantations are relatively poor habitat, but Marsh Tits can sometimes do well in woods with long-rotation coppice-with-standards. The widespread ending of intensive management has seen most British woodlands maturing and increasing in naturalness over the past century, developing more sub-canopy understorey, a taller canopy height and higher structural diversity (Kirby *et al.* 2005, Hopkins & Kirby 2007, Amar *et al.* 2010). The abundance of deer and their browsing activity have also increased, but this has not reduced the woodland understorey in the all-important foraging layers for Marsh Tits (and Willow Tits), particularly above 2m and into the sub-canopy, which have actually expanded (Amar *et al.* 2010, Broughton 2012c). As such, the trends in woodland habitat structure appear to have been moving in the Marsh Tit's favour, meaning that deer impacts or reduced management are unlikely to have driven their widespread decline. Consequently, reintroducing woodland management would seem unlikely to significantly reverse those declines.

For Willow Tits, too, woodland changes do not really explain the species' collapse across Great Britain. Just 40 years ago Willow Tits were still widespread in a range of wooded habitats, from ancient woodlands to farmland copses and pine plantations, essentially as woodland generalists. Although Willow Tits today are primarily associated with wet, early-successional woodlands in Great Britain, this habitat represents a refuge only because Willow Tits have widely been lost from most other types of woodland (Siriwardena 2004, Stewart 2010). Even in wetland habitats, however, Willow Tits continue to decline, and they have already been lost from flagship reserves like in the extensive Dearne Valley in South Yorkshire and Far Ings in Lincolnshire, despite targeted conservation management (Pinder & Carr 2021). As with Marsh Tits, then, the issue of Willow Tit habitat appears to go far beyond vegetation structure and wetness. It therefore seems unlikely that the solution to reversing

Figure 9.1. A Willow Tit in northern England. With no sign of a slowdown in the species' decline, British Willow Tits are at a real risk of national extinction. (© Philip Schofield)

Willow Tit declines is via site-based habitat management, like coppicing, which has been ineffective so far.

What is clear from earlier research is that larger areas of diverse woodland in well-wooded landscapes buffer both species from declines, making them less likely to disappear (Hinsley *et al.* 1995a, 1995b, Broughton *et al.* 2013). This is because bigger patches of woodland offer a greater diversity of habitat niches to support Marsh Tits and Willow Tits throughout the year, and more habitat to support more pairs, with their large territories and even larger home-ranges (see Chapter 5). Well-wooded landscapes also have better connectivity, allowing juveniles to disperse between territories to replace mortality and form new pairs. Expanding and linking together the existing woodlands in regions where both species remain would therefore seem to offer major benefits in terms of population resilience. The pattern of land ownership in Great Britain, and the time needed to create or regenerate diverse wooded landscapes, means that woodland expansion is challenging, and it may not come soon enough for many populations of Marsh Tits or Willow Tits. Nevertheless, targeting grant-funded hedgerow planting and woodland creation towards linking and expanding existing habitat would be a positive strategy. At the very least, protecting the current wooded habitat in and around Marsh Tit and Willow Tit populations should be a priority. Loss of woodland and shrubland, such as commercial or housing development on vulnerable 'brownfield' bushy sites, only exacerbates the fragmentation of habitats and populations, speeding up local declines.

In Fennoscandia there is a very clear link between forestry intensification, increased logging and the rapid decline of Marsh Tit and Willow Tit populations (see Chapter 3). The changes in habitat structure resulting from intensive forestry management, and the increased fragmentation of suitable remaining habitat, have been shown to degrade the birds' territory quality, isolate remaining populations and force them to work harder to make a living in the logged forests. Greater regulation of the forestry industry, improvement of practices and an increase in protected areas are obvious countermeasures, and these have helped to improve the situation in some parts of Fennoscandia, but major issues remain. The situation for Willow Tits in the vast forests of northern Asia is relatively unknown, but major habitat change through rampant logging and climate change is likely to be occurring on a huge scale.

Nest-site provision

As covered in Chapters 3 and 6, provision of nest-sites is a conservation 'red herring' for Marsh Tits and Willow Tits. Only in rare circumstances are nest-sites in short supply in natural forests or semi-natural woodlands, including native woodland in Great Britain. Instead, the major limiting factor is the area of available wooded habitat that can accommodate the birds' large territories. There is now good evidence that providing nest-boxes for either species will have no tangible benefit, and could do more harm than good. This is because the nest-boxes will only rarely be used by the target species and will instead be mostly taken up by Blue Tits, which are even able to excavate boxes filled with a substrate for Willow Tits, like wood chips or deadwood. There is no way of excluding Blue Tits from nest-boxes designed for Marsh Tits or Willow Tits, and so they are best avoided so as not to increase local Blue Tit productivity and abundance, which will only increase future nest competition. If natural nest-sites are genuinely scarce, then providing natural or artificial deadwood snags is the best option, either by ring-barking a small number of living trees or by strapping dead logs to trees. This allows Willow Tits to excavate natural nest-cavities that are better hidden from Blue Tits and predators. For Marsh Tits, by the time that woodland is mature enough to support them then it will already contain suitable nest-sites, and so nest-boxes are unnecessary and will only benefit local Blue Tits.

Bird-feeding and competition

Alongside nest-boxes, the widespread feeding of birds in gardens, woodlands and nature reserves also disproportionately benefits Blue Tits and Great Tits, improving their overwinter survival, inflating their abundance, and so increasing the number of potential competitors for Marsh Tits and Willow Tits (Shutt *et al.* 2021, Shutt & Lees 2021, Broughton *et al.* 2022c). This is a contentious topic, as bird-feeding is a very large industry and preliminary studies have not supported any simple negative relationship between Marsh Tits, Willow Tits and potential competitors (Siriwardena 2004, 2006). However, it is clear from the more detailed evidence outlined in Chapters 3, 4 and 6 that local competition effects are real and widespread, with Marsh Tits and Willow Tits being displaced from nest-sites or foraging areas by dominant species such as Blue Tits and Great Tits. The broad national datasets of tit abundance used in earlier analyses may lack the sensitivity to detect competition effects, given the survey methods that they are based upon, which are known to be relatively poor at representing low-density species like Marsh Tits and Willow Tits (Siriwardena 2004). This limits the scope for more complex analyses, such as non-linear or habitat-specific effects, and so the detailed local studies are probably more informative.

Many people get a lot of pleasure from bird-feeding, and conservation charities have commercial relationships with bird-food businesses that support their work, but the sheer scale of bird-feeding in Great Britain and elsewhere in Europe represents a massive intervention that changes our bird communities (Plummer *et al.* 2019). The implications of this vast experiment in food provision on competition effects have received surprisingly little attention. Regardless, there are good reasons for inferring that bird-feeding significantly benefits dominant and generalist Blue Tits and Great Tits, and also predatory Great Spotted Woodpeckers, as their populations have risen substantially alongside bird-feeding trends (Plummer *et al.* 2019, Massimino *et al.* 2023). It has also been demonstrated at the local scale that greater numbers of these species have negative impacts on Willow Tits and Marsh Tits (Maxwell 2002, Parry & Broughton 2018, Maziarz *et al.* 2023). In particular, the poor nesting success among British Willow Tits is a major red flag, especially with the known bias that underestimates nest losses (see Chapter 6). Consequently, even if national-scale effects have not been confirmed, a precautionary conservation action for Marsh Tits or Willow Tits would be to avoid favouring their potential competitors and predators through supplementary bird-feeding or nest-box provision, particularly in the habitats where these priority species still exist (Figure 9.2). For example, undermining the local quality of remaining Willow Tit habitats by inflating competitor abundance via bird-feeding or nest-boxes is clearly counter-productive.

Figure 9.2. Conserving Marsh Tits (shown) and Willow Tits in Great Britain may involve doing fewer favours for dominant species via bird-feeding and nest-box provision, at least as a precautionary measure until further research clarifies the situation. (© Adam Nicolson)

Knowledge gaps

One of the biggest obstacles to conserving Marsh Tits and Willow Tits is the limited data on which to base actions. Whilst several potential factors can now be discounted, such as nest failure for Marsh Tits or deadwood availability for Willow Tits, we still do not yet know enough to understand the combined drivers of their declines, or how to tackle them. There is a critical lack of long-term, well-resourced, comprehensive research into basic ecology and demography. Indeed, from the late 1940s until the early 2000s there was not a single major study in Great Britain that investigated the territory dynamics, habitat use or breeding success of either species. This is despite the obvious decline of both from the 1970s. By the 2000s, when the conservation urgency had prompted new research, it had become difficult to find stable and functioning populations to study. Severely declining populations do not behave in the same way as successful ones, due to a breakdown in social organisation and shifts in habitat occupation, which eventually become stochastic. For example, in the 1970s British Willow Tits could have been studied in ancient woodland (see Chapter 3), but this is impossible today as they are virtually extinct in this habitat. Instead, the current focus is on young wet woodland where Willow Tits are considered as specialists, as this is the only habitat in which they still occur in sufficient numbers to study (Broughton *et al.* 2020, Pinder & Carr 2021). This demonstrates how the timing of research can influence what answers it produces. It is important to have long-term, comprehensive population data from habitats and regions where populations are stable, as well as where they are declining. For Willow Tits, this may require research projects in countries other than Great Britain, which may still have stable populations.

As discussed in Chapters 4, 5 and 7, we still know relatively little about what Marsh Tits and Willow Tits eat throughout the year, and especially how that food availability might have changed. There is a widespread assumption that invertebrates have declined, but we cannot say whether food availability is a factor in the decline of Marsh Tits and Willow Tits, because the data are lacking. Similarly, there is very little information on the survival rates of Willow Tits in Great Britain, or indeed from most other regions. The Monks Wood study shows that juvenile survival and recruitment seem to be driving the local decline of Marsh Tits, but it is unknown whether that pattern is more widespread, or if it also applies to Willow Tits, and what is really causing it. These are examples of research priorities for which the conservation sector has been quite poor at developing coherent strategies and leadership across a range of partners. Better approaches are needed to understand exactly what is happening with declining woodland birds, such as Marsh Tits and Willow Tits, to harness and support research effort that can tackle species declines more effectively.

Future prospects

In Chapter 1 we saw that the original native range of Marsh Tits and Willow Tits may well have covered almost all of Great Britain, before deforestation, and so this is perhaps the scale at which we should be thinking for national conservation and restoration. Large-scale rewilding and reforestation may provide extensive new wooded habitats that could benefit both species. In Scotland, in particular, rewilding and restoration of native woodland in regions with low urbanisation promises to create large areas of suitable habitat, such as in the Affric Highlands and Carrifran Wildwood. These woodland restoration projects in the Scottish uplands are far from human population centres and their ubiquitous bird-feeding, and their higher elevation and northerly latitude also buffers against the extremes of climate warming seen in southern Great Britain. As they establish and develop, new woodlands like these could provide both Marsh Tits and Willow Tits with the extensively wooded, low-competition environment that they need. Such conditions are likely to become increasingly scarce in England.

A problem with this idea is that the dispersal abilities of Marsh Tits and Willow Tits will not enable them to colonise much of Scotland without assistance, as it is too far from existing populations. A reintroduction through translocation would be the only option, but this has not been attempted for either species, or indeed with any tit species. The skills needed for such an undertaking would have to be learnt, and experience gained, and it would require long-term investment to achieve success. Such measures may be a last resort, but it may be worthwhile formulating a strategy as a contingency, and developing the necessary techniques.

The conservation status of Marsh Tits and Willow Tits in Great Britain, and elsewhere in many parts of Europe, currently looks quite bleak. The population trends of both species are clearly telling us that our woodland ecosystems are under severe strain. This is part of the wider biodiversity crisis, which is closely bound with the climate crisis that we currently face and which overshadows everything. As a society, however, we do have the skills, knowledge and expertise to tackle species declines and restore ecosystems, and to avert the worsening impacts of climate change. There is a large and knowledgeable community of ecologists in scientific institutes, government agencies and NGOs, and among the wider public throughout Great Britain and Europe, who are motivated to conserve our natural heritage. Organising and resourcing that community to research and act upon an effective conservation strategy for our ecosystems is the challenge for society and its governments.

Marsh Tits and Willow Tits are just two of the many sentinels of the modern biodiversity crisis, but the fortunes of these two birds are inextricably linked to the integrity of the woodlands and forests where they live, and reflect our impacts upon them. The growing appreciation of the need for ecosystem restoration, rewilding and large-scale conservation among the public, scientists, landowners and governments offers some hope that Marsh Tits and Willow Tits may continue to find space to exist in future landscapes. Both species are very good indicators of woodland habitat quality, and so if we can ensure that Marsh Tits and Willow Tits are able to thrive at the local, national and regional scales, then it will also safeguard myriad other species and the wider forest ecosystem.

An adult Marsh Tit. (© Burhan Erden/Shutterstock)

APPENDIX 1

Scientific names of species mentioned in the text

Birds

Azure Tit *Cyanistes cyanus*
Azure-winged Magpie *Cyanopica cyanus*
Black-bibbed Tit *Poecile hypermelaenus*
Black-billed Magpie *Pica pica*
Black-capped Chickadee *Poecile atricapillus*
Black Stork *Ciconia nigra*
Blue Tit *Cyanistes caeruleus*
Bohemian Waxwing *Bombycilla garrulus*
Boreal Chickadee *Poecile hudsonicus*
Brambling *Fringilla montifringilla*
Carolina Chickadee *Poecile carolinensis*
Caspian Tit *Poecile hyrcanus*
Chestnut-backed Chickadee *Poecile rufescens*
Coal Tit *Periparus ater*
Collared Flycatcher *Ficedula albicollis*
Common Crane *Grus grus*
Common Nightingale *Luscinia megarhynchos*
Common Pheasant *Phasianus colchicus*
Common Starling *Sturnus vulgaris*
Crested Tit *Lophophanes cristatus*
Eurasian Jay *Garrulus glandarius*
Eurasian Nuthatch *Sitta europaea*
Eurasian Sparrowhawk *Accipiter nisus*
Eurasian Treecreeper *Certhia familiaris*
Eurasian Woodcock *Scolopax rusticola*
Eurasian Wryneck *Jynx torquilla*
Goldcrest *Regulus regulus*
Great Grey Shrike *Lanius excubitor*
Great Spotted Woodpecker *Dendrocopos major*
Great Tit *Parus major*
Japanese Pygmy Woodpecker *Yungipicus kizuki*
Japanese Tit *Parus minor*

Large-billed Crow *Corvus macrorhynchos*
Lesser Spotted Woodpecker *Dryobates minor*
Long-tailed Tit *Aegithalos caudatus*
Marsh Tit *Poecile palustris*
Mountain Chickadee *Poecile gambeli*
Narcissus Flycatcher *Ficedula narcissina*
Northern Hawk Owl *Surnia ulula*
Père David's Tit *Poecile davidi*
Pied Flycatcher *Ficedula hypoleuca*
Pygmy Owl *Glaucidium passerinum*
Siberian Jay *Perisoreus infaustus*
Siberian Tit *Poecile cinctus*
Sichuan Tit *Poecile weigoldicus*
Sombre Tit *Poecile lugubris*
Tawny Owl *Strix aluco*
Varied Tit *Sittiparus varius*
White-browed Tit *Poecile superciliosus*
Willow Tit *Poecile montanus*
Willow Warbler *Phylloscopus trochilus*
Wood Warbler *Phylloscopus sibilatrix*
Yellowhammer *Emberiza citrinella*

Mammals and reptiles

Aurochs *Bos primigenius*
Bank Vole *Myodes glareolus*
Brown Bear *Ursus arctos*
Brown Hare *Lepus europaeus*
Common Weasel *Mustela nivalis*
Eurasian Badger *Meles meles*
Eurasian Beaver *Castor fiber*
Eurasian Elk *Alces alces*
Eurasian Lynx *Lynx lynx*
Eurasian Wolf *Canis lupus*
European Bison *Bison bonasus*
Forest Dormouse *Dryomys nitedula*
Grey Squirrel *Sciurus carolinensis*

Korean Rat-snake *Elaphe anomala*
Muntjac Deer *Muntiacus reevesi*
Pine Marten *Martes martes*
Rabbit *Oryctolagus cuniculus*
Red Deer *Cervus elaphus*
Red Squirrel *Sciurus vulgaris*
Reindeer *Rangifer tarandus*
Roe Deer *Capreolus capreolus*
Steppe Rat-snake *Elaphe dione*
Stoat *Mustela erminea*
Wild Boar *Sus scrofa*
Wood Mouse *Apodemus sylvaticus*
Yellow-necked Mouse *Apodemus flavicollis*

Invertebrates

Barklice Psocoptera
Bumblebees *Bombus* spp.
Caddisflies Trichoptera
Common Wasp *Vespula vulgaris*
Craneflies Tipuloidea
Hen Flea *Ceratophyllus gallinae*
Large Yellow Underwing *Noctua pronuba*
Mayflies Ephemeroptera
Mosquitoes Culicidae
Noctuid moths Noctuidae
Saxon Wasp *Dolichovespula saxonica*
Stoneflies Plecoptera
Tortrix moths Tortricidae
Tree Bumblebee *Bombus hypnorum*
Winter Moth *Operophtera brumata*

Trees and shrubs

Alder *Alnus* spp.
Ash *Fraxinus* spp.
Aspen *Populus* spp.
Beech *Fagus* spp.
Bilberry *Vaccinium myrtillus*
Birch *Betula* spp.
Bird Cherry *Prunus padus*
Black Alder *Alnus glutinosa*
Black Locust *Robinia pseudoacacia*
Black Mulberry *Morus nigra*
Blackthorn *Prunus spinosa*
Bramble *Rubus fruticosus*

Cajander Larch *Larix cajanderi*
Cedar *Cedrus* spp.
Chinese Ash *Fraxinus chinensis*
Chosenia *Chosenia arbutifolia*
Common Ash *Fraxinus excelsior*
Common Beech *Fagus sylvatica*
Common Buckthorn *Rhamnus cathartica*
Common Dogwood *Cornus sanguinea*
Common Elder *Sambucus nigra*
Common Hazel *Corylus avellana*
Common Hornbeam *Carpinus betulus*
Corsican Pine *Pinus nigra*
Cotoneaster *Cotoneaster* spp.
Crab Apple *Malus sylvestris*
Cranberry *Vaccinium oxycoccus*
Currant *Ribes* spp.
Daimyo Oak *Quercus dentata*
Elm *Ulmus* spp.
European Aspen *Populus tremula*
European Hop-hornbeam *Ostrya carpinifolia*
European Larch *Larix decidua*
Eucalyptus *Eucalyptus* spp.
Field Maple *Acer campestre*
Fir *Abies* spp.
Flowering Currant *Ribes sanguineum*
Hawthorn *Crataegus* spp.
Hazel *Corylus* spp.
Hemlock *Tsuga* spp.
Holly *Ilex* spp.
Holm Oak *Quercus ilex*
Honeysuckle *Lonicera periclymenum*
Hornbeam *Carpinus* spp.
Japanese Lime *Tilia japonica*
Japanese Red Pine *Pinus densiflora*
Japanese Sumac *Rhus trichocarpa*
Japanese White Birch *Betula platyphylla*
Jolcham Oak *Quercus serrata*
Juniper *Juniperus communis*
Korean Pine *Pinus koraiensis*
Larch *Larix* spp.
Lime *Tilia* spp.
Manchurian Fir *Abies holophylla*
Manchurian Walnut *Juglans mandshurica*
Maple *Acer* spp.
Mongolian Oak *Quercus mongolica*

Mongolian Poplar *Populus suaveolens*
Norway Maple *Acer platanoides*
Norway Spruce *Picea abies*
Oak *Quercus* spp.
Olive *Olea europaea*
Painted Maple *Acer pictum*
Pear *Pyrus* spp.
Pedunculate Oak *Quercus robur*
Pine *Pinus* spp.
Poplar *Populus* spp.
Raspberry *Rubus idaeus*
Rowan *Sorbus aucuparia*
Sallow *Salix caprea* or *Salix cinerea*
Sawtooth Oak *Quercus acutissima*
Scots Pine *Pinus sylvestris*
Sessile Oak *Quercus petraea*
Siberian Fir *Abies sibirica*
Siberian Spruce *Picea obovata*
Silver Birch *Betula pendula*
Small-leaved Lime *Tilia cordata*
Snowberry *Symphoricarpos albus*
Spindle *Euonymus europaeus*
Spruce *Picea* spp.
Sugar Maple *Acer saccharum*
Sweet Chestnut *Castanea sativa*
Sycamore *Acer pseudoplatanus*
Tree Aralia *Aralia elata*
White Poplar *Populus alba*
Wild Cherry *Prunus avium*
Wild Privet *Ligustrum vulgare*
Wild Service Tree *Sorbus torminalis*
Willow *Salix* spp.
Yew *Taxus baccata*

Other plants

Bamboo Bambusoideae
Barley *Hordeum vulgare*
Black Bryony *Dioscorea communis*
Burdock *Arctium* spp.
Cabbage Thistle *Cirsium oleraceum*
Common Hemp-nettle *Galeopsis tetrahit*
Common Nettle *Urtica dioica*
Common Reed *Phragmites australis*
Cow Parsley *Anthriscus sylvestris*
Duckweed Lemnoideae
Dwarf bamboo *Bambusa* spp.
Flat Neckera moss *Neckera complanata*
Great Reedmace *Typha latifolia*
Hemp *Cannabis sativa*
Japanese Hop *Humulus japonicus*
Knapweed *Centaurea* spp.
Maize *Zea mays*
Millet *Panicum* spp.
Mistletoe *Viscum album*
Nyjer *Guizotia abyssinica*
Oat *Avena sativa*
Peanut *Arachis hypogaea*
Plait-moss *Hypnum cupressiforme*
Pumpkin *Cucurbita* spp.
Scabious *Scabiosa* spp.
Sunflower *Helianthus annuus*
Thistle *Cirsium* spp.
Violet *Viola* spp.
Wheat *Triticum aestivum*
White Bryony *Bryonia dioica*
Wood Sorrel *Oxalis acetosella*
Woody Nightshade *Solanum dulcamara*

APPENDIX 2

Abbreviations and acronyms used in the text

BTO – British Trust for Ornithology

CEDA – Centre for Environmental Data Analysis

DBH – diameter at breast height

EGI – Edward Grey Institute of Field Ornithology

EURING – European Union for Bird Ringing

GIS – geographical information system

Lidar – light detection and ranging

NERC – Natural Environment Research Council

NGO – non-governmental organisation

PAWS – planted ancient woodland site

PIT tag – passive integrated transponder tag

RSPB – Royal Society for the Protection of Birds

UKCEH – UK Centre for Ecology & Hydrology

APPENDIX 3   Biometrics of English Marsh Tits of known age and sex

	Female					Male				
	Mean	s.d.	Median	Range	No. of birds	Mean	s.d.	Median	Range	No. of birds
Wing length (mm)										
First-year	60.8	0.8	61.0	59.0–63.0	152	63.6	0.9	64.0	61.0–66.0	180
Adult	61.6	0.9	62.0	60.0–65.0	120	64.9	0.8	65.0	63.0–67.0	132
All birds	61.2	0.9	61.0	59.0–65.0	272	64.1	1.0	64.0	61.0–67.0	312
Tail length (mm)										
First-year	49.8	1.4	50.0	46.0–55.5	80	52.2	1.4	52.0	49.0–57.5	83
Adult	50.7	1.1	50.8	48.5–53.5	66	53.8	1.2	54.0	50.5–56.0	72
All birds	50.2	1.4	50.0	46.0–55.5	146	52.9	1.5	53.0	49.0–57.5	155
Weight (g)										
First-year	10.4	0.4	10.4	9.1–11.8	249	11.1	0.4	11.1	10.2–12.2	274
Adult	10.4	0.4	10.4	9.3–11.2	167	11.0	0.4	11.0	10.0–12.5	212
All birds	10.4	0.4	10.4	9.1–11.8	416	11.1	0.4	11.1	10.0–12.5	486
Tarsus length (mm)										
All birds	18.7	0.5	18.7	17.8–19.7	35	18.9	0.3	18.9	18.4–19.6	28

Wing length is measured as the maximum chord (straightened and flattened on the ruler). Tail length is measured from the root of the tail, beneath the undertail coverts. Weights are derived from 902 separate measurements of 614 individuals. Tarsus is measured as the maximum length (see figure 18C on p.29 of Svensson 2023).

Source: Broughton et al. 2016, updated with additional Monks Wood data.

APPENDIX 4 Biometrics of English Willow Tits of known age and sex

	Female				Male			
	Mean	s.d.	Median	Range	No. of birds	Mean	s.d.	Median	Range	No. of birds

	Female Mean	Female s.d.	Female Median	Female Range	Female No. of birds	Male Mean	Male s.d.	Male Median	Male Range	Male No. of birds
Wing length (mm)										
First-year	59.3	2.1	59.5	56.0–63.0	8	60.4	1.3	60.5	58.0–62.0	14
Adult	59.0	1.1	59.0	57.0–60.0	25	61.2	1.3	62.0	58.0–63.0	31
All birds	59.0	1.7	59.0	56.0–63.0	33	61.0	1.3	61.0	58.0–63.0	45
Tail length (mm)										
First-year	50.3	1.8	51.0	48.0–52.5	6	52.2	4.0	51.0	49.0–60.0	6
Adult	50.0	2.5	50.0	46.0–53.0	7	52.8	1.9	53.0	49.0–55.0	11
All birds	50.1	2.1	51.0	48.0–53.0	13	52.6	2.7	52.0	49.0–60.0	17
Weight (g)										
First-year	9.8	0.6	9.9	8.9–10.4	7	10.2	0.6	10.4	9.1–11.1	13
Adult	10.1	0.8	10.2	8.5–11.5	21	10.3	0.5	10.2	9.3–11.3	27
All birds	10.0	0.7	10.0	8.5–11.5	28	10.3	0.5	10.3	9.1–11.3	40

Wing length is measured as the maximum chord (straightened and flattened on the ruler). Tail length is measured from the root of the tail, beneath the undertail coverts.

Source: Broughton et al. 2016, updated with additional data.

References

Aalto, P., Ader, A., Baumanis, J., Busse, P., Latja, A., Leivits, A., Miettinen, J., Ojanen, M., Pakkala, H., Tynjälä, M. & Vilbaste, E. 1995. Autumn migration of the Willow Tit (*Parus montanus*). *The Ring* 17: 5–11.

Abe, N. & Kurosawa, O. 1984. Further notes on the morphological differences between *Parus palustris* and *P. montanus*. *Journal of the Yamashina Institute for Ornithology* 16: 142–150.

Alatalo, R. V. 1982. Evidence for interspecific competition among European tits *Parus* spp.: a review. *Annales Zoologici Fennici* 19: 309–317.

Alatalo, R. V. & Lundberg, A. 1983. Laboratory experiments on habitat separation and foraging efficiency in Marsh and Willow Tits. *Ornis Scandinavica* 14: 115–122.

Alatalo, R. V., Gustafsson, L., Lundberg, A. & Ulfstrand, S. 1985. Habitat shift of the Willow Tit *Parus montanus* in the absence of the Marsh Tit *Parus palustris*. *Ornis Scandinavica* 16: 121–128.

Alderman, A., Hinsley, S. A., Broughton, R. K. & Bellamy, P. E. 2011. Local settlement in woodland birds in fragmented habitat: effects of natal territory location and timing of fledging. *Landscape Research* 36: 553–571.

Almond, W. E. & Almond, E. L. 1950. Concealment of food by Marsh Tit. *British Birds* 43: 336–337.

Amann, F. 1980. Alters- und Geschlechtsmerkmale der Nonnenmeise *Parus palustris*. *Der Ornithologische Beobachter* 77: 79–83.

Amann, F. 1997. Ansiedlung und Verhalten der Jungvögel bei der Sumpfmeise *Parus palustris*. *Der Ornithologische Beobachter* 94: 5–18.

Amann, F. 2003. Revierbesetzung und Paarbindung bei der Sumpfmeise *Parus palustris*. *Der Ornithologische Beobachter* 100: 193–210.

Amann, F. 2007. Courtship feeding, diet and hoarding behaviour of Marsh Tits *Parus palustris*. *Der Ornithologische Beobachter* 104: 91–100.

Amar, A., Smith, K. W., Butler, S., Lindsell, J. A., Hewson, C. M., Fuller, R. J. & Charman, E. C. 2010. Recent patterns of change in vegetation structure and tree composition of British broadleaved woodland: evidence from large-scale surveys. *Forestry: an International Journal of Forest Research* 83: 345–356.

Amaral-Rogers, V. 2021. Willow tit survey blog Feb 2021. https://community.rspb.org.uk/ourwork/b/science/posts/willow-tit-survey-blog-feb-2021.

Andreasson, F., Nord, A. & Nilsson, J.-Å. 2023. Variation in breeding phenology in response to climate change in two passerine species. *Oecologia* 201: 279–285.

Andrews, I. J. 2014. The former distribution of Marsh Tits and Willow Tits in Scotland. *Scottish Birds* 34: 218–229.

Anon. 2022. Białowieża forest bird surveys: the end of an era. *Acta Ornithologica* 57: 1–18.

Anvén, B. 1961. Några observationer över entitans (*Parus palustris*) biologi. *Sveriges Ornitologiska Förening* 20: 145–151.

Aparisi, M. P., Schöll, E. M. & Hille, S. M. 2018. Alpine Marsh Tits *Poecile palustris palustris* exhibit no clear sexual dimorphism other than in wing length. *Ringing & Migration* 33: 36–40.

Askeyev, O., Askeyev, A. & Askeyev, I. 2018. Recent climate change has increased forest winter bird densities in East Europe. *Ecological Research* 33: 445–456.

Avery, M. & Leslie, R. 1990. *Birds and Forestry*. T. & A. D. Poyser, London.

Bagg, A. M. 1969. The changing seasons: a summary of the fall migration season, 1968, with special attention to the movements of Black-capped Chickadees. *Audubon Field Notes* 23: 4–12.

Bai, M.-L., Wichmann, F. & Mühlenberg, M. 2005. Nest-site characteristics of hole-nesting birds in a primeval boreal forest of Mongolia. *Acta Ornithologica* 40: 1–14.

Baker, M. C. & Gammon, D. E. 2007. The gargle call of Black-capped Chickadees: ontogeny, acoustic structure, population patterns, function, and processes leading to sharing of call characteristics. In: Otter, K. A. (ed.), *The Ecology and Behavior of Chickadees and Titmice: an Integrated Approach*. Oxford University Press, Oxford. pp. 167–182.

Baker, M. C., Howard, T. M. & Sweet, P. W. 2000. Microgeographic variation and sharing of the gargle vocalization and its component syllables in Black-capped Chickadee (Aves, Paridae, *Poecile atricapillus*) populations. *Ethology* 106: 819–838.

Baker, M. C., Baker, M. S. A. & Gammon, D. E. 2003. Vocal ontogeny of nestling and fledgling Black-capped Chickadees *Poecile atricapilla* in natural populations. *Bioacoustics* 13: 265–296.

Balmer, D. E., Gillings, S., Caffrey, B. J., Swann, R. L., Downie, I. S. & Fuller, R. J. 2013. *Bird Atlas 2007–11: the Breeding and Wintering Birds of Britain and Ireland*. British Trust for Ornithology, Thetford.

Bani, L., Massimino, D., Bottoni, L. & Massa, R. 2006. A multiscale method for selecting indicator species and priority conservation areas: a case study for broadleaved forests in Lombardy, Italy. *Conservation Biology* 20: 512–526.

Bani, L., Massimino, D., Orioli, V., Bottoni, L. & Massa, R. 2009. Assessment of population trends of common breeding birds in Lombardy, Northern Italy, 1992–2007. *Ethology Ecology & Evolution* 21: 27–44.

Banik, M. V., Vysochin, M. O., Atemasov, A. A., Atemasova, T. A. & Devyatko, T. N. 2013. Birds of Dvorichansky National Park and its outskirts (Kharkiv region). *Berkut* 22: 14–24.

Bardin, A. V., Markovets, M. Y. & Mikhaylov, D. V. 1992. Movements of Marsh Tits (*Parus palustris*) along the Courish Spit according to the records of permanent trapping. *Proceedings of the Zoological Institute St Petersburg* 247: 7–17.

Bartosiewicz, L., Zapata, L. & Bonsall, C. 2010. A tale of two shell middens: the natural versus the cultural in 'Obanian' deposits at Carding Mill Bay, Oban, western Scotland. In: VanDerwarker, A. M. & Peres, T. M. (eds.), *Integrating Zooarchaeology and Paleoethnobotany: a Consideration of Issues, Methods, and Cases*. Springer, New York.

Bashta, A.-T. V. 1999. Breeding bird community of monocultural spruce plantation in the Skolivski Beskids (the Ukrainian Carpathians). *Berkut* 8: 9–14.

Beilby, A. 2019. An analysis of Willow Tit (*Poecile montanus*) diet including a comparison of age and seasonal differences. MSc thesis, University of Reading.

Belik, V. P. & Moskalenko, V. M. 1993. Avifaunistic rarities of the Sumy Polesye. 1. Passeriformes. *Berkut* 2: 4–11.

Bellamy, P. 2022. Willow tit habitat usage. Conference presentation, IX International Hole-Nesting Birds Conference, Oxford, UK.

Bellamy, P. E., Hill, R. A., Rothery, P., Hinsley, S. A., Fuller, R. J. & Broughton, R. K. 2009. Willow Warbler *Phylloscopus trochilus* habitat in woods with different structure and management in southern England. *Bird Study* 56: 338–348.

Bellamy, P. E., Charman, E. C., Riddle, N., Kirby, W. B., Broome, A. C., Siriwardena, G. M., Grice, P. V., Peach, W. K. & Gregory, R. D. 2022. Impact of woodland agri-environment management on woodland structure and target bird species. *Journal of Environmental Management* 316: 115221.

Berndt, R. & Winkel, W. 1987. Abundance dynamic and breeding data of Marsh Tits (*Parus palustris*). Results collected in south-east Lower Saxony. *Die Vogelwelt* 108: 121–131.

Beskardes, V., Keten, A., Arslangundogdu, Z. & Anderson, J. T. 2018. Habitat use by tit species in the Yuvacik watershed, Turkey. *Fresenius Environmental Bulletin* 27: 9033–9039.

Betts, M. M. 1955. The food of titmice in oak woodland. *Journal of Animal Ecology* 24: 282–323.

BirdLife International 2023. IUCN Red List for birds. Downloaded from http://datazone.birdlife.org.

Black, R. & Twydell, M. 2022. SOS Marsh Tit survey 2020/21. *Sussex Bird Report* 2021: 253–258.

Blumgart, D., Botham, M. S., Menéndez, R. & Bell, J. R. 2022. Moth declines are most severe in broadleaf woodlands despite a net gain in habitat availability. *Insect Conservation and Diversity* 15: 496–509.

Blunsden, T. & Goodenough, A. E. 2023. Influence of nest box design and nesting material on ectoparasite load for four woodland passerines. *Bird Study* 70: 25–36.

Bobiec, A. 2002. Living stands and dead wood in the Białowieża forest: suggestions for restoration management. *Forest Ecology and Management* 165: 125–140.

Bock, C. E. & Lepthien, L. W. 1976. Synchronous eruptions of boreal seed-eating birds. *The American Naturalist* 110: 559–571.

Bonderud, E. S., Otter, K. A., Burg, T. M., Marini, K. L. D. & Reudink, M. W. 2018. Patterns of extra-pair paternity in mountain chickadees. *Ethology* 124: 378–386.

Brewer, R. 1961. Comparative notes on the life history of the Carolina Chickadee. *The Wilson Bulletin* 73: 348–373.

Brodin, A. 1992. Cache dispersion affects retrieval time in hoarding Willow Tits. *Ornis Scandinavica* 23: 7–12.

Brodin, A. 1993. Low rate of loss of Willow Tit caches may increase adaptiveness of long-term hoarding. *The Auk* 110: 642–645.

Brodin, A. 1994a. The role of naturally stored food supplies in the winter diet of the boreal Willow Tit *Parus montanus. Ornis Svecica* 4: 31–40.

Brodin, A. 1994b. Separation of caches between individual willow tits hoarding under natural conditions. *Animal Behaviour* 47: 1031–1035.

Brodin, A. 2005. Mechanisms of cache retrieval in long-term hoarding birds. *Journal of Ethology* 23: 77–83.

Brodin, A. & Lundborg, K. 2003. Is hippocampal volume affected by specialization for food hoarding in birds? *Proceedings of the Royal Society B* 270: 1555–1563.

Brodin, A. & Urhan, A. U. 2014. Interspecific observational memory in a non-caching *Parus* species, the great tit *Parus major. Behavioral Ecology and Sociobiology* 68: 649–656.

Broggi, J., Hohtola, E. & Koivula, K. 2021. Winter feeding influences the cost of living in boreal passerines. *Ibis* 163: 260–267.

Broggi, J., Watson, H., Nilsson, J. & Nilsson, J.-Å. 2022. Carry-over effects on reproduction in food-supplemented wintering great tits. *Journal of Avian Biology* 2022: e02969.

Brotons, L. 2000. Winter spacing and non-breeding social system of the Coal Tit *Parus ater* in a subalpine forest. *Ibis* 142: 657–667.

Broughton, R. K. 2002. *Birds of the Hull Area.* Kingston Press, Hull.

Broughton, R. K. 2005. Hissing display of incubating Marsh Tit *Parus palustris* and anti-predator response of young. *British Birds* 98: 267–268.

Broughton, R. K. 2006. An example of polygyny in the Marsh Tit. *British Birds* 99: 211–212.

Broughton, R. K. 2008. Singing by female Marsh Tits: frequency and function. *British Birds* 101: 155–156.

Broughton, R. K. 2009. Separation of Willow Tit and Marsh Tit in Britain: a review. *British Birds* 102: 604–616.

Broughton, R. 2012a. XC99087 Marsh Tit female 'begging' call. www.xeno-canto.org/99087.

Broughton, R. K. 2012b. Nest defence behaviour by marsh tits in response to a stuffed weasel. *British Birds* 105: 39–40.

Broughton, R. K. 2012c. Habitat modelling and the ecology of the marsh tit (*Poecile palustris*). PhD thesis, Bournemouth University.

Broughton, R. K. 2015. Low incidence of leg and foot injuries in colour-ringed Marsh Tits *Poecile palustris*. *Ringing & Migration* 30: 37–42.

Broughton, R. K. 2019. Calls of nestling and fledgling Marsh Tits *Poecile palustris*. *Ringing & Migration* 34: 95–102.

Broughton, R. K. 2020. Current and future impacts of nest predation and nest-site competition by invasive eastern grey squirrels *Sciurus carolinensis* on European birds. *Mammal Review* 50: 38–51.

Broughton, R. K. & Alker, P. J. 2017. Separating British Marsh Tits *Poecile palustris* and Willow Tits *P. montana* using a new feature trialled in an online survey. *Ringing & Migration* 32: 43–49.

Broughton, R. K. & Hinsley, S. A. 2014. A nestbox trial for British Marsh Tits *Poecile palustris*. *Ringing & Migration* 29: 77–80.

Broughton, R. K. & Hinsley, S. A. 2015. The ecology and conservation of the Marsh Tit in Britain. *British Birds* 108: 12–29.

Broughton, R. K., Hinsley, S. A., Bellamy, P. E., Hill, R. A. & Rothery, P. 2006. Marsh Tit territories in a British broadleaved wood. *Ibis* 148: 744–752.

Broughton, R. K., Hinsley, S. A. & Bellamy, P. E. 2008a. Separation of Marsh Tit *Poecile palustris* from Willow Tit *Poecile montana* using a bill criterion. *Ringing & Migration* 24: 101–103.

Broughton, R. K., Hinsley, S. A., Bellamy, P. E., Carpenter, J. E. & Rothery, P. 2008b. Ageing and sexing Marsh Tits *Poecile*

palustris using wing length and moult. *Ringing & Migration* 24: 88–94.

Broughton, R. K., Hill, R. A., Bellamy, P. E. & Hinsley, S. A. 2010. Dispersal, ranging and settling behaviour of Marsh Tits *Poecile palustris* in a fragmented landscape in lowland England. *Bird Study* 57: 458–472.

Broughton, R. K., Hill, R. A., Bellamy, P. E. & Hinsley, S. A. 2011. Nest-sites, breeding failure and causes of non-breeding in a population of British Marsh Tits *Poecile palustris*. *Bird Study* 58: 229–237.

Broughton, R. K., Hill, R. A., Freeman, S. N., Bellamy, P. E. & Hinsley, S. A. 2012a. Describing habitat occupation by woodland birds with territory mapping and remotely sensed data: an example using the marsh tit (*Poecile palustris*). *The Condor* 114: 812–822.

Broughton, R. K., Hill, R. A., Henderson, L. J., Bellamy, P. E. & Hinsley, S. A. 2012b. Patterns of nest placement in a population of Marsh Tits *Poecile palustris*. *Journal of Ornithology* 153: 735–746.

Broughton, R. K., Hill, R. A. & Hinsley, S. A. 2013. Relationships between patterns of habitat cover and the historical distribution of the Marsh Tit, Willow Tit and Lesser Spotted Woodpecker in Britain. *Ecological Informatics* 14: 25–30.

Broughton, R. K., Bellamy, P. E., Hill, R. A. & Hinsley, S. A. 2014. Winter habitat selection by Marsh Tits *Poecile palustris* in a British woodland. *Bird Study* 61: 404–412.

Broughton, R. K., Bellamy, P. E., Hill, R. A. & Hinsley, S. A. 2015a. Winter social organisation of Marsh Tits *Poecile palustris* in Britain. *Acta Ornithologica* 50: 11–21.

Broughton, R. K., Hebda, G., Maziarz, M., Smith, K. W., Smith, L. & Hinsley, S. A. 2015b. Nest-site competition between bumblebees (Bombidae), social wasps (Vespidae) and cavity-nesting birds in Britain and the Western Palearctic. *Bird Study* 62: 427–437.

Broughton, R. K., Alker, P. J., Bellamy, P. E., Britton, S., Dadam, D., Day, J. C., Miles,

M. & Hinsley, S. A. 2016. Comparative biometrics of British Marsh Tits *Poecile palustris* and Willow Tits *P. montana*. *Ringing & Migration* 31: 30–40.

Broughton, R. K., Dadam, D., Maziarz, M., Bellamy, P. E. & Hinsley, S. A. 2018a. An efficient survey method for estimating populations of Marsh Tits *Poecile palustris*, a low-density woodland passerine. *Bird Study* 65: 299–305.

Broughton, R. K., Day, J. C., Carpenter, J. E., Gosler, A. G. & Hinsley, S. A. 2018b. Offspring sex ratio of a woodland songbird is unrelated to habitat fragmentation. *Journal of Ornithology* 159: 593–596.

Broughton, R. K., Maziarz, M. & Hinsley, S. A. 2019. Social structure of Coal Tits *Periparus ater* in temperate deciduous forest. *Journal of Ornithology* 160: 117–126.

Broughton, R. K., Parry, W. & Maziarz, M. 2020. Wilding of a post-industrial site provides a habitat refuge for an endangered woodland songbird, the British Willow Tit *Poecile montanus kleinschmidti*. *Bird Study* 67: 269–278.

Broughton, R. K., Bullock, J. M., George, C., Hill, R. A., Hinsley, S. A., Maziarz, M., Melin, M., Mountford, J. O., Sparks, T. H. & Pywell, R. F. 2021a. Long-term woodland restoration on lowland farmland through passive rewilding. *PLoS ONE* 16: e0252466.

Broughton, R. K., Chetcuti, J., Burgess, M. D., Gerard, F. F. & Pywell, R. F. 2021b. A regional-scale study of associations between farmland birds and linear woody networks of hedgerows and trees. *Agriculture, Ecosystems & Environment* 310: 107300.

Broughton, R. K., Karpińska, O., Kamionka-Kanclerska, K. & Maziarz, M. 2022a. Do large herbivores have an important role in initiating tree cavities for hole-nesting birds in European forests? *Acta Ornithologica* 57: 107–121.

Broughton, R. K., Bullock, J. M., George, C., Gerard, F., Maziarz, M., Payne, W. E., Scholefield, P. A., Wade, D. & Pywell, R. F. 2022b. Slow development of woodland vegetation and bird communities during 33 years of passive rewilding in open farmland. *PLoS ONE* 17: e0277545.

Broughton, R., Shutt, J. & Lees, A. 2022c. Rethinking bird feeding. *British Birds* 115: 2–6.

Brunelli, M. & Fraticelli, F. 2020. Sulla presenza della Cincia alpestre *Poecile montanus* in Appennino centrale [On the presence of Willow Tit *Poecile montanus* in the Central Apennines]. *Rivista Italiana di Ornitologia* 89 (2). https://doi.org/10.4081/rio.2019.442

Buckley, P. 2020. Coppice restoration and conservation: a European perspective. *Journal of Forest Research.* 25: 125–133.

Cambridge Bird Club 1996. *Cambridgeshire Bird Report 1995*. CBC, Cambridge.

Cambridge Bird Club 1998. *Cambridgeshire Bird Report 1997*. CBC, Cambridge.

Carpenter, J. 2008. An investigation of causes of population decline in the Marsh Tit *Poecile palustris* in Britain. DPhil thesis, University of Oxford.

Carpenter, J., Smart, J., Amar, A., Gosler, A., Hinsley, S. & Charman, E. 2010. National-scale analyses of habitat associations of Marsh Tits *Poecile palustris* and Blue Tits *Cyanistes caeruleus*: two species with opposing population trends in Britain. *Bird Study* 57: 31–43.

Carr, G. & Lunn, J. 2017. Thriving Willow Tits in a post-industrial landscape. *British Birds* 110: 330–340.

Catsadorakis, G. & Källander, H. 1999. Densities, habitat and breeding parameters of the Sombre Tit *Parus lugubris* in Prespa National Park, Greece. *Bird Study* 46: 373–375.

Chalmers, M. 2017. The last stronghold of the Willow Tit in Hampshire. *Hampshire Bird Report* 2016: 224–232.

Cholewa, M. & Wesołowski, T. 2011. Nestling food of European hole-nesting passerines: do we know enough to test the adaptive hypotheses on breeding seasons? *Acta Ornithologica* 46: 105–116.

Clarke, A. L., Sæther, B.-E. & Røskaft, E. 1997. Sex biases in avian dispersal: a reappraisal. *Oikos* 79: 429–438.

Clayton, N. S. 1992. The ontogeny of food-storing and retrieval in Marsh Tits. *Behaviour* 122: 11–25.

Clemmons, J. R. 1995. Vocalizations and other stimuli that elicit gaping in nestling Black-capped Chickadees (*Parus atricapillus*). *The Auk* 112: 603–612.

Clemmons, J. & Howitz, J. L. 1990. Development of early vocalizations and the chick-a-dee call in the black-capped chickadee, *Parus atricapillus. Ethology* 86: 203–223.

Cleverley, R., Shore, M., Smith, L. & Smith, K. 2019. Observation of a Lesser Spotted Woodpecker losing its nest hole to Marsh Tits. *British Birds* 112: 684–685.

Cole, E. F., Regan, C. E. & Sheldon, B. C. 2021. Spatial variation in avian phenological response to climate change linked to tree health. *Nature Climate Change* 11: 872–878.

Cowie, R. J., Krebs, J. R. & Sherry, D. F. 1981. Food storing by marsh tits. *Animal Behaviour* 29: 1252–1259.

Cramp, S. & Perrins, C. M. 1993. *The Birds of the Western Palearctic. Vol. VII: Flycatchers to Shrikes*. Oxford University Press, Oxford.

Crane, E. 2022. Can wildwoods help us avert climate disaster? *British Wildlife* 33: 490–497.

Cromack, D. 2018. *Nestboxes: Your Complete Guide*. British Trust for Ornithology, Thetford.

Curry, R. L. 2005. Hybridization in chickadees: much to learn from familiar birds. *The Auk* 122: 747–758.

Czyż, B., Rowiński, P. & Wesołowski, T. 2012. No evidence for offspring sex ratio adjustment in Marsh Tits *Poecile palustris* breeding in a primeval forest. *Acta Ornithologica* 47: 111–118.

Dadam, D., Broome, A., Conway, G., Riddle, N. & Siriwardena, G. 2020. *Resurvey of AES Woodland Creation for Woodland Birds*. Defra report LM0486. Defra, London.

Dale, S. 2001. Female-biased dispersal, low female recruitment, unpaired males, and the extinction of small and isolated bird populations. *Oikos* 92: 344–356.

Deeming, D. C. & du Feu, C. R. 2008. Measurement of brood patch temperature of British passerines using an infrared thermometer. *Bird Study* 55: 139–143.

Delmée, E., Dachy, P. & Simon, P. 1972. Contribution à la biologie des Mésanges (Paridae) en milieu forestier. *Aves* 9: 1–80.

Desrochers, A. & Hannon, S. J. 1997. Gap crossing decisions by forest songbirds during the post-fledging period. *Conservation Biology* 11; 1204–1210.

Desrochers, A., Hannon, S. J. & Nordin, K. E. 1988. Winter survival and territory acquisition in a northern population of Black-capped Chickadees. *The Auk* 105: 727–736.

Dewolf, P. 1987. Un nouveau critère de distinction entre la Mésange Nonnette et la Mésange Boréale. *Bulletin à l'Usage du Bagueur Ornithologue* 2: 10–11.

Dhondt, A. 2007. What drives differences between North American and Eurasian tit studies? In: Otter, K. A. (ed.), *The Ecology and Behavior of Chickadees and Titmice: an Integrated Approach*. Oxford University Press, Oxford, pp. 299–310.

Dhondt, A. A. 2012. *Interspecific Competition in Birds*. Oxford University Press, Oxford.

Dhondt, A. A. & Adriaensen, F. 1999. Experiments on competition between Great and Blue Tit: Effects on Blue Tit reproductive success and population processes. *Ostrich* 70: 39–48.

Dhondt, A. A. & Hublé, J. 1969. Een geval van hybridisatie tussen glanskopmees ♀ (*Parus palustris*) en een matkopmees ♂ (*Parus montanus*) te Gent. *Gerfaut* 59: 374–377.

Dietrich, V. C. J., Schmoll, T., Winkel, W. & Lubjuhn, T. 2003. Survival to first breeding is not sex-specific in the Coal Tit (*Parus ater*). *Journal für Ornithologie* 144: 148–156.

Dingemanse, N. J., Both, C., van Noordwijk, A. J., Rutten, A. L. & Drent, P. J. 2003. Natal dispersal and personalities in great tits (*Parus major*). *Proceedings of the Royal Society B* 270: 741–747.

Doherty, P. F. & Grubb, T. C. 2002. Survivorship of permanent-resident birds in a fragmented forested landscape. *Ecology* 83: 844–857.

Dolenec, Z. 2006. Laying date of marsh tits *Parus palustris* in relation to climate change. *Biologia* 61: 635–637.

Drocić, N. & Drocić, S. 2013. Prilog poznavanju prirodnih hibrida ptica u Bosni i Hercegovini (hibrid *Parus major* × *Poecile palustris* u srednjoj Bosni 2013. godine). *Bilten Mreže posmatrača ptica u Bosni i Hercegovini* 9: 83–90.

Dunn, E. K. 1976. Laying dates of four species of tit in Wytham Wood, Oxfordshire. *British Birds* 69: 45–50.

Dunn, E. 1977. Predation by Weasels (*Mustela nivalis*) on breeding tits (*Parus* spp.) in relation to the density of tits and rodents. *Journal of Animal Ecology* 46: 633–652.

Duquet, M. 1995. Un hybride probable Mésange nonette *Parus palustris* × M. charbonnière *Parus major*. *Ornithos* 2: 93.

East, M. L. & Perrins, C. M. 1988. The effect of nestboxes on breeding populations of birds in broadleaved temperate woodlands. *Ibis* 130: 393–401.

Eck, S. 1980. Intraspezifische evolution bei Graumeisen (Aves, Paridae: *Parus*, subgenus *Poecile*). *Zoologische Abhandlungen* 36: 135–219.

Eggers, S. & Low, M. 2014. Differential demographic responses of sympatric Parids to vegetation management in boreal forest. *Forest Ecology and Management* 319: 169–175.

Ehrenroth, B. 1973. Studies on migratory movements of the Willow Tit *Parus montanus borealis* Selys-Longchamps. *Ornis Scandinavica* 4: 87–96.

Ekman, J. 1979a. Coherence, composition and territories of winter social groups of the Willow Tit *Parus montanus* and the Crested Tit *P. cristatus*. *Ornis Scandinavica* 10: 56–68.

Ekman, J. 1979b. Non-territorial Willow Tits *Parus montanus* in late summer and early autumn. *Ornis Scandinavica* 10: 262–267.

Ekman, J. 1984. Density-dependent seasonal mortality and population fluctuations of the temperate-zone Willow Tit (*Parus montanus*). *Journal of Animal Ecology* 53: 119–134.

Ekman, J. 1986. Tree use and predator vulnerability of wintering passerines. *Ornis Scandinavica* 17: 261–267.

Ekman, J. 1989. Ecology of non-breeding social systems of *Parus*. *The Wilson Bulletin* 101: 263–288.

Ekman, J. & Askenmo, C. 1986. Reproductive cost, age-specific survival and a comparison of the reproductive strategy in two European tits (genus *Parus*). *Evolution* 40: 159–168.

Enoksson, B., Angelstam, P. & Larsson, K. 1995. Deciduous forest and resident birds: the problem of fragmentation within a coniferous forest landscape. *Landscape Ecology* 10: 267–275.

EURING 2023. The EURING databank. https://euring.org/data-and-codes/euring-databank.

Facey, S. L., Botham, M. S., Heard, M. S., Pywell, R. F. & Staley, J. T. 2014. Moth communities and agri-environment schemes: Examining the effects of hedgerow cutting regime on diversity, abundance, and parasitism. *Insect Conservation and Diversity* 7: 543–552.

Farina, A. 1983. Habitat preferences of breeding tits. *Monitore Zoologico Italiano* 17: 121–131.

Farine, D. R., Aplin, L. M., Sheldon, B. C. & Hoppitt, W. 2015. Interspecific social networks promote information transmission in wild songbirds. *Proceedings of the Royal Society B* 282: 20142804.

Farjon, A. 2022. Wild oakwoods: can they exist in lowland England? *British Wildlife* 34: 178–188.

Ficken, M. S. & Witkin, S. R. 1977. Responses of Black-capped Chickadee flocks to predators. *The Auk* 94: 156–157.

Ficken, M. S., Ficken, R. W. & Witkin, S. R. 1978. Vocal repertoire of the Black-capped Chickadee. *The Auk* 95: 34–48.

Fijen, T. P. M. 2015. Singing Willow Tits *Poecile montanus*: Sino-Japanese song type recorded in the southern and western Altai, Kazakhstan, June–July 2013. *Sandgrouse* 37: 94–96.

Fitzpatrick, S. 1994. Nectar-feeding by suburban Blue Tits: contribution to the diet in spring. *Bird Study* 41: 136–145.

Forestry Commission 2001. *National Inventory of Woodland and Trees: England.* Forestry Commission, Edinburgh.

Forrester, R. W. & Andrews, I. J. (eds.) 2007. *The Birds of Scotland.* Scottish Ornithologists' Club, Aberlady.

Foster, J. & Godfrey, C. 1950. A study of the British Willow-Tit. *British Birds* 43: 351–361.

Fransson, T., Kolehmainen, T., Moss, D. & Robinson, R. 2023. EURING list of longevity records for European birds. https://euring.org/files/documents/EURING_longevity_list_20230901.pdf.

Frederiksen, K. S., Jensen, M., Larsen, E. H. & Larsen, V. H. 1972. Nogle data til belysning af yngletidspunkt og kuldstørrelse hos mejser (Paridae). *Dansk Ornitologisk Forenings Tidsskrift* 66: 73–85.

Fujita, K. & Takahashi, T. 2009. Ecological role of the Great Tit *Parus major* as a seed disperser during winter. *Ornithological Science* 8: 157–161.

Fuller, R. J. 2022. Population density and stability of breeding birds in English oak woodland over a 32-year period in relation to habitat structure and edges. *Acta Ornithologica* 57: 49–70.

Fuller, R. J. & Henderson, A. C. B. 1992. Distribution of breeding songbirds in Bradfield Woods, Suffolk, in relation to vegetation and coppice management. *Bird Study* 39: 73–88.

Fuller, R. J., Noble, D. G., Smith, K. W. & Vanhinsbergh, D. 2005. Recent declines in populations of woodland birds in Britain: a review of possible causes. *British Birds* 98: 116–143.

Gauthier, S., Bernier, P., Kuuluvainen, T., Shvidenko, A. Z. & Schepaschenko, D. G. 2015. Boreal forest health and global change. *Science* 349: 819–822.

Geer, T. A. 1978. Effects of nesting Sparrowhawks on nesting tits. *The Condor* 80: 419–422.

Gibb, J. 1954a. Feeding ecology of tits, with notes on treecreeper and goldcrest. *Ibis* 96: 513–543.

Gibb, J. 1954b. Population changes of titmice, 1947–1951. *Bird Study* 1: 40–48.

Gibb, J. A. 1960. Populations of tits and goldcrests and their food supply in pine plantations. *Ibis* 102: 163–208.

Gibb, J. A. & Betts, M. M. 1963. Food and food supply of nestling tits (Paridae) in Breckland pine. *Journal of Animal Ecology* 32: 489–533.

Gill, F. B., Slikas, B. & Sheldon, F. H. 2005. Phylogeny of titmice (Paridae): II. Species relationships based on sequences of the mitochondrial cytochrome-*b* gene. *The Auk* 122: 121–143.

Ginn, H. B. & Melville, D. S. 1983. *Moult in Birds.* BTO Guide 19. BTO, Tring.

Glutz von Blotzheim, U. N. & Bauer, K. M. 1993. *Handbuch der Vögel Mitteleuropas, Vol. 13.* Aula-Berlag, Wiebelsheim.

Gold, C. S. & Dahlsten, D. L. 1983. Effects of parasitic flies (*Protocalliphora* spp.) on nestlings of Mountain and Chestnut-backed Chickadees. *The Wilson Bulletin* 95: 560–572.

Gosler, A. 1993. *The Great Tit.* Hamlyn, London.

Gosler, A., Clement, P. & Garcia, E. F. J. 2020a. Willow Tit (*Poecile montanus*), version 1.0. In: Billerman, S. M., Keeney, B. K., Rodewald, P. G. & Schulenberg, T. S. (eds.), *Birds of the World.* Cornell Lab of Ornithology, Ithaca, NY.

Gosler, A., Clement, P., Garcia, E. F. J., Christie, D. A. & Kirwan, G. M. 2020b. Marsh Tit (*Poecile palustris*), version 1.0.

In: del Hoyo, J., Elliott, A., Sargatal, J., Christie, D. A. & de Juana, E. (eds.), *Birds of the World*. Cornell Lab of Ornithology, Ithaca, NY.

Gove, B., Ghazoul, J., Power, S. & Buckley, P. 2004. *The Impacts of Pesticide Spray Drift and Fertiliser Over-spread on the Ground Flora of Ancient Woodland*. English Nature Research Report 614. English Nature, Peterborough.

Greenwood, P. J., Harvey, P. H. & Perrins, C. M. 1979. The role of dispersal in the Great Tit (*Parus major*): the causes, consequences and heritability of natal dispersal. *Journal of Animal Ecology* 48: 123–142.

Groom, J. D. & Grubb, T. C. 2006. Patch colonization dynamics in Carolina Chickadees (*Poecile carolinensis*) in a fragmented landscape: a manipulative study. *The Auk* 123, 1149–1160.

Grubb, T. C. & Bronson, C. L. 1995. Artificial snags as nesting sites for chickadees. *The Condor* 97: 1067–1070.

Guzy, A. I. 1994a. Peculiarities of the breeding and autumn migrational bird communities in oak–beech forests of Bukovina. *Berkut* 3: 3–9.

Guzy, A. I. 1994b. Structure and age successions of ornithocenosises in hornbeam–beech and pure beech forests of the Ukrainian Carpathians. *Berkut* 3: 79–88.

Guzy, A. I. 1995. Birds of pure beech and hornbeam–beech old forests of the Ukrainian Carpathians. *Berkut* 4: 18–24.

Haftorn, S. 1973. Lappmeisa *Parus cinctus* i hekketiden. Forplantning, stemmeregister og hamstring av naering. *Sterna* 12: 91–155.

Haftorn, S. 1979. Incubation and regulation of egg temperature in the Willow Tit *Parus montanus*. *Ornis Scandinavica* 10: 220–234.

Haftorn, S. 1988. Incubating female passerines do not let the egg temperature fall below the 'physiological zero temperature' during their absences from the nest. *Ornis Scandinavica* 19: 97–110.

Haftorn, S. 1993a. Ontogeny of the vocal repertoire of the Willow Tit *Parus montanus*. *Ornis Scandinavica* 24: 267–289.

Haftorn, S. 1993b. Is the Coal Tit *Parus ater* really the most subordinate of the Scandinavian tits? *Ornis Scandinavica* 24: 335–338.

Haftorn, S. 1994. A case of polygyny in the Willow Tit *Parus montanus*. *Ornis Fennica* 71: 68–71.

Haftorn, S. 1995. A case of extra-pair copulation in the Willow Tit *Parus montanus*. *Ornis Fennica* 72: 180–182.

Haftorn, S. 1996. Egg-laying behaviour in tits. *The Condor* 98: 863–865.

Haftorn, S. 1997a. One Norwegian territory of the Marsh Tit *Parus palustris* during 35 years. *Ibis* 139: 379–381.

Haftorn, S. 1997b. Natal dispersal and winter flock formation in the Willow Tit *Parus montanus*. *Fauna norvegica* Series C Cinclus 20: 17–35.

Haftorn, S. 1999. Initial winter flock formation in the Willow Tit *Parus montanus*. Do immigrating juveniles assess the quality of territorial adults? *Ibis* 141: 109–114.

Hailman, J. P. 1989. The organization of major vocalizations in the Paridae. *The Wilson Bulletin* 101: 305–343.

Hailman, J. P., Ficken, M. S. & Ficken, R. W. 1985. The 'chick-a-dee' calls of *Parus atricapillus*: a recombinant system of animal communication compared with written English. *Semiotica* 56: 191–224.

Halley, D. J. & Gjershaug, J. O. 1998. Inter- and intra-specific dominance relationships and feeding behaviour of Golden Eagles *Aquila chrysaetos* and Sea Eagles *Haliatetus albicilla* at carcasses. *Ibis* 140: 295–301.

Hanmer, H. J., Dadam, D. & Siriwardena, G. M. 2022. Evidence that rural wintering bird populations supplement suburban breeding populations. *Bird Study* 69: 12–27.

Harrap, S. & Quinn, D. 1995. *Chickadees, Tits, Nuthatches and Treecreepers.* Princeton University Press, Princeton, NJ.

Harris, S. J., Massimino, D., Balmer, D. E., Kelly, L., Noble, D. G., Pearce-Higgins, J. W., Woodcock, P., Wotton, S. & Gillings, S. 2022. *The Breeding Bird Survey 2021.* BTO Research Report 745. British Trust for Ornithology, Thetford.

Hartel, T., Plieninger, T. & Varga, A. 2015. Wood-pastures in Europe. In: Kirby, K. J. & Watkins, C. (eds.), *Europe's Changing Woods and Forests.* CABI, Wallingford, pp. 61–76.

Hartert, E. 1910. *Die Vögel der paläarktischen Fauna systematische Übersicht der in Europa, Nord-Asien und der Mittelmeerregion vorkommenden Vögel.* R. Friedländer & Sohn, Berlin.

Hasselquist, D. & Kempenaers, B. 2002. Parental care and adaptive brood sex ratio manipulation in birds. *Proceedings of the Royal Society B* 357: 363–372.

Healy, S. D. & Suhonen, J. 1996. Memory for locations of stored food in Willow Tits and Marsh Tits. *Behaviour* 133: 71–80.

Hellmayr, C. E. 1900. Einige Bemerkungen über Graumeisen. *Ornithologisches Jahrbuch* 11: 201–217.

Herbert, S. Hotchkiss, A., Reid, C. & Hornigold, K. 2022. *Woodland Creation Guide.* Woodland Trust, Grantham.

Hewson, C. M. & Fuller, R. J. 2006. Little evidence of temporal changes in edge-use by woodland birds in southern England. *Bird Study* 53: 323–327.

Hewson, C. M., Amar, A., Lindsell, J. A., Thewlis, R. M., Butler, S., Smith, K. & Fuller, R. J. 2007. Recent changes in bird populations in British broadleaved woodland. *Ibis* 149: 14–28.

Hildén, O. 1971. Activities of Finnish bird stations in 1969. *Ornis Fennica* 48: 125–130.

Hildén, O. 1983. A hybrid *Parus ater × P. montanus* found in Finland. *Ornis Fennica* 60: 58–61.

Hildén, O. & Ketola, H. 1985. A mixed pair of *Parus cinctus* and *P. montanus* nesting in Kuusamo. *Ornis Fennica* 62: 26.

Hill, R. A. & Broughton, R. K. 2009. Mapping the understorey of deciduous woodland from leaf-on and leaf-off airborne LiDAR data: a case study in lowland Britain. *ISPRS Journal of Photogrammetry and Remote Sensing* 64: 223–233.

Hinde, R. A. 1952. The behaviour of the Great Tit (*Parus major*) and some other related species. *Behaviour* Supplement 2: III, V–X, 1–201.

Hinks, A. E., Cole, E. F., Daniels, K. J., Wilkin, T. A., Nakagawa, S. & Sheldon, B. C. 2015. Scale-dependent phenological synchrony between songbirds and their caterpillar food source. *The American Naturalist* 186: 84–97.

Hinsley, S. A., Bellamy, P. E., Newton, I. & Sparks, T. H. 1995a. Habitat and landscape factors influencing the presence of individual breeding bird species in woodland fragments. *Journal of Avian Biology* 26: 94–104.

Hinsley, S. A., Bellamy, P. E. & Newton, I. 1995b. Bird species turnover and stochastic extinction in woodland fragments. *Ecography* 18: 41–50.

Hinsley, S. A., Bellamy, P. E., Newton, I. & Sparks, T. H. 1996. Influences of population size and woodland area on species distributions in small woods. *Oecologia* 105: 100–106.

Hinsley, S. A., Bellamy, P. E. & Wyllie, I. 2005. The Monks Wood avifauna. In: Gardiner, C. & Sparks, T. (eds.), *Ten Years of Change: Woodland Research at Monks Wood NNR, 1993–2003. Proceedings of the 50th Anniversary Symposium, December 2003.* English Nature, Peterborough, pp. 75–85.

Hinsley, S. A., Carpenter, J. E., Broughton, R. K., Bellamy, P. E., Rothery, P., Amar, A., Hewson, C. M. & Gosler, A. G. 2007. Habitat selection by Marsh Tits *Poecile palustris* in the UK. *Ibis* 149: 224–233.

Hinsley, S. A., Hill, R. A., Fuller, R. J., Bellamy, P. E. & Rothery, P. 2009. Bird species distributions across woodland canopy structure gradients. *Community Ecology* 10: 99–110.

Hogstad, O. 1978. Differentiation of foraging niche among tits, *Parus* spp., in Norway during winter. *Ibis* 120: 139–146.

Hogstad, O. 1987. Social rank in winter flocks of willow tits *Parus montanus*. *Ibis* 129: 1–9.

Hogstad, O. 1988. Advantages of social foraging of Willow Tits *Parus montanus*. *Ibis* 130: 275–283.

Hogstad, O. 1990a. Winter floaters in Willow Tits *Parus montanus* a matter of choice or making the best of a bad situation? In: Blondel, J., Gosler, A., Lebreton, J. D. & McCleery, R. (eds.), *Population Biology of Passerine Birds*. NATO ASI Series, volume 24. Springer, Berlin, Heidelberg, pp. 415–421.

Hogstad, O. 1990b. Dispersal date and settlement of juvenile Willow Tits *Parus montanus* in winter flocks. *Fauna norvegica Series C Cinclus* 13: 49–55.

Hogstad, O. 2011. Wing length as a predictor of body size in the Willow Tit *Poecile montanus*. *Ornis Norvegica* 34: 24–27.

Hogstad, O. 2014. Ecology and behaviour of winter floaters in a subalpine population of Willow Tits, *Poecile montanus*. *Ornis Fennica* 91: 29–38.

Hogstad, O. & Kroglund, R. T. 1993. The throat badge as a status signal in juvenile male willow tits *Parus montanus*. *Journal für Ornithologie* 134: 413–423.

Hogstad, O. & Slagsvold, T. 2018. Survival of Willow Tits *Poecile montanus*: the significance of flock membership, social rank and body size. *Ornis Norvegica* 41: 13–18.

Holleback, M. 1974. Interactions and the dispersal of the family in Black-capped Chickadees. *The Wilson Bulletin* 86: 466–488.

Holzinger-Umlauf, H. A.-M., Marschang, R. E., Gravendyck, M. & Kaleta, E. F. 1997. Investigation on the frequency of *Chlamydia* sp. infections in tits (Paridae). *Avian Pathology* 26: 779–789.

Hong, S.-H. & Kwak, J.-I. 2011. Characteristics of appearance by vegetation type of Paridae in urban forest of Korea. *Korean Journal of Environment and Ecology* 25: 760–766.

Hopkins, J. J. & Kirby, K. J. 2007. Ecological change in British broadleaved woodland since 1947. *Ibis* 149: 29–40.

Hubbs, C. L. 1955. Hybridization between fish species in nature. *Systematic Biology* 4: 1–20.

Hughes, S., Kunin, W., Watts, K. & Ziv, G. 2023. New woodlands created adjacent to existing woodlands grow faster, taller and have higher structural diversity than isolated counterparts. *Restoration Ecology* 31: e13889.

Iwasa, M., Hori, K. & Aoki, N. 1995. Fly fauna of bird nests in Hokkaido, Japan (Diptera). *The Canadian Entomologist* 127: 613–621.

Jabłoński, P. G. & Lee, S. D. 2002. Foraging niche shifts in mixed-species flocks of tits in Korea. *Journal of Field Ornithology* 73: 246–252.

Jaroszewicz, B., Cholewińska, O., Gutowski, J. M., Samojlik, T., Zimny, M. & Latałowa, M. 2019. Białowieża Forest: a relic of the high naturalness of European forests. *Forests* 10: 849.

Järvinen, A. 1982. Ecology of the Siberian Tit *Parus cinctus* in NW Finnish Lapland. *Ornis Scandinavica* 13: 47–55.

Järvinen, A. 1997. Interspecific hybridization between the Siberian Tit *Parus cinctus* and the Willow Tit *Parus montanus* produces fertile offspring. *Ornis Fennica* 74: 149–152.

Johansson, H. 1972. Clutch size and breeding success in some hole-nesting passerines in Central Sweden. *Ornis Fennica* 49: 1–6.

Johansson, U. S., Ekman, J., Bowie, R. C. K., Halvarsson, P., Ohlson, J. I., Price, T. D. & Ericson, P. G. P. 2013. A complete multilocus species phylogeny of the tits and chickadees (Aves: Paridae). *Molecular Phylogenetics and Evolution* 69: 852–860.

Johnston, T. L. 1936. Nesting habits of the Willow-Tit in Cumberland. *British Birds* 29: 378–380.

Kay, Q. O. N. 1985. Nectar from willow catkins as a food source for Blue Tits. *Bird Study* 32: 40–44.

Kirby, K. 2020a. *Woodland Flowers.* Bloomsbury Publishing, London.

Kirby, K. 2020b. Tree and shrub regeneration across the Knepp Estate in Sussex, Southern England. *Quarterly Journal of Forestry* 114: 230–236.

Kirby, K. J. & Watkins, C. 2015. The Forest Landscape Before Farming. In: Kirby, K. J. & Watkins, C. (eds.), *Europe's Changing Woods and Forests.* CABI, Wallingford, pp. 33–45.

Kirby, K. J., Smart, S. M., Black, H. I. J., Bunce, R. G. H., Corney, P. M. & Smithers, R. J. 2005. *Long Term Ecological Change in British Woodland (1971–2001).* English Nature Research Reports 653. English Nature, Peterborough.

Kirby, K. J., Goldberg, E. A., Isted, R., Perry, S. C. & Thomas, R. C. 2016. Long-term changes in the tree and shrub layers of a British nature reserve and their relevance for woodland conservation management. *Journal for Nature Conservation* 31: 51–60.

Knaus, P., Antoniazza, S., Wechsler, S., Guélat, J., Kéry, M., Strebel, N. & Sattler, T. 2018. *Swiss Breeding Bird Atlas 2013–2016. Distribution and population trends of birds in Switzerland and Liechtenstein.* Swiss Ornithological Institute, Sempach.

Kniprath, E. 1967. Untersuchungen zur Variation der Rückenfärbung der beiden Meisen *Parus montanus* und *Parus palustris. Journal für Ornithologie* 108: 1–46.

Koenig, W. D. 2001. Synchrony and periodicity of eruptions by boreal birds. *The Condor* 103: 725–735.

Koivula, K. & Orell, M. 1988. Social rank and winter survival in the Willow Tit *Parus montanus. Ornis Fennica* 65: 114–120.

Koivula, K., Orell, M. & Rytkönen, S. 1991. Mate guarding in forest-living, territorial Willow Tits. *Ornis Fennica* 68: 105–113.

Koivula, K., Lahti, K., Orell, M. & Rytkönen, S. 1993. Prior residency as a key determinant of social dominance in the willow tit (*Parus montanus*). *Behavioral Ecology and Sociobiology* 33: 283–287.

Konno, S. 2018. Examination of the identification criteria of Marsh Tit *Poecile palustris* and Willow Tit *P. montanus* in Hokkaido. *The Bulletin of the Japanese Bird Banding Association* 30: 1–13.

Kopij, G. 2019. The effect of urbanization on population densities of forest passerine species in a central European city. *Ornis Hungarica* 27: 207–220.

Krams, I. A. 1996. Predation risk and shifts of foraging sites in mixed Willow and Crested Tit flocks. *Journal of Avian Biology* 27: 153–156.

Krams, I. A., Luoto, S., Krama, T., Krams, R., Sieving, K., Trakimas, G., Elferts, D., Rantala, M. J. & Goodale, E. 2020. Egalitarian mixed-species bird groups enhance winter survival of subordinate group members but only in high-quality forests. *Scientific Reports* 10: 4005.

Kumpula, S., Vatka, E., Orell, M. & Rytkönen, S. 2023. Effects of forest management on the spatial distribution of the willow tit (*Poecile montanus*). *Forest Ecology and Management.* 529: 120694.

Kurosawa, R. & Askins, R. A. 2003. Effects of habitat fragmentation on birds in deciduous forests in Japan. *Conservation Biology* 17: 695–707.

Kvist, L., Martens, J., Ahola, A. & Orell, M. 2001. Phylogeography of a Palaearctic sedentary passerine, the willow tit (*Parus montanus*). *Journal of Evolutionary Biology* 14: 930–941.

Laaksonen, M. & Lehikoinen, E. 1976. Age determination of Willow and Crested Tits *Parus montanus* and *P. cristatus. Ornis Fennica* 53: 9–14.

Lack, D. 1971. *Ecological Isolation in Birds.* Blackwell Scientific Publications, London.

Lahti, K. 1998. Social dominance and survival in flocking passerine birds: a review with an emphasis on the Willow Tit *Parus montanus. Ornis Fennica* 75: 1–17.

Lahti, K., Koivula, K., Orell, M. & Rytkönen,

S. 1996. Social dominance in free-living Willow Tits *Parus montanus*: determinants and some implications of hierarchy. *Ibis* 138: 539–544.

Laiolo, P, Rolando, A. & Valsania, V. 2004. Responses of birds to the natural re-establishment of wilderness in montane beechwoods of North-western Italy. *Acta Oecologica* 25: 129–136.

La Mantia, T., Bonaviri, L. & Massa, B. 2014. Ornithological communities as indicators of recent transformations on a regional scale: Sicily's case. *Avocetta* 38: 67–81.

Lampila, S., Orell, M., Belda, E. & Koivula, K. 2006. Importance of adult survival, local recruitment and immigration in a declining boreal forest passerine, the willow tit *Parus montanus*. *Oecologia* 148: 405–413.

Lampila, S., Orell, M. & Kvist, L. 2011. Willow tit *Parus montanus* extrapair offspring are more heterozygous than their maternal half-siblings. *Journal of Avian Biology* 42: 355–362.

Last, J. & Burgess, M. 2015. Nestboxes and fieldcraft for monitoring Willow Tits. *British Birds* 108: 30–36.

Latałowa, M., Zimny, M., Jędrzejewska, B. & Samojlik, T. 2018. Białowieża Primeval Forest: a 2000-year interplay of environmental and cultural forces in Europe's best preserved temperate woodland. In: Kirby, K. J. & Watkins, C. (eds.), *Europe's Changing Woods and Forests*. CABI, Wallingford, pp. 243–264.

Lawson, B., Lachish, S., Colvile, K. M., Durrant, C., Peck, K. M., Toms, M. P., Sheldon, B. C. & Cunningham, A. A. 2012. Emergence of a novel avian pox disease in British tit species. *PLoS ONE* 7: e40176.

Lehikoinen, A. & Virkkala, R. 2018. Population trends and conservation status of forest birds. In: Mikusiński, G., Roberge, J.-M. & Fuller, R. J. (eds.), *Ecology and Conservation of Forest Birds*. Cambridge University Press, Cambridge, pp. 391–426.

Lemmon, D., Withiam, M. L. & Barkan, C. P. L. 1997. Mate protection and winter pair-bonds in Black-capped Chickadees. *The Condor* 99: 424–433.

Lens, L. 1996. Wind stress affects foraging site competition between Crested Tits and Willow Tits. *Journal of Avian Biology* 27: 41–46.

Lewis, A. J. G., Amar, A., Cordi-Piec, D. & Thewlis, R. M. 2007. Factors influencing Willow Tit *Poecile montanus* site occupancy: a comparison of abandoned and occupied woods. *Ibis* 149: 205–213.

Lewis, A. J. G., Amar, A., Daniells, L., Charman, E. C., Grice, P. & Smith, K. 2009a. Factors influencing patch occupancy and within-patch habitat use in an apparently stable population of Willow Tits *Poecile montanus kleinschmidti* in Britain. *Bird Study* 56: 326–337.

Lewis, A. J. G., Amar, A., Charman, E. C. & Stewart, F. R. P. 2009b. The decline of the Willow Tit in Britain. *British Birds* 102: 386–393.

Liu, N., Li, Y. & Liu, J. 1989. Studies of interspecific relationship between Great Tit and Willow Tit. *Zoological Research* 10: 277–284.

Lõhmus, A. 2022. Ecological sustainability at the forest landscape level: a bird assemblage perspective. *Land* 11: 1965.

Löhrl, H. 1950. Beobachtungen zur Soziologie und Verhaltensweise von Sumpfmeisen (*Parus palustris communis*) im Winter. *Zeitschrift für Tierpsychologie* 7: 417–424.

Löhrl, H. 1987. Bastardierung von Weiden- und Sumpfmeise *Parus montanus* × *P. palustris* im Nordschwarzwald. *Journal für Ornithologie* 128: 248–251.

Londi, G., Tellini Florenzano, G., Campedelli, T., Cutini, S. & Massa, B. 2012. Le zone ornitologiche della Sicilia: un metodo per l'individuazione oggettiva di ecoregioni. *Il Naturalista siciliano* 36: 459–493.

London Natural History Society. 2014. *London Bird Report 2012*. London Natural History Society, London.

Lucas, J. R. & Freeberg, T. M. 2007. 'Information' and the *chick-a-dee* call: Communicating with a complex vocal system. In: Otter, K. A. (ed.), *The Ecology and Behavior of Chickadees and Titmice: an Integrated Approach*. Oxford University Press, Oxford, pp. 199–213.

Ludescher, F. B. 1973. Sumpfmeise (*Parus p. palustris* L.) und Weidenmeise (*P. montanus salicarius* Br.) als sympatrische Zwillingsarten. *Journal für Ornithologie* 114: 3–56.

MacColl, A. D. C., du Feu, C. R. & Wain, S. P. 2014. Significant effects of season and bird age on use of coppice woodland by songbirds. *Ibis* 156: 561–575.

Macdonald, B. 2022. *Cornerstones: Wild Forces That Can Change Our World*. Bloomsbury Publishing, London.

MacGregor-Fors, I. 2022. Winter thriving: on the role of a boreal city on bird communities. *Journal of Urban Ecology* 8: juac010.

Madin, D. F. 1979. Birds of Hayley Wood. Cambridge Bird Club's survey of CAMBIENT Reserves. *Nature in Cambridgeshire* 22: 34–38.

Malengreau, A. 2020. XC544648 Willow Tit *Poecile montanus*. www.xeno-canto. org/544648.

Markovets, M. Y. 1992. Sexing of willow tits (*Parus montanus*). *Russian Journal of Ornithology* 1: 111–113.

Markovets, M. Y. 2001. *Population ecology of the Marsh Tit (Parus palustris)*. PhD Thesis, Zoological Institute, St. Petersburg.

Markovets, M. Yu. & Visotsky, V. G. 1993. Breeding biology of the marsh tit (*Parus palustris*) on the Courish Spit. *Journal of Ornithology* 2: 61–69.

Martens, J. & Nazarenko, A. A. 1993. Microevolution of eastern palaearctic Grey tits as indicated by their vocalizations (*Parus [Poecile]*: Paridae, Aves) I. *Parus montanus*: contributions to the Fauna of the Far East, No. 2. *Journal of Zoological Systematics and Evolutionary Research* 31: 127–143.

Martens, J., Ernst, S. & Petri, B. 1995. Reviergesänge ostasiatischer Weidenmeisen *Parus montanus* und ihre mikroevolutive Ableitung. *Journal für Ornithologie* 136: 367–388.

Martini, I., Galipò, G., Foderi, C., Tocci, R. & Sargentini, C. 2021. Ornithical community of Vallombrosa Biogenetic National Nature Reserve (Italy). *European Zoological Journal* 88: 254–268.

Mason, W. L. 2007. Changes in the management of British forests between 1945 and 2000 and possible future trends. *Ibis* 149: 41–52.

Massimino, D., Orioli, V., Pizzardi, F., Massa, R. & Bani, L. 2010. Usefulness of coarse grain data on forest management to improve bird abundance models. *Italian Journal of Zoology* 77: 71–80.

Massimino, D., Woodward, I. D., Hammond, M. J., Barber, L., Barimore, C., Harris, S. J., Leech, D. I., Noble, D. G., Walker, R. H., Baillie, S. R. & Robinson, R. A. 2023. *BirdTrends 2022: Trends in Numbers, Breeding Success and Survival for UK Breeding Birds*. Research Report 753. British Trust for Ornithology, Thetford. www.bto.org/birdtrends.

Matsuoka, S. & Kudo, M. (2009). Imprisonment by glaze ice may have caused the death of a cavity-roosting marsh tit (*Parus palustris*). *Bulletin of the Forestry and Forest Products Research Institute* 8: 149–155.

Matthysen, E. 1990. Nonbreeding social organization in *Parus*. *Current Ornithology* 7: 209–249.

Matthysen, E. 1998. *The Nuthatches*. T. & A. D. Poyser, London.

Maxwell, J. 2001. And these are the birds, that live in the house that Jack built! *Scottish Bird News* 62: 1–3.

Maxwell, J. 2002. Nest-site competition with Blue Tits and Great Tits as a possible cause of declines in Willow Tit numbers: observations in the Clyde area. *Glasgow Naturalist* 24: 47–50.

Maziarz, M. 2019. Breeding birds actively modify the initial microclimate of occupied tree cavities. *International Journal of Biometeorology* 63: 247–257.

Maziarz, M. & Broughton, R. K. 2015. Breeding microhabitat selection by Great Tits *Parus major* in a deciduous primeval forest (Białowieża National Park, Poland). *Bird Study* 62: 358–367.

Maziarz, M., Wesołowski, T., Hebda, G. & Cholewa, M. 2015. Natural nest-sites of Great Tits (*Parus major*) in a primeval temperate forest (Białowieża National Park, Poland). *Journal of Ornithology* 156: 613–623.

Maziarz, M., Broughton, R. K. & Wesołowski, T. 2017. Microclimate in tree cavities and nest-boxes: implications for hole-nesting birds. *Forest Ecology and Management* 389: 306–313.

Maziarz, M., Grendelmeier, A., Wesołowski, T., Arlettaz, R., Broughton, R. K. & Pasinelli, G. 2019. Patterns of predator behaviour and Wood Warbler *Phylloscopus sibilatrix* nest survival in a primaeval forest. *Ibis* 161: 854–866.

Maziarz, M., Broughton, R. K., Casacci, L. P., Dubiec, A., Maák, I. & Witek, M. 2020. Thermal ecosystem engineering by songbirds promotes a symbiotic relationship with ants. *Scientific Reports* 10: 20330.

Maziarz, M., Broughton, R. K., Beck, K. B., Robinson, R. A. & Sheldon, B. C. 2023. Temporal avoidance as a means of reducing competition between sympatric species. *Royal Society Open Science* 10: 230521.

Melin, M., Hinsley, S. A., Broughton, R. K., Bellamy, P. E. & Hill, R. A. 2018. Living on the edge: utilising lidar data to assess the importance of vegetation structure for avian diversity in fragmented woodlands and their edges. *Landscape Ecology* 33: 895–910.

Mennill, D. J., Ramsay, S. M., Boag, P. T. & Ratcliffe, L. M. 2004. Patterns of extrapair mating in relation to male dominance status and female nest placement in black-capped chickadees. *Behavioral Ecology* 15: 757–765.

Merne, O. J. 1993. Marsh Tit in County Wicklow: a species new to Ireland. *Irish Birds* 5: 74–75.

Mikusiński, G., Roberge, J.-M. & Fuller, R. J. 2018. *Ecology and Conservation of Forest Birds*. Cambridge University Press, Cambridge.

Mildenberger, H. 1984. *Die Vögel des Rheinlandes. Vol. 2*. Ges. Rheinischer Ornithologen, Düsseldorf.

Millett, M., McGrail, S., Creighton, J. D., Gregson, C. W., Heal, S. V. E., Hillam, J., Holdridge, L., Jordan, D., Spencer, P. J., Stallibrass, S., Stevens, D. & Turner, J. 1987. The Archaeology of the Hasholme Logboat. *Archaeological Journal* 144: 69–155.

Mishima, T. 1969. An example of *Parus varius × P atricapillus* and a pale phase of *Emberiza spodocephala*. *Miscellaneous Reports of the Yamashina Institute of Ornithology and Zoology* 5: 676–678.

Møller, A. P. 1989. Parasites, predators and nest boxes: facts and artefacts in nest box studies of birds? *Oikos* 56: 421–423.

Moores, N. 2017. Birds and their conservation in Rason Special Economic Zone, Democratic People's Republic of Korea. *Forktail* 33: 124–133.

Morley, A. 1949. Observations on courtship-feeding and coition of the Marsh-Tit. *British Birds* 42: 233–239.

Morley, A. 1950. The formation and persistence of pairs in the Marsh-Tit. *British Birds* 43: 387–393.

Morley, A. 1953. Field observation on the biology of the Marsh Tit. *British Birds* 46: 233–238, 273–287, 332–346.

Morris, I. 2021. The population status of the Willow Tit in Ceredigion. *Birds in Wales* 18: 31–35.

Morse, D. H. 1978. Structure and foraging patterns of flocks of flocks of tits and associated species in an English woodland during the winter. *Ibis* 120: 298–312.

Mörtberg, U. M. 2001. Resident bird species in urban forest remnants: landscape and habitat perspectives. *Landscape Ecology* 16: 193–203.

Murakami, M. 2002. Foraging mode shifts of four insectivorous bird species under temporally varying resource distribution in a Japanese deciduous forest. *Ornithological Science* 1: 63–69.

Nakamura, T. 1970. A study of Paridae community in Japan. II. Ecological separation of feeding sites and foods. *Journal of the Yamashina Institute for Ornithology* 6: 141–169.

Nakamura, T. 1975. A study of Paridae community in Japan. III. Ecological separation in social structure and distribution. *Journal of the Yamashina Institute for Ornithology* 7: 603–636.

Nakamura, H. & Wako, Y. 1988. Food storing behaviour of Willow Tit *Parus montanus*. *Journal of the Yamashina Institute for Ornithology* 20: 21–36.

Naylor, K. A. 2021 Northern Willow Tit *Poecile montanus borealis*. Historical Rare Birds. www.historicalrarebirds.info/cat-ac/northern-willow-tit.

Newton, I. 1986. *The Sparrowhawk*. T. & A. D. Poyser, London.

Newton, I. 1998. *Population Limitation in Birds*. Academic Press, London.

Newton, I. 2010. *Bird Migration*. Collins, London.

Newton, I. 2017. *Farming and Birds*. Collins, London.

Nilsson, J.-Å. 1989a. Causes and consequences of natal dispersal in the marsh tit, *Parus palustris*. *Journal of Animal Ecology* 58: 619–636.

Nilsson, J.-Å. 1989b. Establishment of juvenile marsh tits in winter flocks: an experimental study. *Animal Behaviour* 38: 586–595.

Nilsson, J.-Å. 1991. Clutch size determination in the Marsh Tit (*Parus palustris*). *Ecology* 72: 1757–1762.

Nilsson, J.-Å. 1992. Variation in wing length in relation to sex and age of Marsh Tits *Parus palustris*. *Ornis Svecica* 2: 7–12.

Nilsson, J.-Å. 1993. Energetic constraints on hatching asynchrony. *The American Naturalist* 141: 158–166.

Nilsson, J.-Å. 2003. Ectoparasitism in marsh tits: costs and functional explanations. *Behavioral Ecology* 14: 175–181.

Nilsson, J.-Å. & Nord, A. 2017. The use of the nest for parental roosting and thermal consequences of the nest for nestlings and parents. *Behavioral Ecology and Sociobiology* 71: 171.

Nilsson, J.-Å. & Smith, H. G. 1985. Early fledgling mortality and the timing of juvenile dispersal in the marsh tit *Parus palustris*. *Ornis Scandinavica* 16: 293–198.

Nilsson, J.-Å. & Smith, H. G. 1988a. Effects of dispersal date on winter flock establishment and social dominance in marsh tits *Parus palustris*. *Journal of Animal Ecology* 57: 917–928.

Nilsson, J.-Å. & Smith, H. G. 1988b. Incubation feeding as a male tactic for early hatching. *Animal Behaviour* 36: 641–647.

Nilsson, J.-Å. & Svensson, M. 1996. Sibling competition affects nestling growth strategies in Marsh Tits. *Journal of Animal Ecology* 65: 825–836.

Nilsson, S. G. 1984. The evolution of nest-site selection among hole-nesting birds: the importance of nest predation and competition. *Ornis Scandinavica* 15: 167–175.

Nomi, D., Yuta, T. & Koizumi, I. 2017. Breeding biology of four sympatric tits in northern Japan. *The Wilson Journal of Ornithology* 129: 294–300.

Oddmund, K. Rudolfsen, R. & Schmoll, T. 2020. Extra-pair paternity in the boreal, socially monogamous Grey-headed Chickadee (*Poecile cinctus*). *Ornis Fennica* 97: 28–44.

Opdam, P., Rijsdijk, G. & Hustings, F. 1985. Bird communities in small woods in an agricultural landscape: effects of area and isolation. *Biological Conservation* 34, 333–352.

Orell, M. 1983. Nestling growth in the Great Tit *Parus major* and the Willow Tit *P. montanus*. *Ornis Scandinavica* 11: 173–178.

Orell, M. 1989. Population fluctuations and survival of Great Tits *Parus major* dependent on food supplied by man in winter. *Ibis* 131: 112–127.

Orell, M. & Koivula, K. 1988. Cost of reproduction: parental survival and production of recruits in the Willow Tit *Parus montanus. Oecologia* 77: 423–432.

Orell, M. & Ojanen, M. 1980. Overlap between breeding and moulting in the Great Tit *Parus major* and Willow Tit *P. montanus* in northern Finland. *Ornis Scandinavica* 11: 43–49.

Orell, M. & Ojanen, M. 1983. Breeding biology and population dynamics of the Willow Tit *Parus montanus. Annales Zoologici Fennici* 20: 99–114.

Orell, M., Rytkönen, S. & Koivula, K. 1994a. Causes of divorce in the monogamous willow tit, *Parus montanus*, and consequences for reproductive success. *Animal Behaviour* 48: 1143–1154.

Orell, M., Koivula, K., Rytkönen, S. & Lahti, K. 1994b. To breed or not to breed: causes and implications of non-breeding habit in the willow tit *Parus montanus. Oecologia* 100: 339–346.

Orell, M., Rytkönen, S., Koivula, K., Ronkainen, M. & Rahiala, M. 1996. Brood size manipulations within the natural range did not reveal intragenerational cost of reproduction in the Willow Tit *Parus montanus. Ibis* 138: 630–637.

Orell, M., Lahti, K. & Matero, J. 1999a. High survival rate and site fidelity in the Siberian Tit *Parus cinctus*, a focal species of the taiga. *Ibis* 141: 460–468.

Orell, M., Lahti, K., Koivula, K., Rytkönen, S. & Welling, P. 1999b. Immigration and gene flow in a northern Willow Tit (*Parus montanus*) population. *Journal of Evolutionary Biology* 12: 283–295.

Orłowski, G. & Ławniczak, D. 2009. Changes in breeding bird populations in farmland of south-western Poland between 1977–1979 and 2001. *Folia Zoologica* 58: 228–239.

Östlund, L. Zackrisson, O & Axelsson, A.-L. 1997. The history and transformation of a Scandinavian boreal forest landscape since the 19th century. *Canadian Journal of Forest Research* 27: 1198–1206.

Ottvall, R., Edenius, L., Elmberg, J., Engström, H., Green, M., Holmqvist, N., Lindström, Åke, Pärt, T. & Tjernberg, M. 2009. Population trends for Swedish breeding birds. *Ornis Svecica* 19: 117–192.

Pakanen, V. M., Koivula, K., Orell, M., Rytkönen, S. & Lahti, K. 2016a. Sex-specific mortality costs of dispersal during the post-settlement stage promote male philopatry in a resident passerine. *Behavioral Ecology and Sociobiology* 70: 1727–1733.

Pakanen, V. M., Orell, M., Vatka, E., Rytkönen, S. & Broggi, J. 2016b. Different ultimate factors define timing of breeding in two related species. *PLoS ONE* 11: e0162643.

Pakanen, V. M., Ahonen, E., Hohtola, E & Rytkönen, S. 2018. Northward expanding resident species benefit from warming winters through increased foraging rates and predator vigilance. *Oecologia* 188: 991–999.

Palmgren, P. 1932. Zur Biologie von *Regulus r. regulus* (L.) und *Parus atricapillus borealis* Selys; eine vergleichend-ökologische Untersuchung. *Acta Zoologica Fennica* 14: 1–113.

Panayotopoulou, M. 2005. The winter social system of sombre tits *Parus lugubris*, in Greece. Thesis, Lund University, Sweden.

Parry, W. 2017. Hooked on Willow Tits. *LifeCycle* 6: 12–13.

Parry, W. & Broughton, R. K. 2018. Nesting behaviour and breeding success of Willow Tits *Poecile montanus* in north-west England. *Ringing & Migration* 33: 75–85.

Pavlova, A., Rohwer, S., Drovetski, S. V. & Zink, R. M. 2006. Different post-Pleistocene histories of Eurasian parids. *Journal of Heredity* 97: 389–402.

Perrins, C. M. 1965. Population fluctuations and clutch-size in the Great Tit, *Parus major* L. *Journal of Animal Ecology* 34: 601–647.

Perrins, C. 1979. *British Tits*. Collins, London.

Perrins, C. M. & Gosler, A. G. 2010. Birds. In: Savill, P. S., Perrins, C. M., Kirby, K. J. & Fisher, N. (eds.), *Wytham Woods: Oxford's Ecological Laboratory*. Oxford University Press, Oxford, pp. 145–171.

Peterken, G. 2019. Defining 'natural woodland'. *British Wildlife* 30: 157–159.

Peterken, G. 2022. A long-term perspective on rewilding woodland. *British Wildlife* 33: 584–589.

Peterken, G. 2023. *Trees and Woodlands*. Bloomsbury Publishing, London.

Peterken, G. & Mountford, E. 2017. Woodland development: a long-term study of Lady Park Wood. CABI, Wallingford.

Pettorelli, N. & Bullock, J. M. 2023. Restore or rewild? Implementing complementary approaches to bend the curve on biodiversity loss. *Ecological Solutions and Evidence* 4: e12244.

Piaskowski, V. D., Weise, C. M. & Ficken, M. S. 1991. The body ruffling display of the Black-capped Chickadee. *The Wilson Bulletin* 103: 426–434.

Pinder, S. 2021. *SP12 Final Report: Willow Tit*. Back From the Brink. https://naturebftb. co.uk/wp-content/uploads/2022/01/SP12-Willow-Tit-Final-Report-BftB-Website.pdf.

Pinder, S. & Carr, G. 2021. *Willow Tit Conservation Handbook*. Back From the Brink, naturebftb.co.uk.

Plummer, K. E., Risely, K., Toms, M. P. & Siriwardena, G. M. 2019. The composition of British bird communities is associated with long-term garden bird feeding. *Nature Communications* 10: 2088.

Popov, V. A. 1978. *Birds of the Volga and Kama region. Passerines*. Nauka, Moscow.

Porro, Z., Chiantante, G. & Bogliani, G. 2020. Associations between forest specialist birds and composition of woodland habitats in a highly modified landscape. *Forest Ecology and Management* 458: 117732.

Pravosudov, V. V. & Pravosudova, E. V. 1996. The breeding biology of the Willow Tit in northeastern Siberia. *The Wilson Bulletin* 101: 80–93.

Pritchard, R., Hughes, J., Spence, I. M., Haycock, B. & Brenchley, A. (eds.) 2021. *The Birds of Wales*. Liverpool University Press, Liverpool.

Prokofjeva, I. W. 1990. Seasonable variability of feeding Willow Tit in the north-west USSR conditions. *Trudy Zoologicheskogo Instituta, Akademiya Nauk SSSR* 210: 89–94.

Purevdorj, Z., Munkhbayar, M., Paek, W. K., Ganbold, O., Jargalsaikhan, A., Purevee, E., Amartuvshin, T., Genenjamba, U., Nyam B. & Lee, J. W. 2022. Relationships between bird assemblages and habitat variables in a boreal forest of the Khentii Mountain, northern Mongolia. *Forests* 13: 1037.

Pusey, A. E. 1987. Sex-biased dispersal and inbreeding avoidance in birds and mammals. *Trends in Ecology and Evolution* 2: 295–299.

Rackham, O. 1976. *Trees and Woodland in the British Landscape*. J. M. Dent, London.

Rackham, O. 2006. *Woodlands*. HarperCollins, London.

Ram, D., Axelsson, A.-L., Green, M., Smith, H. G. & Lindström, Å. 2017. What drives current population trends in forest birds: forest quantity, quality or climate? A large-scale analysis from northern Europe. *Forest Ecology and Management* 385: 177–188.

Ramsay, S. M., Mennill, D. J., Otter, K. A., Ratcliffe, L. M. & Boag, P. T. 2003. Sex allocation in black-capped chickadees *Poecile atricapilla*. *Journal of Avian Biology* 34: 134–139.

Reinertsen, R. E. (1983) Nocturnal hypothermia and its energetic significance for small birds living in the Arctic and subarctic regions: a review. *Polar Research* 1: 269–284.

Revill, J., Rustell, A., & Last, J. 2010. What WILTIs want (and how to provide it). *BTO Nest Record News* 26: 12–13.

Rheinwald, G. 1975. Gewichtsentwicklung

einiger nestjunger Höhlenbrüter. *Journal für Ornithologie* 116: 55–64.

Rhim, S. J., Hur, W. H. & Lee, W. S. 2003. Characteristics of bird communities between slope and valley in natural deciduous forest, South Korea. *Journal of Forestry Research* 14: 221–224.

Rhim, S. J., Son, S. H. & Kim, K. J. 2011. Breeding ecology of tits *Parus* spp. using artificial nest boxes in a coniferous forest over a five-year period. *Forest Science and Technology* 7: 141–144.

Richards, T. J. 1958. Concealment and recovery of food by birds, with some relevant observations on squirrels. *British Birds* 51: 497–508.

Rintamäki, P. T., Ojanen, M., Pakkala, H., Tynjälä, M., & Lundberg, A. 2000. Blood parasites of juvenile Willow Tits *Parus montanus* during autumn migration in northern Finland. *Ornis Fennica* 77: 83–87.

Robb, G. N., McDonald, R. A., Chamberlain, D. E., Reynolds, S. J., Harrison, T. J. E. & Bearhop, S. 2008. Winter feeding of birds increases productivity in the subsequent breeding season. *Biology Letters* 4: 220–223.

Roberge, J.-M., Virkkala, R. & Mönkkönen, M. 2018. Boreal forest bird assemblages and their conservation. In: Mikusiński, G., Roberge, J.-M. & Fuller, R. J. (eds.), *Ecology and Conservation of Forest Birds*. Cambridge University Press, Cambridge, pp. 183–230.

Robinson, M. G. 1950. Concealment of food by Marsh-Tit. *British Birds* 43: 336.

Robinson, R. A., Leech, D. I. & Clark, J. A. 2023. *The Online Demography Report: Bird Ringing and Nest Recording in Britain and Ireland in 2022*. British Trust for Ornithology, Thetford. www.bto.org/ringing-report.

Rolando, A. 1983. Ecological segregation of tits and associated species in two coniferous woods of northern Italy. *Italian Journal of Zoology* 17: 1–18.

Romanowski, E. 1978. Der gesang von Sumpf- und Weidenmeise (*Parus palustris*

und *Parus montanus*): variation und funktion. *Die Vogelwarte* 29: 235–253.

Romanowski, E. 1979. Der Gesang von Sumpf- und Weidenmeise (*Parus palustris* und *Parus montanus*): reaktionsauslösende Parameter. *Die Vogelwarte* 30: 48–65.

Rustell, A. 2015. The effects of avian nest predation and competition on the Willow Tit in Britain. *British Birds* 108: 37–41.

Rytkönen, S., Koivula, K. & Orell, M. 1990. Temporal increase in nest defence intensity of the willow tit (*Parus montanus*): parental investment or methodological artifact? *Behavioral Ecology and Sociobiology* 27: 283–286.

Rytkönen, S., Koivula, K. & Orell, M. 1996. Patterns of per-brood and per-offspring provisioning efforts in the Willow Tit *Parus montanus*. *Journal of Avian Biology* 27: 21–30.

Sage, R. B., Hoodless, A. N., Hewson, C. M., Wilson, S., Le Clare, C., Marchant, J. H., Draycott, R. A. H. & Fuller, R. J. 2011. Assessing breeding success in common woodland birds using a novel method. *Bird Study* 58: 409–420.

Sahvon, V. V. 2007. Structure of bird communities of floodplain forests of the black alder of Belarussian Polesye. *Branta* 10: 27–36.

Salzburger, W., Martens, J., Nazarenko, A. A., Sun, Y.-H., Dallinger, R. & Sturmbauer, C. 2002. Phylogeography of the Eurasian Willow Tit (*Parus montanus*) based on DNA sequences of the mitochondrial cytochrome b gene. *Molecular Phylogenetics and Evolution* 24: 26–34.

Saurola, P. 1981. Recoveries of irruptive willow tits (*Parus montanus*) ringed in Finland. *Lintumies* 16: 106–110.

Schaub, M. & Amann, F. 2001. Saisonale Variation der Überlebensraten von Sumpfmeisen *Parus palustris*. *Der Ornithologische Beobachter* 98: 223–235.

Scott, G. W. 1999. Separation of Marsh Tits *Parus palustris* and Willow Tits *Parus montanus*. *Ringing & Migration* 19: 323–326.

Sellers, R. M. 1984. Movements of Coal, Marsh and Willow Tits in Britain. *Ringing & Migration* 5: 79–89.

Sells, J. D. 1998. Coal Tits, Marsh Tits and the Batsford nestbox scheme. *The Gloucestershire Naturalist* 11: 74–79.

Selva, N., Jędrzejewska, B., Jędrzejewski, W. & Wajrak, A. 2011. Factors affecting carcass use by a guild of scavengers in European temperate woodland. *Canadian Journal of Zoology* 83: 1590–1601.

Sherry, D. F. 1989. Food storing in the Paridae. *The Wilson Bulletin* 101: 289–304.

Shimazaki, A., Yamaura, Y., Senzaki, M., Yabuhara, Y., Akasaka, T. & Nakamura, F. 2016. Urban permeability for birds: an approach combining mobbing-call experiments and circuit theory. *Urban Forestry and Urban Greening* 19: 167–175.

Shrubsole, G. 2022. *The Lost Rainforests of Britain*. William Collins, London.

Shutt, J. D. & Lees, A. C. 2021. Killing with kindness: Does widespread generalised provisioning of wildlife help or hinder biodiversity conservation efforts? *Biological Conservation* 261: 109295.

Shutt, J. D., Trivedi, U. H. & Nicholls, J. A. 2021. Faecal metabarcoding reveals pervasive long-distance impacts of garden bird feeding. *Proceedings of the Royal Society B* 288: 20210480.

Sibley, C. G. 1955. Behavioral mimicry in the titmice (Paridae) and certain other birds. *The Wilson Bulletin* 67: 128–132.

Siffczyk, C., Brotons, L., Kangas, K. & Orell, M. 2003. Home range size of willow tits: a response to winter habitat loss. *Oecologia* 136: 635–642.

Sikora, D. J. 2021. The frequency method in studying habitat preferences of common forest birds in south-east Poland. *Baltic Forestry* 27: 180–188.

Simms, E. 1971. *Woodland Birds*. Collins, London.

Siriwardena, G. M. 2004. Possible roles of habitat, competition and avian nest predation in the decline of the Willow Tit *Parus montanus* in Britain. *Bird Study* 51: 193–202.

Siriwardena, G. M. 2006. Avian nest predation, competition and the decline of British Marsh Tits *Parus palustris*. *Ibis* 148: 255–265.

Skilsky, I. V. 1998. Structure and peculiarities of forming of the park ornithocomplex in Chernivtsi. *Berkut* 7: 3–11.

Skliar, O. Yu. & Knysh, N. P. 2016. New data about rare and insufficiently known bird species of the Hetmansky National Park and its environs (Sumy region, NE Ukraine). *Berkut* 25: 15–24.

Skórka, P. & Wójcik, J. D. 2003. Winter bird communities in a managed mixed oak–pine forest (Niepołomice Forest, southern Poland). *Acta Zoologica Cracoviensia* 46: 29–41.

Smith, H. G. 1993a. Parental age and reproduction in the Marsh Tit *Parus palustris*. *Ibis* 135: 196–201.

Smith, H. G. 1993b. Seasonal decline in clutch size of the Marsh Tit (*Parus palustris*) in relation to date-specific survival of offspring. *The Auk* 110: 889–899.

Smith, K. W., Burges, D. J. & Parks, R. A. 1992. Breeding bird communities of broadleaved plantation and ancient pasture woodlands of the New Forest. *Bird Study* 39: 132–141.

Smith, K. W., Smith, L., Charman, E., Briggs, K., Burgess, M., Dennis, C., Harding, M., Isherwood, C., Isherwood, I. & Mallord, J. 2011. Large-scale variation in the temporal patterns of the frass fall of defoliating caterpillars in oak woodlands in Britain: implications for nesting woodland birds. *Bird Study* 58: 506–511.

Smith, S. M. 1967. Seasonal changes in the survival of the Black-capped Chickadee. *The Condor* 69: 344–359.

Smith, S. M. 1984. Flock switching in chickadees: why be a winter floater? *The American Naturalist* 123 81–98.

Smith, S. M. 1991. *The Black-capped Chickadee: Behavioural Ecology and Natural History*. Cornell University Press, New York.

Smith, S. M. 1992. Pairbond persistence and 'divorce' in Black-capped Chickadees. *The Wilson Bulletin* 104: 338–342.

Smith, S. M. 1994. Social influences on the dynamics of a northeastern Black-capped Chickadee population. *Ecology* 75: 2043–2051.

Smith, S. T. 1972. Communication and other social behaviour in *Parus carolinensis*. *Publications of the Nuttall Ornithological Club* 11: 1–125.

Smout, T. C., MacDonald, A. R. & Watson, F. 2007. *A History of the Native Woodlands of Scotland, 1500–1920*. Edinburgh University Press, Edinburgh.

Snow, D. W. 1954. The habitats of Eurasian tits (*Parus* spp.). *Ibis* 96: 565–585.

Snow, D. W. 1956. The specific status of the Willow Tit. *Bulletin of the British Ornithologists' Club* 76: 29–31.

Snow, D. & Snow, B. 1988. *Birds and Berries*. T. & A. D. Poyser, London.

Söderström, B. 2009. Effects of different levels of green- and dead-tree retention on hemi-boreal forest bird communities in Sweden. *Forest Ecology and Management* 257: 215–222.

Sokolov, L. V., Markovets, M. Y., Yefremov, V. D. & Shapoval, A. P. 2002. Irregular migrations (irruptions) in six bird species on the Courish Spit in the Baltic Sea in 1957–2002. *Avian Ecology and Behaviour* 9: 39–53.

Southern, H. N. & Morley A. 1950. Marsh-tit territories over six years. *British Birds* 43: 33–45.

Southwood, T. R. E. 1961. The number of species of insect associated with various trees. *Journal of Animal Ecology* 30: 1–8.

Spencer, J. W. & Kirby, K. J. 1992. An inventory of ancient woodland for England and Wales. *Biological Conservation* 62: 77–93.

Spina, F., Baillie, S. R., Bairlein, F., Fiedler, W. & Thorup, K. (eds.) 2022. *The Eurasian African Bird Migration Atlas: Willow Tit*. https://migrationatlas.org/node/1808.

Staley, J. T., Sparks, T. H., Croxton, P. J.,

Baldock, K. C. R., Heard, M. S., Hulmes, S., Hulmes, L., Peyton, J., Amy, S. R. & Pywell, R. F. 2012. Long-term effects of hedgerow management policies on resource provision for wildlife. *Biological Conservation* 145: 24–29.

Stanković, D., Jönsson, J. & Raković, M. 2019. Diversity of avian blood parasites in wild passerines in Serbia with special reference to two new lineages. *Journal of Ornithology* 160: 545–555.

Steele, R. C. & Welch, R. C. 1973. *Monks Wood: a Nature Reserve Record*. Nature Conservancy, Huntingdon.

Stefanski, R. A. 1967. Utilization of the breeding territory in the Black-capped Chickadee. *The Condor* 69: 259–267.

Stejneger, L. 1886. The British marsh-tit. *Proceedings of the United States National Museum* 9: 200–201.

Stenning, M. 2018. *The Blue Tit*. T. & A. D. Poyser, London.

Stewart, F. 2010. Ecology and conservation of the Willow Tit (*Poecile montanus*) in Britain. PhD thesis, University of Nottingham.

Suzuki, T. N. 2012. Long-distance calling by the Willow Tit, *Poecile montanus*, facilitates formation of mixed-species foraging flocks. *Ethology* 118: 10–16.

Svensson, L. 2023. *Identification Guide to European Passerines*, 5th edition. Lullula Förlag, Stockholm.

Szulkin, M. & Sheldon, B. C. 2008. Dispersal as a means of inbreeding avoidance in a wild bird population. *Proceedings of the Royal Society B* 275: 703–711.

Tang, L., Shao, G., Piao, Z., Dai, L., Jenkins, M. A., Wang, S., Wu, G., Wu, J. & Zhao, J. 2010. Forest degradation deepens around and within protected areas in East Asia. *Biological Conservation* 143: 1295–1298.

Tansley, A. 1939. *The British Islands and their Vegetation*. Cambridge University Press, Cambridge.

Temperley, G. W. 1934. Some breeding-habits of the British Willow-Tit. *British Birds* 28: 165–169.

Thessing, A. 1999. Growth and heritability of size traits of Willow Tit *Parus montanus* nestlings: a supplementary feeding experiment. *Ornis Fennica* 76: 107–114.

Thönen, W. 1962. Stimmgeographische, ökologische und verbreitungsgeschictliche Studien über die Mönchsmeise (*Parus montanus* Conrad). *Der Ornithologische Beobachter* 59: 101–172.

Tiainen, J. 1980. Adaptedness of the Willow Tit *Parus montanus* to the migratory habit. *Ornis Fennica* 57: 77–81.

Tojo, H. 2009. Breeding bird community of the Ogawa Forest Reserve, an old-growth deciduous forest in central Japan. *Ornithological Science* 8: 105–115.

Tomiałojć, L. 2011. Changes in breeding bird communities of two urban parks in Wrocław across 40 years (1970–2010): before and after colonization by important predators. *Ornis Polonica* 52: 1–25.

Tomiałojć, L. & Wesołowski, T. 2005. The avifauna of Białowieża Forest: a window into the past. *British Birds* 98: 174–193.

Tritsch, C., Martens, J., Sun, Y.-H., Heim, W., Strutzenberger, P. & Päckert, M. 2017. Improved sampling at the subspecies level solves a taxonomic dilemma: a case study of two enigmatic Chinese tit species (Aves, Passeriformes, Paridae, *Poecile*). *Molecular Phylogenetics and Evolution* 107: 538–550.

Ulfstrand, S. & Nilsson, S. G. 1976. Quantitative composition and foraging niches of a passerine bird guild in pine plantations in Denmark during winter. *Ornis Scandinavica* 7: 171–178.

Unno, A. 2002. Tree species preferences of insectivorous birds in a Japanese deciduous forest: the effect of different foraging techniques and seasonal change of food resources. *Ornithological Science* 1: 133–142.

Urhan, A. U., Emilsson, E. & Brodin, A. 2017. Evidence against observational spatial memory for cache locations of conspecifics in marsh tits *Poecile palustris*. *Behavioral Ecology and Sociobiology* 71: 34.

Ustaoglu, E. & Collier, M. J. 2018. Farmland abandonment in Europe: an overview of drivers, consequences, and assessment of the sustainability implications. *Environmental Reviews* 26: 396–416.

Van Balen, J. H., Booy, C. J. H., van Franeker, J. A. & Osieck, E. R. 1982. Studies on hole-nesting birds in natural nest sites. 1. Availability and occupation of natural nest sites. *Ardea* 70: 1–24.

Vandergert, P. & Newell, J. 2003. Illegal logging in the Russian Far East and Siberia. *International Forestry Review* 5: 303–306.

van Dorp, D. & Opdam, P. F. M. 1987. Effects of patch size, isolation and regional abundance on forest bird communities. *Landscape Ecology* 1: 59–73.

van Overveld, T., Careau, V., Adriaensen, F. & Matthysen, E. 2014. Seasonal- and sex-specific correlations between dispersal and exploratory behaviour in the great tit. *Oecologia* 174: 109–120.

Van Tienderen, P. H. & Van Noordwijk, A. J. 1988. Dispersal, kinship and inbreeding in an island population of the Great Tit. *Journal of Evolutionary Biology* 1: 117–137.

Vatka, E., Orell, M. & Rytkönen, S. 2011. Warming climate advances breeding and improves synchrony of food demand and food availability in a boreal passerine. *Global Change Biology* 17: 3002–3009.

Vatka, E., Kangas, K., Orell, M., Lampila, S., Nikula, A. & Nivala, V. 2014. Nest site selection of a primary hole-nesting passerine reveals means to developing sustainable forestry. *Journal of Avian Biology* 45: 187–196.

Vaurie, C. & Snow, D. 1957. Systematic notes on Palearctic birds. No. 27. Paridae: the genera *Parus* and *Sylviparus*. *American Museum Novitates* 1852: 1–43.

Vera, F. W. M. 2000. *Grazing Ecology and Forest History*. CABI, Wallingford.

Verhulst, S., Perrins, C. M. & Riddington, R. 1997. Natal dispersal of Great Tits in a patchy environment. *Ecology* 78: 864–872.

Virkkala, R. 1990. Ecology of the Siberian Tit

Parus cinctus in relation to habitat quality: effects of forest management. *Ornis Scandinavica* 21: 139–146.

Virkkala, R. 2016. Long-term decline of southern boreal forest birds: consequence of habitat alteration or climate change? *Biodiversity and Conservation* 25: 151–167.

Visser, M. E. & Gienapp, P. 2019. Evolutionary and demographic consequences of phenological mismatches. *Nature Ecology & Evolution* 3: 879–885.

Volchanetsky, I. B. 1950. Birds of the edges of deciduous forests of Kharkiv and Sumy regions. *Proceedings of the Scientific Research Institute of Biology, Kharkiv State University A. M. Gorky* 14–15: 193–223.

von Brömssen, A. & Jansson, C. 1980. Effects of food addition to Willow Tit *Parus montanus* and Crested Tit *P. cristatus* at the time of breeding. *Ornis Scandinavica* 11: 173–178.

Wang, H., Gao, W., Wang, D., Liu, D. & Deng, W. 2003. Nest-site characteristics and reproductive success of five species of birds breeding in natural cavities. *Acta Ecologica Sinica* 23: 1377–1385.

Wang, J., Wei, Y., Zhang, L., Jiang, Y., Li, K. & Wan, D. 2021. High level of extra-pair paternity in the socially monogamous Marsh Tits (*Poecile palustris*). *Avian Research* 12: 69.

Waterman, J., Desrochers, A. & Hannon, S. 1989. A case of polyandry in the Black-capped Chickadee. *The Wilson Bulletin* 101: 351–353.

Weise, C. M. & Meyer, J. R. 1979. Juvenile dispersal and development of site-fidelity in the Black-capped Chickadee. *The Auk* 96: 40–55.

Weiser, E. L. 2021. Fully accounting for nest age reduces bias when quantifying nest survival. *Ornithological Applications* 123: duab030.

Welling, P., Koivula, K. & Lahti, K. 1995. The dawn chorus is linked with female fertility in the Willow Tit *Parus montanus*. *Journal of Avian Biology* 26: 241–246.

Welling, P., Orell, M., Koivula, K. & Lahti, K. 1997. The frequency and timing of copulations in the Willow Tit *Parus montanus*. *Ornis Fennica* 74: 51–58.

Wernham, C. V., Toms, M. P., Marchant, J. H., Clark, J. A., Siriwardena, G. M. & Baillie, S. R. (eds.) 2002. *The Migration Atlas: Movements of the Birds of Britain and Ireland.* T. & A. D. Poyser, London.

Wesołowski, T. 1989. Nest-sites of hole-nesters in a primaeval temperate forest (Białowieża National Park, Poland). *Acta Ornithologica* 25: 321–351.

Wesołowski, T. 1995a. Birds from a primaeval temperate forest hardly use feeders in winter. *Ornis Fennica* 72: 132–134.

Wesołowski, T. 1995b. The loss of avian cavities by injury compartmentalization in a primeval European forest. *The Condor* 97: 256–257.

Wesołowski, T. 1996. Natural nest sites of Marsh Tit (*Parus palustris*) in a primeval forest (Białowieża National Park, Poland). *Die Vogelwarte* 38: 235–249.

Wesołowski, T. 1998. Timing and synchronisation of breeding in a Marsh Tit *Parus palustris* population from a primeval forest. *Ardea* 86: 89–100.

Wesołowski, T. 2000. Time-saving mechanisms in the reproduction of Marsh Tits (*Parus palustris*). *Journal of Ornithology* 141: 309–318.

Wesołowski, T. 2001. Host–parasite interactions in natural holes: marsh tits (*Parus palustris*) and blow flies (*Protocalliphora falcozi*). *Journal of Zoology* 255: 495–503.

Wesołowski, T. 2002. Anti-predator adaptations in nesting Marsh Tits *Parus palustris*: the role of nest-site security. *Ibis* 144: 593–601.

Wesołowski, T. 2003. Clutch size and breeding performance of Marsh Tits *Parus palustris* in relation to hole size in a primeval forest. *Acta Ornithologica* 38: 65–72.

Wesołowski, T. 2006. Nest site re-use: Marsh Tit *Poecile palustris* decisions in a primeval forest. *Bird Study* 53: 199–204.

Wesołowski, T. 2007a. Primeval conditions: what can we learn from them? *Ibis* 149: 64–77.

Wesołowski, T. 2007b. Lessons from long-term hole-nester studies in a primeval temperate forest. *Journal of Ornithology* 148: 395–405.

Wesołowski, T. 2012. 'Lifespan' of non-excavated holes in a primeval temperate forest: a 30 year study. *Biological Conservation* 153: 118–126.

Wesołowski, T. 2013. Timing and stages of nest-building by Marsh Tits (*Poecile palustris*) in a primeval forest. *Avian Biology Research* 6: 31–38.

Wesołowski, T. 2015. Dispersal in an extensive continuous forest habitat: Marsh Tit *Poecile palustris* in the Białowieża National Park. *Journal of Ornithology* 156: 349–361.

Wesołowski, T. 2017. Failed predator attacks: a direct test of security of tree cavities used by nesting Marsh Tits (*Poecile palustris*). *The Auk* 134: 802–810.

Wesołowski, T. 2023. Weather, food and predation shape the timing of Marsh Tit breeding in primeval conditions: a long-term study. *Journal of Ornithology* 164: 253–274.

Wesołowski, T. & Maziarz, M. 2012. Dark tree cavities: a challenge for hole nesting birds? *Journal of Avian Biology* 43: 454–460.

Wesołowski, T. & Neubauer, G. 2017. Diet of Marsh Tit *Poecile palustris* nestlings in a primeval forest in relation to food supply and age of young. *Acta Ornithologica* 52: 105–118.

Wesołowski, T. & Rowiński, P. 2012. The breeding performance of Blue Tits *Cyanistes caeruleus* in relation to the attributes of natural holes in a primeval forest. *Bird Study* 59: 437–448.

Wesołowski, T. & Stańska, M. 2001. High ectoparasite loads in hole nesting birds: a nest box bias? *Journal of Avian Biology* 32: 281–285.

Wesołowski, T. & Wierzcholska, S. 2018. Tits as bryologists: patterns of bryophyte use in nests of three species cohabiting a primeval forest. *Journal of Ornithology* 159: 733–745.

Wesołowski, T., Czeszczewik, D., Rowiński, P. & Walankiewicz, W. 2002. Nest soaking in natural holes: a serious cause of breeding failure? *Ornis Fennica* 79: 132–138.

Wesołowski, T., Cholewa, M., Hebda, G., Maziarz, M. & Rowiński, P. 2016. Immense plasticity of timing of breeding in a sedentary forest passerine, *Poecile palustris*. *Journal of Avian Biology* 47: 129–133.

Wesołowski, T, Fuller, R. J. & Clade, M. 2018. Temperate forests: a European perspective on variation and dynamics in bird assemblages. In: Mikusiński, G., Roberge, J.-M. & Fuller, R. J. (eds.), *Ecology and Conservation of Forest Birds*. Cambridge University Press, Cambridge, pp. 253–317.

Wesołowski, T., Czeszczewik, D., Hebda, G., Maziarz, M., Mitrus, C., Rowiński, P. & Neubauer, G. 2022. Long-term changes in breeding bird community of a primeval temperate forest: 45 years of censuses in the Białowieża National Park (Poland). *Acta Ornithologica* 57: 71–100.

Wilkin, T. A., Garant, D., Gosler, A. G. & Sheldon, B. C. 2006. Density effects on life-history traits in a wild population of the great tit *Parus major*: analyses of long-term data with GIS techniques. *Journal of Animal Ecology* 75: 604–615.

Witherby, H. F. 1934. The Willow-Tit's method of boring its nesting-hole. *British Birds* 27: 320–324.

Woodward, I., Aebischer, N., Burnell, D., Eaton, M., Frost, T., Hall, C., Stroud, D. A. & Noble, D. 2020. Population estimates of birds in Great Britain and the United Kingdom. *British Birds* 113: 69–104.

Xirouchakis, S. 2005. The avifauna of the western Rodopi forests (N. Greece). *Belgian Journal of Zoology* 135: 261–269.

Yapp, W. B. 1962. *Birds and Woods*. Oxford University Press, London.

Yatsiuk, Y. 2015. Occurrence of Willow Tit (*Parus montanus*) in Oak forests of Kharkiv region: species expansion or omission by researchers? In: Regional aspects of Floristic and Faunistic Studies: Materials of the Second International Workshop and Conference, Chernivtsi, Ukraine, pp. 152–154.

Yoon, J., Jung, J.-S., Joo, E.-J., Kim, B.-S. & Park, S.-R. 2017. Parent birds assess nest predation risk: influence of cavity condition and avian nest predator activity. *Journal of Avian Biology* 48: 691–699.

Zakkak, S., Radovic, A., Nikolov, S. C., Shumka, S., Kakalis, L. & Kati, V. 2015. Assessing the effect of agricultural land abandonment on bird communities in southern-eastern Europe. *Journal of Environmental Management* 164: 171–179.

Zhang, L., Liu, J., Gao, Z., Zhang, L., Wan, D., Liang, W. & Møller, A. P. 2020. Comparative analysis of hissing calls in five tit species. *Behavioural Processes* 171: 104029.

Zhang, L., Bai, L., Wang, J., Wan, D. & Liang, W. 2021. Occupation rates of artificial nest boxes by secondary cavity-nesting birds: the influence of nest site characteristics. *Journal for Nature Conservation* 63: 126045.

Zub, K., Czeszczewik, D., Ruczyński, I., Kapusta, A. & Walankiewicz, W. 2017. Silence is not golden: the hissing calls of tits affect the behaviour of a nest predator. *Behavioral Ecology and Sociobiology* 71: 1–7.

Zurfluh, K., Albini, S., Mattmann, P., Kindle, P., Nüesch-Inderbinen, M., Stephan, R. & Vogler, B. R. 2019. Antimicrobial resistant and extended-spectrum β-lactamase producing *Escherichia coli* in common wild bird species in Switzerland. *MicrobiologyOpen* 8: e845.

Acknowledgements

First, I am very grateful for the patient guidance of the Bloomsbury team, notably Katy Roper and Laura Browning-Brant, and particularly Hugh Brazier for his skilled and insightful editing. I am honoured that the hugely talented Darren Woodhead created the vividly energetic cover art.

On joining the UKCEH and its predecessors at the Monks Wood research station I was able to begin a research career with woodland birds, which I owe to the tremendous support of Shelley Hinsley and Paul Bellamy in developing the Marsh Tit project. I was also supported by other valued colleagues, including Richard Pywell, Helen Roy, Jane Hall, John Day, Stephen Freeman, Owen Mountford, Peter Rothery, Tim Sparks and some excellent students, namely Jane Mason, Deborah Marchant and Lindsay Henderson. I have had great luck in working with Ross Hill at Monks Wood and Bournemouth University on using remote sensing in woodland research.

At the University of Oxford's EGI I have enjoyed rewarding collaborations and support from Ben Sheldon, Jane Carpenter, Ross Crates, Andrew Gosler, Keith McMahon and Matt Wood. From the BTO, I thank present and former staff, especially Lee Barber, Carl Barimore, Jez Blackburn, Jacquie Clark, Daria Dadam, Samantha Franks, Rob Fuller, Ellie Leech, Dorian Moss and Rob Robinson. From the RSPB, I thank Arjun Amar, Malcolm Burgess, Ken Smith, Nigel Butcher, Graham Elliott and especially Paul Bellamy again. The Natural History Museum at Tring provided valuable access to collections, and for access to study sites I am also very grateful to the Abbots Ripton Estate, Milton Estate, Forestry Commission, John Chilvers and the Wildlife Trusts. I am especially thankful to Natural England for access to Monks Wood and support for the research, particularly to Katy Smith and Tony Juniper. Much of the Monks Wood work was made possible thanks to Sarah Caesar, Geoff Leach and Douglas Hall for fieldwork support and curry evenings.

Many people have contributed to projects, shared information or images, or helped to publicise the research. Particular thanks go to Dan Alder, Peter Alker, Ian Andrews, Ashley Banwell, Patrick Barkham, Keith Betton, Philip Bone, Simon Brebner, Alan Brimmell, Stewart Britton, Ken Davies, Tony Davis, Chris du Feu, Hugh Ford, the late Jim Fowler, Robert Fredagsvik, John Gibson, Jon Groom, Iain Hamlin, Philip Hanmer, Peter Hendry, Ken Hindmarch, Harri Hölttä, Julian Hughes, Alex Inzani, Anne Laine, John Last, Alexander Lees, J. W. Ross MacKinnon, Mikhail Markovets, Giulia Masoero, Geoff Mawson, Jill Mead, Markus Melin, Stephen Menzie and *British Birds* magazine, Michael Miles, Ash Murray, David Neal, Adam Nicolson, Jan-Åke Nilsson, Johan Nilsson, Han Onderwater, Markku Orell, Steve Ormerod, Garth Peacock, Wayne Parry, Matt Prior, Samuele Ramellini, Alasdair Reid, Roger Riddington, Peter Roseveare, Allan Rustell, Philip Schofield, John Sells, Jack Shutt, the BBC *Springwatch* team, Henry Stanier, Finn Stewart, Peter Stronach, Andrew Tongue, the Treswell Wood Integrated Population Monitoring Group, John Walder, Phoebe Weston and Yehor Yatsiuk. Thanks are also due to the many contributors and staff involved in data collection, preparation and availability for the BTO and RSPB surveys, and to the European Union for Bird Ringing (EURING) for making data available through the EURING databank.

Finally, I am deeply indebted to the research teams in the Białowieża Forest who have greatly enriched my studies and wider experience, particularly the late Tomasz Wesołowski, and also Marta Cholewa, Dorota Czeszczewik, Monika Czuchra, Grzegorz Hebda, Cezary Mitrus, Grzegorz Neubauer, Patryk Rowiński, the late Wiesław Walankiewicz, Wanda Wesołowska and especially Marta Maziarz. I am further grateful to Marta for her support while I was writing this book, and for long-term collaborations at Białowieża, Monks Wood, Oxford and beyond.

Index